PLANT BREEDING
AT A GLANCE

PLANT BREEDING

AT A GLANCE

Dr. PHUNDAN SINGH
Former Director
Central Institute for Cotton Research
NAGPUR-440010

KALYANI PUBLISHERS
LUDHIANA – NEW DELHI – NOIDA (U.P.) – HYDERABAD – CHENNAI
KOLKATA – CUTTACK – GUWAHATI – KOCHI – BANGALORE

KALYANI PUBLISHERS

Head Office :
1/1, Rajinder Nagar, **Ludhiana**-141 008
E-mail: kalyanibooks@yahoo.co.in

Administrative Office :
4863/2B, Bharat Ram Road,
24, Daryaganj, **New Delhi-110 002**
E-mail: kalyani_delhi@yahoo.co.in

Works :
B-16, Sector 8, **Noida (U.P.)**

Branch Offices :
No. 1, Mahalakshmi Street, T. Nagar, **Chennai**-600 017
110/111, Bharatia Towers, Badambadi, **Cuttack**-753 009 (Orissa)
3-5-1108, Narayanaguda, **Hyderabad**-500 029
No. 10.2B, Ramanath Mazumdar Street, **Kolkata**-700 009
Arunalaya, First Floor, Saraswati Road
Pan Bazar, **Guwahati**-781 001 (Phone No. 2731274)
Convent Road, **Kochi**-682 035
No. 24 & 25 1st floor, Hameed Shah Complex
Cubbonpet Main Road, **Bangalore**–560 002

KPPJ 13529 07 N-22505 05

Second Revised and Enlarged Edition 2010
Third Revised Edition 2013

Reprinted, 2017

TYPESETTING AT
(GEN–13–PBG–GEN)
Gautam Graphic Printers, Near Mai Hiran Gate, Jalandhar

ISBN 978-93-272-3478-7

PRINTED IN INDIA
At Process World India C-90 Sec-63 Noida
and published by Mrs Usha Raj Kumar for
Kalyani Publishers, New Delhi-110 002

Preface

Third Edition

It is a matter of great pleasure to write preface to the third edition of the book "**Plant Breeding At A glance**". The earlier editions of this book were very well received by the students, teachers and researchers. As a result, the book went out of stock and was reprinted many times. In this edition several changes have been made to make it more useful to the readers. The important changes are as follows.

1. Two chapters one on Introduction to Computers and another on Introduction to Genomics and Bioinformatics have been rewritten and thoroughly revised to provide up to date information.
2. The lists of ICAR Crop Breeding Institutes and State and Central Agricultural Universities have been appended.

The cooperation extended by my wife Mrs. Jaswanti Singh during preparation of this revised manuscript is highly appreciable. Hope, this Volume would be more useful to the students, teachers and other readers. Constructive suggestions of students and teachers are invited for further improvement of this book.

Nagpur **Phundan Singh**

2013

Preface

First Edition

Plant breeding, which is also called Crop Improvement Technology, plays a key role in improving genetic makeup of crop plants in relation to their economic use for human beings. Plant breeding is offered as a separate subject at graduate and postgraduate levels in all Agricultural Universities. At university level, students have to attempt descriptive, objective and short answer type of questions. There are several books of foreign and Indian origin on descriptive aspect of plant breeding. Some books are also available on objective aspect. However, there is no book which can provide solution to short answer type of questions.

The present book "Plant Breeding at a Glance" has been designed to fulfil this long felt need of Indian students. This volume covers the courses offered at graduate and postgraduate levels in various Agricultural Universities and also the syllabi of various competitive examinations.

This book contains 25 chapters. Each chapter has been dealt with in question answer form. Most of the questions are short answer type. Such approach helps students in clarity of concepts and better understanding of subject. I have cited few examples. Teachers and students can frame a lot more such questions for exercise. At the end glossary has been presented as a ready reference for various technical terms which are commonly used in Plant Breeding.

I am extremely thankful to my wife Mrs. Jaswanti Singh for her fullest cooperation and understanding during preparation of this manuscript. Infact I could not attend her properly for which I feel guilty.

Hope this book would be useful to the students and teachers of Plant breeding and also for those who are preparing for various competitive examinations such as ARSE, SET, NET and civil services examinations.

Constructive suggestions of students and teaching faculty are invited for further improvement of this book.

Phundan Singh

2004

KALYANI BOOKS BY THE SAME AUTHOR

SN.	Title of Book	SN.	Title of Book
A.	**Plant Breeding**	23.	Principles of Genetics (For UG Students)
1.	Essentials of Plant Breeding	24.	Plant Genetics
2.	Plant Breeding: Molecular and New Approaches	25.	Cotton Genetics
3.	Principles of Seed Technology	26.	Objective Genetics
4.	Practical in Crop Breeding	27.	Genetics At a Glance
5.	Practical and Numerical in Plant Breeding	28.	Objective Genetics and Plant Breeding
6.	Numerical Problems in Plant Breeding and Genetics	29.	Objective Genetics and Plant Breeding (Hindi Edn)
7.	Objective Science of Plant Breeding	30.	Glossary cum Dictionary of Genetics and Plant Breeding
8.	Plant Breeding At a Glance	31.	Molecular Genetics (Objective)
9.	Plant Breeding (For Under Graduate Students)	32.	Molecular genetics (At a Glance)
10.	Principles of Plant Breeding	33.	Molecular Genetics (Subjective)
11.	Cotton Breeding	34.	Genetics and Man
12.	Heterosis Breeding in Cotton	35	Genetics Today
13.	Breeding Hybrid Cotton	C.	**Quantitative Genetics**
14.	Breeding Transgenic Bt. Cotton	36.	Biometrical Techniques in Plant Breeding
15.	Cotton Improvement in India	37.	Application of Biometrical Techniques in Plant Breeding (Hindi Edn)
16.	Glossary of Plant Breeding and Genetics	38.	Quantitative Genetics
17.	IPR and Plant Breeders' Rights (Subjective)	39.	Objective Quantitative Genetics
18.	IPR and Plant Breeders' Rights (Objective)	40.	Quantitative Genetics At a Glance
19.	IPR and Plant Breeders' Rights (At a Glance)	D.	**Biotechnology**
20.	Elements of Baby Corn	41.	Introduction to Biotechnology
B.	**Genetics**	42.	Plant Biotechnology
21.	Elements of Genetics	43.	Principles of Plant Biotechnology
22.	Genetics		

Contents

SECTION-I
INTRODUCTORY TOPICS

INTRODUCTION TO PLANT BREEDING

Q. 1. What is plant breeding ?

Ans. Plant breeding is defined as the art, science and technology of improving genetic make up of crop plants in relation to their economic use for mankind.

Q. 2. Who defined plant breeding as current phase of crop evolution?

Ans. N.W. Simmonds (1979) defined plant breeding as the current phase of crop evolution.

Q. 3. Who used the term technology for plant breeding ?

Ans. Riley (1978) defined plant breeding as a technology of developing superior crop plants for various purposes.

Q. 4. What is plant breeding as per Frankel (1968) ?

Ans. According to Frankel (1968), plant breeding is the genetic adjustment of plants to social, cultural, economic and technological aspects of the environment.

Q. 5. How Poehlman defined plant breeding ?

Ans. According to J.M. Poehlman (1959), plant breeding is the art and the science of improving the heredity of crop plants for the benefit of human kind.

Q. 6. What are different branches of plant breeding ?

Ans. Various branches of plant breeding are given below.

(i) Agricultural Plant Breeding
(ii) Horticultural Plant Breeding
(iii) Tree Breeding
(iv) Medicinal Plant Breeding
(v) Stress Breeding
(vi) Quality Breeding
(vii) Mutation Breeding
(viii) Polyploidy Breeding
(ix) Physiological crop Breeding

(x) Introgressive Breeding
(xi) Transgenic Breeding
(xii) Maintenance Breeding
(xiii) Molecular breeding

Q. 7. What are important areas of plant breeding ?

Ans. There are three main areas of plant breeding as given below.

(i) Plant genetic resources or germplasm
(ii) Plant breeding techniques
(iii) Seed production techniques.

Q. 8. What are objectives of plant breeding ?

Ans. Important objectives of plant breeding are given below.

(i) Higher yield
(ii) Improved quality
(iii) Resistance to biotic stresses
(iv) Resistance to abiotic stresses
(v) Earliness
(vi) Synchronous maturity
(vii) Photo insensitivity
(viii) Thermo insensitivity
(ix) Wider adaptability
(x) Low toxic substances
(xi) Short stature etc.

Q. 9. What are disciplines which are involved in plant breeding ?

Ans. Involvement of following disciplines is required in crop improvement work.

(i) Morphology and Taxonomy
(ii) Genetics and cytogenetics
(iii) Plant physiology
(iv) Plant pathology
(v) Entomology
(vi) Agronomy
(vii) Soil science
(viii) Biochemistry
(ix) Agricultural engineering
(x) Statistics
(xi) Biometrics
(xii) Computers
(xiii) Biotechnology.

Q. 10. What is the role of taxonomy in plant breeding ?

Ans. Knowledge of taxonomy is useful in the identification of wild species and wild relatives of crop plants.

Q. 11. How knowledge of genetics is useful in plant breeding ?

Ans. It is useful in the study of inheritance of various aligogenic characters, linkage studies and gene transfer.

Q. 12. What is the role of plant physiology in plant breeding ?

Ans. Knowledge of plant physiology is useful in breeding crop cultivars resistant to drought, heat, cold, frost etc. and also for developing crop ideotypes.

Q. 13. How knowledge of plant pathology and entomology is useful in plant breeding ?

Ans. Pathologist helps in breeding disease resistant varieties and entomologist helps in evolving insect resistant cultivars.

Q. 14. What is the role of agronomy in plant breeding ?

Ans. Knowledge of agronomy is useful in developing crop cultivars resistant to herbicides and parasitic weeds.

Q. 15. How knowledge of soil science is useful in plant breeding ?

Ans. Knowledge of soil science is useful in evolving crop cultivars resistant to soil salinity, alkalinity and acidity.

Q. 16. What is the role of biochemistry in plant breeding ?

Ans. Biochemist helps breeder in developing crop cultivars with : (i) Better quality and (ii) Low level of toxic substances.

Q. 17. What is the role of agricultural engineer in plant breeding ?

Ans. Agricultural engineer helps breeder in evolving crop cultivars suitable for machine harvesting.

Q. 18. How knowledge of statistics and biometrics is useful to the plant breeder ?

Ans. Knowledge of statistics and biometrics is useful for proper conduct of breeding experiments and for various biometrical analyses.

Q. 19. How computers are useful in plant breeding ?

Ans. In plant breeding, computers are useful in biometrical and statistical analyses, storage of information, report writing, diagrammatic representation, updating of information, plant modelling etc.

Q. 20. What is the role of biotechnology in crop improvement ?

Ans. Biotechnology plays important role in following ways :

(i) In developing transgenic varieties with herbicide resistance, good quality and resistance to biotic and abiotic stresses.

(ii) In marker aided selection (molecular breeding)

(iii) In overcoming barriers of cross incompatibility through somatic hybridization.

(iv) In developing transgenic male sterility for use in hybrid seed production of different crops.

Q. 21. What is transgenic breeding ?

Ans. A branch of plant breeding which deals with genetic improvement of crop plants for various economic characters through plant biotechnology (genetic engineering) is known as transgenic breeding.

Q. 22. What is molecular breeding ?

Ans. A branch of plant breeding which deals with genetic improvement of crop plants for various economic characters through indirect selection for linked molecular markers (DNA markers) is called molecular breeding. It utilizes molecular techniques such as RFLP, AFLP, RAPD, CAPS, SSR etc. for selection of superior plants.

Q. 23. What is maintenance breeding ?

Ans. A branch of plant breeding which deals with principles of pure seed production is known as maintenance breeding or seed production technology. It deals with ways and means of maintaining genetic and physical purity of released and notified varieties and parents of commercial hybrids.

Q. 24. What is physiological crop breeding ?

Ans. It is a branch of plant breeding which deals with genetic improvement of crop plants in relation to following physiological parameters.

(i) CO_2 fixation efficiency

(ii) Nutrient uptake capacity

(iii) Harvest index

(iv) Reduction in photorespiration and transpiration rate

(v) Development of ideal plant types

(vi) Development of thermo and photoinsensitive genotypes etc.

Q. 25. What is introgressive breeding ?

Ans. It deals with genetic improvement of crop plants by transferring desirable genes from wild species into cultivated species of crop plants.

Q. 26. Who discovered sex in plants ?

Ans. The sex in plants was first discovered in 1694 by Camerarius, a German researcher.

Q. 27. Who made first interspecific cross ?

Ans. The first interspecific cross was made by Thomas Fairchild in 1717 in England between two species of **Dianthus** *i.e. Dianthus barbatus* X D. Caryophyllus.

Q. 28. Who identified self and cross pollinated plants ?

Ans. In plants, self pollination and cross pollination were first identified by Sprengel in 18th century.

Q. 29. Who developed semi-dwarf varieties of wheat ?

Ans. Semi dwarf varieties of wheat were first developed by Norman E. Borlaug in 1965. He developed semidwarf varieties of wheat at CIMMYT in Mexico.

Q. 30. When was Borlaug awarded Nobel Prize ?

Ans. Norman E. Borlaug was awarded Nobel Prize in 1970 for developing semidwarf varieties of wheat. His varieties brought revolution in wheat production the world over.

Q. 31. Who is known as the father of green revolution ?

Ans. Norman E. Borlaug is know as the father of green revolution.

Q. 32. What is the source of dwarfing gene in wheat ?

Ans. In wheat, Norin 10 line is the source of dwarfing gene which has been used in developing semidwarf varieties of wheat.

Q. 33. Who is known as the father of Indian green revolution ?

Ans. M.S. Swaminathan, an eminent plant breeder, is knwon as the father of green revolution in India.

Q. 34. What is Tift 23A ?

Ans. Tift 23A is an important source of cytoplasmic genic male sterility in pearl millet.

Q. 35. What is Kafir 60 ?

Ans. Kafir 60 is an important source of cytoplasmic genic male sterility in **Sorghum**.

Q. 36. What is Gregg 399 ?

Ans. Gregg 399 is an important source of genetic male sterility in cotton. It has been used in developing commercial hybrids in cotton.

Q. 37. What is the significance of Gossypium harknessii in cotton ?

Ans. In cotton, wild diploid species *Gossypium harknessii* is an important source of cytoplasmic genic male sterility. This CGMS source is being used in developing commercial hybrids of cotton.

Q. 38. What is source of dwarfing gene in rice ?

Ans. In rice, line Dee-Geo-Woogen is the source of dwarfing gene. It has been used in developing semi dwarf varieties of rice.

Q. 39. Who developed first cotton hybrid ?

Ans. The world's first cotton hybrid (H4) was developed by C.T. Patel in 1970 for commercial cultivation in Gujarat (India). For this outstanding achievement, CT Palel was awarded Padamshree.

Q. 40. Who is known as the father of hybrid cotton ?

Ans. C.T. Patel is known as the father of hybrid cotton and Gujarat is known as home of the hybrid cotton.

Q. 41. Who developed first rice hybrid ?

Ans. The world's first rice hybrid was developed in China in 1976 by Y.L. Ping et al. using cytoplasmic genic male sterility.

Q. 42. Who developed first pigeonpea hybrid ?

Ans. In pigeon pea, world's first hybrid (ICPH8) was developed in India in 1991 from International Crop Research Institute for Semiarid Tropics (ICRISAT), Hyderabad. It was developed using genetic male sterility.

Q. 43. Who developed first transgenic plant ?

Ans. The first transgenic plant was developed by Fraley et al. in 1983 in USA in Tobacco.

Q. 44. Who developed first transgenic cotton ?

Ans. The transgenic cotton plant was first developed by Monsanto Company of USA in 1997.

Q. 45. What is terminator gene ?

Ans. In transgenic plants, a gene which allows germination of seed in one generation only is called terminator gene. This gene leads to abortion of embryo resulting in termination of seed viability.

Q. 46. Who identified terminator gene ?

Ans. The terminator gene was first identified by Monsanto Company of USA in 1997.

Q. 47. What is traitor gene ?

Ans. In transgenic plants, a gene which responds to specific brand of fertilizers and insecticides only is known as traitor gene. Use of such gene will lead to monopoly of a company for sale of seed, fertilizers and insecticides.

Q. 48. Who identified traitor gene ?

Ans. The traitor gene was first identified by Monsanto Seed Company of USA in 1998. Use of this gene has not been permitted in plant breeding.

Q. 49. Who wrote famous book "Principles of Plant Breeding" ?

Ans. The famous book Principles of Plant Breeding was written by RW Allard of USA in 1960.

Q. 50. Who wrote the book Methods of Plant Breeding ?

Ans. The book Methods of Plant Breeding was written by Hays, Immer and Smith in 1955.

Q. 51. Who wrote famous book Breeding Field Crops ?

Ans. The book Breeding Field Crops was written by John Milton Poehlman in 1959 in USA.

Q. 52. Who wrote famous book Breeding Asian Field Crops ?

Ans. The book Breeding Asian Field Crops was written by JM Poehlman and Dhirendra Borthakur.

Q. 53. Who wrote the Book Principles of Crop Improvement ?

Ans. The book Principles of Crop Improvement was written by NW Simmonds, an English Plant breeder, in 1979.

Q. 54. Who wrote the book Principles of Cultivar Development ?

Ans. The book Principles of Cultivar Development was written by Walter Fehr, an American Plant Breeder, in 1987.

Q. 55. Who was TS Venkatraman ?

Ans. TS Venkatraman was a famous Indian Sugarcane breeder. He was Director of Sugarcane Breeding Institute, Coimbatore. He developed several high yielding varieties of sugarcane.

Q. 56. Who was BP Paul ?

Ans. BP Paul was an eminent wheat breeder. He developed several varieties of wheat. He was first Directer General of ICAR.

Q. 57. Who was Puskarnath ?

Ans. Puskarnath was a famous potato breeder. He developed several varieties of potato. He was Director of Central Potato Research Institute Simla.

Q. 58. Who was K. Ramiah ?

Ans. K. Ramiah was a renowned rice breeder. He developed several varieties of rice. He was Director of Central Rice Research Institute, Cuttack.

Q. 59. Who was Bosi Sen ?

Ans. Bosi sen was a famous maize breeder. He developed several maize varieties at Vivekanand Parvatiya Krishi Anusandhan Shala, Almora (Uttranchal).

Q. 60. Who was D.S. Athwal ?

Ans. D.S. Athwal was a pearlmillet breeder at Punjab Agricultural University, Ludhiana. He developed several varieties of pearl millet.

Q. 61. Who is M.S. Swaminathan ?

Ans. M.S. Swaminathan is a famous plant breeder who is responsible for green revolution in India. He was second Director General of ICAR. He was also Director General of International Rice Research Institute, Manila, Phillipines.

Q. 62. Who is N.G.P. Rao ?

Ans. He is an eminent sorghum breeder. He developed several hybrids in sorghum. He was vice chancellor of Marathwada Agricultural University, Parbhani and Chairman of Agricultural Scientist Recruitment Board, New Delhi.

Q. 63. Who was Dharam Pal Singh ?

Ans. Dharam Pal Singh was a famous oil seed breeder at Kanpur. He was vice chancellor of G.B. Pant University of Agriculture and Technology. Pant Nagar.

Q. 64. Who was C.T. Patel ?

Ans. C.T. Patel was a famous cotton breeder at Surat. He developed world's first cotton hybrid (H4) in 1970 for which he was awarded Padamshree. He is known as father of hybrid cotton.

Q. 65. What are major achievements of plant breeding in India ?

Ans. Major achievements of plant breeding in India are given below :

(i) Improvement in yield (ii) Improvement in quality (iii) Resistance to biotic and abiotic stresses (iv) Earliness (v) Wide adaptability (vi) Dwarfness, etc.

Q. 66. What are undesirable effects of plant breeding ?

Ans. Undersirable effects of plant breeding are given below.

(i) Reduction in diversity (ii) Narrow genetic base (iii) Danger of uniformity (iv) Sometimes undesirable combinations.

2

CROP GENETIC RESOURCES AND CENTRES OF DIVERSITY

Q. 1. What do you mean by plant genetic resources ?

Ans. Genetic material of plants which is of value as a resource for present and future generations of people is referred to as plant genetic resources.

Q. 2. What do you mean by gene pool ?

Ans. The whole library of different alleles of a species or sum total of genes in a species is known as gene pool, also called germplasm, genetic stock and genetic resources.

Q. 3. Who coined the term gene pool ?

Ans. The term gene pool was coined by Dobzhansky in 1951.

Q. 4. Who coined the term germplasm ?

Ans. The term germplasm was first used by Weismann in 1883.

Q. 5. What are different kinds of germplasm ?

Ans. Germplasm includes following plant materials.

(i) Land races (ii) Obsolete cultivars (iii) Modern cultivars (iv) Homozygous breeding materials (v) Wild races of cultivated species (vi) Wild species (vii) Mutants, etc.

Q. 6. What do you mean by land races ?

Ans. Primitive cultivars which were selected and cultivated by farmers for many generations are called land races. Main features of land races are given below. They have :

(i) More genetic diversity.
(ii) Wider adaptability, and broad genetic base.
(iii) High degree of resistance to biotic and abiotic stresses.

Q. 7. What are main drawbacks of land races ?

Ans. There are two main drawbacks of land races as follows.

(i) *Produce of land races is less uniform.*
(ii) Land races are low yielders.

Q. 8. What are obsolete cultivars ?

Ans. Improved cultivars of the recent past are known as obsolete cultivars.

Q. 9. What are modern cultivars ?

Ans. Currently cultivated high yielding varieties are known as modern cultivars. Modern cultivars are also known as improved cultivars or advanced cultivars. Modern cultivars constitute major part of working collectors and are extensively used in breeding programmes for developing productive cultivars.

Q. 10. What are different types of gene pool ?

Ans. Based on degree of relationship, Harlan and de Wet (1971) divided the gene pool into following three groups.

(i) Primary gene pool
(ii) Secondary gene pool
(iii) Tertiary gene pool.

Q. 11. What is primary gene pool ?

Ans. The gene pool in which intermating is easy and leads to production of fertile hybrids is known as primary gene pool. It includes genotypes (collections) of the same species and is designated as GP_1.

Q. 12. What is secondary gene pool ?

Ans. The genetic material which leads to partial fertility on crossing with primary gene pool is called secondary gene pool. It includes genotypes of related species and is designated as GP_2.

Q. 13. What is tertiary gene pool ?

Ans. The genetic material which leads to production of sterile hybrids on crossing with primary gene poal is called tertiary gene pool. It includes wild species of a crop and is designated as GP_3.

Q. 14. What are different types of seed collections ?

Ans. Based on use in crop improvement, seed collections are of following three types.

(i) Base Collections
(ii) Active Collections, and
(iii) Working Collections.

Q. 15. What do you mean by base collections ?

Ans. Total accessions available in a crop species are known as base collections. These are used for long term storage. Seeds are stored at −18 or −20°C. These are also known as principal collections.

Q. 16. What do you mean by active collections ?

Ans. Collections which are actively utilized in the breeding programmes are called active collections. These are used for medium term (8-10 years) storage. Seeds are stored at 0°C.

Q. 17. What do you mean by working collections ?

Ans. Collections which are frequently utilized by breeders in their crop improvement programmes are known as working collections. These are used for short term storage (3-5 years) and are stored at 5–10°C.

Q. 18. What are core collections ?

Ans. A set of accessions derived from base collections to represent the genetic spectrum of variability in the whole collections are called core collections.

Q. 19. Who proposed idea of core collections ?

Ans. The concept of core collections was proposed by Frankel in 1984.

Q. 20. What are various germplasm activities ?

Ans. There are six important germplasm activities as given below.

(i) Exporation and Collection, (ii) Conservation, (iii) Evaluation, (iv) Documentation, (v) Distribution, and (vi) Utilization.

Q. 21. What is in-situ conservation of germplasm ?

Ans. Protection of genetic diversity of crop plants from genetic erosion is called conservation. It is of two types, viz. in-situ conservation and ex-situ conservation. Conservation of germplasm under natural habitat is called in-situ conservation. It requires establishment of biosphere reserve and national parks for protection of endangered species.

Q. 22. What is ex-situ conservation ?

Ans. Conservation of germplanes in the gene banks is called ex-situ conservations. It has following advantages.

(i) It helps in conserving biodiversity of a species at one place.
(ii) Handling of germplasm is easy.
(iii) It is cheap method of germplasm conservation.

Q. 23. Who classified seed into orthodox and recalcitrant groups ?

Ans. Roberts (1973) classified seeds into two groups, viz. orthodox and recalcitrant for storage purposes.

Q. 24. What are orthodox seeds ?

Ans. Seeds which can be dried to low moisture contents and stored at low temperature without losing their viability are known as orthodox seeds. This group includes seeds of following crops.

Maize	Soybean	Cotton
Wheat	Greengram	Sunflower

Rice	Blackgram	Beans
Barley	Cowpea	Egg plant
Chickpea	Field Pea	Brassicas
Lentil		

Q. 25. What are recalcitrant seeds ?

Ans. Seeds which show very drastic loss in viability with decrease in moisture content below 12 or 13% are called recalcitrant seeds. Seed of such crops can not be stored in seed banks. This group includes following crops.

Coconut	Jack fruit
Tea	Rubber
Coffee	Oil palm etc.

Q. 26. When was plant introduction work started in India ?

Ans. In India, plant introduction work was started in 1946 in the Division of Botany, IARI, New Dlehi.

Q. 27. When was Division of Plant Introduction established in India ?

Ans. In India, Division of Plant Introduction was establishes in 1961 under the leadership of Dr. Har Bhajan Singh at IARI, New Delhi.

Q. 28. What is NBPGR ?

Ans. NBPGR refers to National Bureau of Plant Genetic Resources, New Delhi.

Q. 29. When was NBPGR established ?

Ans. NBPGR was established in 1976 at IARI, New Delhi. Infact, in 1976, the Division of Plant Introduction was elevated to the status of Independent Institute known as National Bureau of Plant Genetic Resources.

Q. 30. What are main functions of NBPGR ?

Ans. Main functions of NBPGR are given below :

(i) To coordinate germplasm collection and conservation work.

(ii) It deals with export and import of germplasm of agricultural and horticultural crops.

(iii) To organize short term training course on various activities of germplasm.

(iv) *To provide consultancy and guidance for creating cold storage facilities.*

(v) To take decision for setting up gene sanctuaries of endangered species.

Q. 31. What is IBPGR ?

Ans. IBPGR refers to International Board for Plant Genetic Resources, Rome, Italy. It was established in 1974 to coordinate global work of germplasm collection and conservation.

Q. 32. What is IPGRI ?

Ans. IPGRI refers to International Plant Genetic Resources Institute, Rome Italy. IPGRI was established in October 1991. However, it started functioning in January 1994. In fact IBPGR, which was established in 1974 was renamed as IPGRI in January 1994. Now IPGRI coordinates the work of germplasm collection, conservation, exchange etc. at global level.

Q. 33. Who was N.I. Vavilov ?

Ans. N.I. Vavilov was a famous Russian Plant Explorer.

Q. 34. Who was Har Bhajan Singh ?

Ans. Dr. Har Bhajan Singh was a famous India Plant Explorer/ germplasm curator.

Q. 35. What is Plant Quarantine ?

Ans. Quarantine refers to the prophylactic measures which are used to prevent the entry of new diseases, insects and weeds from other countries.

Q. 36. What are quaratine organisations in India ?

Ans. Quarantine inspection offices are located at all seaports and International Airports. In India quarantine inspection is done by following organisations.

(i) **NBPGR**. It deals with quarantine Inspection of Agricultural and Horticultural crops.

(ii) **Forest, Research Institute Dehradun.** It deals with quarantine inspection of propagules of forest trees.

(iii) **Botanieal Survey of India, Kolkata**. It deals with quarantine inspection of medicinal plants and other plant species.

(iv) **Directorate of Plant Protection, Quarantine and Storage, Faridabad.** It deals with quarantine inspection of food grains and other products imported for human consumption.

Q. 37. What is genetic diversity ?

Ans. Genetic diversity refers to total amount of genetic variation present in a population or species. In other words, variety of genes and genotypes found in a particular crop species is known as genetic diversity.

Q. 38. What is genetc erosion ?

Ans. Gradual reduction in genetic variability due to elimination of various genotypes from a population is called genetic erosion. It results due to replacement of land races with improved cultivars, modernization of agriculture and various developmental activities.

Q. 39. What is extinction ?

Ans. Extinction refers to permanent loss of a crop species due to various reasons.

Q. 40. What is Biodiversity ?

Ans. Biodiversity refers to total variability present within and among species of all living organisms and their habitats. In other words, biodiversity is sum of genes, species and ecosystem in a region. Thus biodiversity is of following three types.

(i) genetic diversity (ii) species diversity, and (iii) ecosystem diversity

Q. 41. What is significance of species biodiversity ?

Ans. Diversity of species offers opportunities for people to select species of their choices for food, cloth, shelter and medicinal uses. Diversity in the species has offered items of food, fuel, cloth, timber and medicines to human beings. Crop diversity provides an insurance against crop losses caused due to diseases and insects.

Q. 42. What is CBD ?

Ans. CBD refers to convention on Biological Diversity. The CBD was signed in 1992 and became operational in December 1993 with following objectives.

(i) Conservation of biodiversity

(ii) Sustainable use of biodiversity, and

(iii) Exchange of genetic resources for benefits of all countries.

Q. 43. What is centre of diversity ?

Ans. A place, region or area where maximum variability of crop plants is observed is called centre of diversity, also called centres of origin. Centres of diversity are of three types, viz. primary centres, secondary centres and microcentres.

Q. 44. What are primary centres of diversity ?

Ans. Original homes of crop plants which are generally uncultivated areas such as mountains, hills, river valleys, forests etc. are called primary centres of diversity. Main features of these centres are given below.

(i) They have wide genetic diversity

(ii) Have large number of dominant genes

(iii) Mostly have wild characters

(iv) Exhibit less crossing over

(v) Natural selection operates

Q. 45. What are secondary centres of diversity ?

Ans. Cultivated areas with vast genetic diversity are called secondary centres of diversity. Main features of these centres are given below.

(i) Have lesser genetic diversity than primary centres.

(ii) Have large number of recessive genes.

(iii) Mostly have desirable characters.

(iv) Exhibit more crossing over.
(v) Both natural and artificial selections operate.

Q. 46. What are micro-centres of genetic diversity ?

Ans. Small areas within the centre of diversity that exhibit tremendous genetic diversity of crop plants are known as micro centres. Main features of micro-centres are given below.

(i) *They represent small areas within the centres of diversity.*
(ii) Exhibit tremendous genetic diversity.
(iii) *The rate of evolution in these centres in faster than larger areas.*
(iv) They are important sites for study of crop evolution.

Q. 47. Who coined the term micro-centre ?

Ans. The term micro-centre was first used by Harland in 1948.

Q. 48. What are differences between centres of origin and centres of diversity ?

Ans. Main differences between centres of origin and centres of diversity are given below.

(i) *Centres of origin are geographical areas where crop plants have originated.*
(ii) Centre of diversity refers to a location where vast genetic variability for a crop and its variability is found.

Thus, centre of origin and centre of diversity for a crop may be the same or different.

Q. 49. What is law of parallel variation ?

Ans. This law states that a particular variation observed in a crop species is also expected to be available in another related species; also called law of homologous series of variation.

Q. 50. Who proposed law of parallel variation ?

Ans. The concept of parallel variation was developed by N.I. Vavilov in 1951.

Q. 51. What are Vavilovian Centres of diversity ?

Ans. Based on his studies of global exploration and collection, Vovilov proposed eight main centres and three subsidiary centres of diversity as follows.

(i) Main Centres. Include China, Hindustan, Central Asia, Persia, Mediterranea, Abyssinia, Mexico and South America.

(ii) Subsidiary Centres. Indo-Malaysia, Chile, Brazil and Paraguya.

Q. 52. What are gene sanctuaries ?

Ans. Protected areas of great genetic diversity under natural conditions (*i.e.*

in-situ conservation) are called gene sanctuaries. Gene sanctuary is also known as natural park or biosphere reserve. India has set up its first gene sanctuary in the Garo Hills of Assam for wild relatives of citrus. Efforts are also being made to set up gene sanctuaries for banana, sugarcane, rice and mango.

Q. 53. What are advantages of gene sanctuaries ?

Ans. Main advantages of gene sanctuaries are given below.

(i) *It protects the loss of genetic diversity caused by human intervention.*

(ii) It allows natural selection and evolution to operate.

Q. 54. What are drawbacks of gene sanctuaries ?

Ans. There are two main drawbacks of gene sanctuaries as given below.

(i) *Entire variability of a crop species can not be conserved.*

(ii) Its maintenance and establishment is a difficult task.

Q. 55. What do you mean by gene banks ?

Ans. A place or organisation where germplasm is conserved in living state is called gene bank; also known as germplasm bank. Gene banks are of two types, viz. seed gene banks and field gene banks.

Q. 56. What is seed gene bank ?

Ans. A place where germplasm is conserved in the form of seeds is called seed gene bank. Main advantages of seed gene banks are given below.

(i) *Whole variability can be conserved in a small space.*

(ii) Handling of germplasm is easy.

(iii) *Germplasm is conserved under pathogen and insect free environment.*

Q. 57. What are demerits of seed gene bank ?

***Ans.** Main drawbacks of seed gene bank are given below.*

(i) Seeds of recalcitrant species can not be stored.

(ii) *It requires constant power supply.*

(iii) It requires periodical evaluation of viability and fresh seeds for storage.

Q. 58. What are field gene banks ?

Ans. Those areas of land in which germplasm of recalcitrant crop species is maintained in the form of plants are called field gene banks; also called plant gene banks. Field gene banks are maintained for crops like coconut, Rubber, Mango, Cassava, Yam, Cocoa etc.

Q. 59. What are advantages of field gene banks ?

Ans. Main advantages of field gene banks are given below.

(i) Field gene banks provide opportunities for continuous evaluation of germplasm for various economic characters.

(ii) Also provide opportunities for utilization of germplasm in breeding programmes.

Q. 60. What are drawbacks/limitations of field gene banks ?

Ans. Main demerits of field gene banks are given below.

(i) The entire variability can not be conserved.

(ii) The germplasm is exposed to pathogens and insects. Also exposed to natural disasters such as bush fire, cyclones, floods etc.

(iii) Maintenance is a costly affair.

Q. 61. What is cryopreservations ?

Ans. Long term conservation of plant tissues (meristems) at extremely low temperature *i.e.*–196°C in liquid nitrogen is called cryopreservation. This method is widely used for conservation of horticultural species.

Q. 62. What is characterization ?

Ans. Recording of highly heritable phenotypic characters in germplasm collections is called characterization. Evaluation refers to recording of both oligogenic and polygenic characters.

3

CROP BIODIVERSITY AND INDIAN LEGISLATION

Q.1. What Biodiversity ?

Ans. The biodiversity refers to the variety of life in all forms, levels and combinations. The term biodiversity includes genetic diversity, species diversity , and ecosystem diversity. In other words, the variety of life forms -the different plants, animals and micro organisms, the genes they contain and ecosystems they may form is referred to as biodiversity. It is usually considered at three levels, viz., genetic diversity, species diversity and ecosystem diversity. Crop biodiversity refers to the variety of genes and genotypes found in crop plants. Biodiversity is a modern term which simply means" the variety of life on earth". This variety of life can be measured in several ways.

Q.2. What are different levels of Biodiversity ?

Ans. There are three levels of biodiversity, viz., genetic diversity, species diversity , and ecological diversity. These levels are all interrelated yet distinct enough that they can be studied as three separate components.

Q.3. Define Genetic Diversity.

Ans. Genetic diversity refers to the variation between individuals of the same species. This includes genetic variation between individuals in a single population, as well as variations between different populations of the same species. Genetic differences can now be measured using increasingly sophisticated techniques. These differences are the raw material of evolution. Genetic diversity is the variety present at the level of genes. Genes are made of DNA which determines expression of a trait in an individual. This level of diversity can differ by alleles (different variants of the same gene, such as yellow and white flower colour), by entire genes or by several genes.

Genetic diversity can be measured at many different levels such as population, species and community. The genetic diversity is important at each of these levels. The main points related to genetic diversity are listed below.

(i) It represents variation between individuals of the same species.
(ii) It represents the variation at gene level.

(iii) It can be measured at population, species and community levels.
(iv) Sophisticated techniques are required for its measurement.
(v) It represents the raw material for evolution and adaptation.
(vi) It is effected by the environment and competition with other species.

Q.4. What is the relationship between Genetic diversity and Adaptation ?

Ans. The amount of diversity at the genetic level is important because it represents the raw material for evolution and adaptation. More genetic diversity in a species or population means a greater ability for some of the individuals in it to adapt to changes in the environment. Less diversity leads to uniformity resulting in poor adaptation. For example, modern varieties of wheat and rice are high yielding and highly uniform.We can harvest a good crop from such varieties, but can be a problem when a disease or parasite attacks the field, as every plant in the field will be susceptible having the identical genetic constitution. Thus monocultures are unable to deal well with changing conditions.

Q.5. What the effect of environment on Genetic Diversity ?

Ans. Within species, genetic diversity often increases with environmental variability, which can be expected. If the environment often changes, different genes will have an advantage at different times or places. In this situaition genetic diversity remains high because many genes are in the population at any given time. If the environment didn't change, then the small number of genes that had an advantage in that unchanging environment would spread at the cost of the others, causing a drop in genetic diversity.

Q.6. What is the relationship between Genetic Diversity and Communities ?

Ans. In communities, it can increase with the diversity of species. How much it increases depends not only on the number of species, but also on how closely related the species are. Species that are closely related *(e.g.* two species of wheat) have similar genetic structures and makeup and therefore. do not contribute much additional genetic diversity. These closely related species will contribute to genetic diversity in the community less than more remotely-related species *(e.g.* a wheat and a barley) would.

An increase in species diversity can also affect the genetic diversity. If there are many species, the genetic diversity at that level will be larger than when there are fewer species. On the other hand, genetic diversity within each species can decrease. This can happen if the large number of species means so much competition that each species must be extremely specialized which will lead to little genetic diversity within any of the species.

Q.7. What is Species Diversity ?

Ans. Species diversity refers to the variety of species in a given region or area.

This can either be determined by counting the number of different species present, or by determining taxonomic diversity. Taxonomic diversity is more precise and considers the relationship of species to each other. *For* example, an area containing three species of wheat and two species of barley, is more diverse than an area containing five species of wheat, even though they both contain the same number of species. High species biodiversity is not always necessarily a good thing. For example, a habitat may have high species biodiversity because many common and widespread species are invading it at the expense of species restricted to that habitat.

The species level crop biodiversity is the easiest to work on both from practical and theoretical point of view. Species are relatively easy to identify by eye in the field, whereas genetic diversity requires laboratories, time and resources to identify. Species are well known and are distinct units of diversity. Each species can be considered to have a particular "role" in the ecosystem, so the addition or loss of single species may have consequences for the system as a whole. The main points related to species diversity are listed below.

(i) It represents variety of species in a given region or area.
(ii) It identification is easy by naked eye in the field.
(iii) It can be determined simple by counting the number of different species.
(iv) Species are distinct units of diversity.

Q.8. What is Ecosystem Diversity ?

Ans. Ecosystem diversity refers to communities of plants and animals, together with the physical characteristics of their environment (e.g. geology, soil and climate). Ecosystem diversity is more difficult to measure because there are rarely clear cut boundaries between different ecosystems because they overlap into one another. However, if consistent criteria are chosen to define the limits of an ecosystem, then their number and distribution can also be measured.

Ecosystem-level diversity deals with species distributions and community patterns, the role and function of key species, and combines species functions and interactions. This is the least-understood level of the three types of biodiversity due to the complexity of the interactions.

One of the difficulties in examining communities is that the transitions between them are usually not very sharp. A lake may have a very sharp boundary between it and the deciduous forest it is in, but the deciduous forest will shift much more gradually to grasslands or to a coniferous forest. This lack of sharp boundaries is known as "open communities" (as opposed to "closed communities," which would have sudden transitions) and makes studying ecosystems difficult. The main points related to ecosystem biodiversity are given below.

(i) It represents communities of plants and animals together with physical structures of their environment

(ii) The measurement of ecosystem diversity is more difficult than genetic diversity and species diversity.

(iii) The assessment of ecosystem diversity needs many complex measurements to be taken over a long period of time

(iv) It is a time consuming task and completion of such work requires adequate staff and financial support.

Q.9. What are the differences between genetic diversity and species diversity?

Ans. There are several differences between genetic diversity and species diversity which are presented in Table 3.1

TABLE 3.1 Comparison of genetic biodiversity and species biodiversity

SN	Particulars	Genetic Diversity	Species Diversity
1.	Represents	Variation between individuals of same species.	Variety of species in a given region.
2.	Identification	Difficult	Easy
3.	Assessment	Requires sophisticated laboratory techniques.	Can be assessed by counting the species.
4.	Conservation	Possible in gene banks	Possible in gene banks.
5.	Environmental effects	Present	Present

Q.10. What are Advantages of Crop Biodiversity ?

Ans. There are several advantages of crop biodiversity. Some important advantages or benefits of crop biodiversity are briefly presented as follows. The crop biodiversity provides:

(i) Protection from biotic stresses such as diseases and insects.

(ii) Protection from abiotic stresses such as drought, soil salinity , soil alkalinity , soil acidity, cold temperature, heat etc.

(iii) Choice of quality for food and other items,

(iv) Choice of items for food, fibre, oil, fuel, fodder to animals, timber and medicines.

(v) Wider adaptation to environmental changes.

(vi) Broad genetic base to the populations.

Q.11. What are the reasons for Loss of crop biodiversity ?

Ans. There are two main reasons of depletion of biodiversity in crop plants, viz., extinction and genetic erosion. Extinction refers to permanent loss of a crop species due to various reasons. Genetic erosion refers to the gradual reduction in genetic variability, in population of a species, due to elimination of various genotypes. There are five main reasons of genetic erosion as follows

(1) Replacement of Land Races by Modern Cultivars
(2) Industrial agriculture
(3) Farming into wild habitat.
(4) Clean Cultivation.
(5) Developmental Activities such as development of national high ways, towns, cities, airports, seaports etc.

Q.12. What is the need for Conservation of Biodiversity ?

Ans. Conservation refers to protection of genetic diversity of crop plants from genetic erosion. The loss of biodiversity poses a serious threat to agriculture and the livelihoods of millions of people. Conserving biodiversity and using it wisely is a global necessity. Biodiversity provides the foundation for our agricultural systems. It provides the sources of traits to improve yield, quality , resistance to pests and diseases and adapt to changing environmental conditions, such as global warming. The protection of plant diversity is essential for food security and ecological well-being. Biodiversity is also a direct source of food for many people and is a essential part of our life support system. Without biodiversity our ecosystems, the planet's entire biosphere, cannot function.

Q.13. What are approaches of conserving biodiversity ?

Ans. There are different approaches to conserving biodiversity and different ways of using genetic resources. These are (i) on farm management, (ii) In–situ conservation, (iii) Ex–situ conservation and (iv) complementary conservation.

Q.14. Describe on farm management of biodiversity.

On farm management involves the maintenance of crop species on the farm or in home gardens. The effectiveness of strategies to maintain and use crop or livestock diversity on farms depends on the extent to which local varieties continue to meet the needs of farmers and communities. Many plant genetic resources, especially those of minor crops, are managed as part of agricultural production systems. This type of biodiversity conservation has been termed 'conservation through use'. There are important reasons for supporting on farm maintenance of crop and livestock diversity:

It ensures the ongoing processes of evolution and adaptation of crops to their environments.

1. It allows for the continued selection of superior material by farmers that meets their needs and preferences.

2. It helps preserve indigenous knowledge, strengthens local institutions and promotes farmers' participation in national biodiversity conservation programmes.
3. It provides a necessary backup to gene bank collection.
4. It provides natural laboratories for agricultural research.

The biodiversity of only those species can be maintained on the cultivators' farms in home gardens which have greater use to the farmers and rural communities.

Q.15. What is In situ *conservation* ?

Ans. *In situ* (=on-site) conservation and use refers to the maintenance and use of wild plant populations in the habitats where they naturally occur and have evolved without the help of human beings. In other words, it refers to conservation of biodiversity under natural habitats. In this method, the wild species and the complete ecosystems are preserved together.

The wild populations regenerate naturally, and are dispersed naturally by wild animals, winds and in water courses. The evolution is purely driven by environmental pressures and any changes in one component affect the other.

The main focus of Biodiversity's work is on the maintenance of the genetic diversity of the wild species and wild relatives of crop plants in the protected areas (national parks and nature reserves).

Q.16. What are advantages and disadvantages of in-situ conservation of biodiversity ?

Ans. The advantages and disadvantages of in-situ conservation of biodiversity are presented below.

Advantages

1. The biodiversity is conserved under natural habitats.
2. It improves of the populations.
3. It conserve complete ecosystem means several species together.
4. It protects endangered species in their original habitats.

Disadvantages

1. It requires establishment of gene sanctuaries and national parks..
2. Several areas have to be conserved to protect a single species.
3. The management of protected areas poses several problems.
4. It is costly method of biodiversity conservation.

Q.17. What is Ex *situ* conservation of biodiversity ?

Ans. *Ex situ* (= off-site) conservation of germplasm takes place outside the natural habitat or outside the production system, in facilities specifically created for this purpose. Depending on the type of species to be conserved, different *ex situ* conservation methods may be used. The gene banks are used for ex situ conservation of biodiversity. In gene banks, biodiversity is conserved for its sustainable use by breeders, farmers and researchers for agricultural development.

To make the genetic resources useful to farmers, breeders and researchers, gene bank managers must carefully document the collected materials, make the information available and establish a transparent and safe system for its distribution. Such material should be used by breeders and other researchers for agricultural development.

Q.18. What are the advantages and disadvantages of Ex situ conservation of biodiversity ?

Ans. The advantages and disadvantages of Ex situ conservation of biodiversity are given below.

Advantages

1. One species can be conserved at one place.
2. The biodiversity is conserved either in seed gene banks or in field gene banks.
3. Handling of material is easy.
4. It requires less space for conservation.

Disadvantages

1. It requires establishment of seed gene banks and field gene banks.
2. It require regular funding for maintenance of gene banks.
3. It involves high technical man power.
4. It requires periodical rejuvenation in case of seed crops.

Q.19. What are differences between In-situ and Ex-situ conservation of biodiversity ?

Ans. There are several differences-between In-situ and Ex-situ conservation of biodiversity which are presented in Table 3.2

Table 3.2 Comparison of In situ and Ex situ conservation

SN	Particulars	In Situ Conservation	Ex situ Conservation
1.	Conservation place	Natural habitats	Gene banks
2.	Basic requirements	Gene sanctuaries and national parks	Seed Gene banks and field gene banks
3.	Skill required	Moderate	High
4.	Species conserved	Several	Few at one place

5.	Initial cost involved	Moderate	High
6.	Management	Difficult	Easy
7.	Ecosystem conservation	Possible	Not possible

Q.20. Describe briefly Complementary conservation of biodiversity

Ans. It involves a proper combination of different conservation approaches for the sustainable use (in present and future) of genetic diversity existing in a target gene pool.

A proper combination of different methods depends on the species being conserved, the local infrastructure and human resources, the number of accessions in a given collection, its geographic site and intended use of the conserved germplasm. It does not advocate a particular method. A good complementary conservation strategy does not categorize crops or species into definitive classes. It is dynamic, and lends itself to meet the challenges of changes that are occurring in the field of genetic resources as it is open to new technologies and new needs.

Q.21. What is Convention on Biological Diversity (CBD) ?

Ans. The Convention on Biological Diversity is also known as the Biodiversity Convention. It is an international treaty that was adopted in Rio de Janeiro in June 1992. The Convention was opened for signature at the Earth Summit in Rio de Janeiro on 5 June 1992 and entered into force on 29 December 1993. The biodiversity agreement has been signed by 189 countries.

Q.22. What are the main goals of CBD ?

Ans. The Convention has following three main goals:

1. conservation of biological diversity (or biodiversity):
2. sustainable use of its components; and
3. fair and equitable sharing of benefits arising from genetic resources.

Q.23. What is Cartagena Protocol ?

Ans. The Cartagena Protocol also known as the Biosafety Protocol, was adopted in January 2000. The Biosafety Protocol seeks to protect biological diversity from the potential risks posed by living modified organisms resulting from modern biotechnology.

The Biosafety Protocol makes clear that products from new technologies must be based on the precautionary principle and allow developing nations to balance public health against economic benefits. It will allow countries to ban imports of a genetically modified organism if or product which is not safe for use.

Q.24. What are the main functions of the Conference of the Parties (COP) ?

Ans. The convention's governing body is the **Conference of the Parties** *(COP)*, consisting of all governments (and regional economic integration organizations) that have ratified the treaty. The main functions of *COP* are to:

(i) review progress under the Convention,
(ii) identifies new priorities,.
(iii) sets work plans for members,
(iv) make amendments to the Convention,
(v) create expert advisory bodies,
(vi) review progress reports by member nations, and
(vii) collaborate with other international organizations and agreements.

The Conference of the Parties uses expertise and support from several other bodies that are established by the convention. In addition to committees or mechanisms established on an ad hoc basis, two main organs are Secretariat and SBSTT A which are discussed below. :

Q.25. What are the function of the Secretariat of CBD ?

Ans. The CBD Secretariat is based in Montreal (Canada). It operates under the United Nations Environment Programme. The main functions of the secretariat are to:

(i) organize meetings,
(ii) draft documents,
(iii) assist member governments in the implementation of the programme of work,
(iv) coordinate with other international organizations, and
(v) collect and disseminate information.

Q.26. What is the role SBSTTA ?

Ans. The Subsidiary Body on Scientific, Technical and Technological Advice (SBSTTA) is a committee composed of experts from member governments competent in relevant fields. It plays a key role in making recommendations to the *COP* on scientific and technical issues. Thirteenth Meeting of the Subsidiary Body on Scientific, Technical and Technological Advice(SBSTTA-13) was held from 18 to 22 February 2008 in the FAO at Rome, Italy. SBSTTA-13 delegates met in the Committee of the Whole in the morning to finalize and adopt recommendations on the in-depth reviews of the work programmes on agricultural and forest biodiversity and SBSTT A' s modus operandi for the consideration of new and emerging issues.

Q.27. What is Indian Biodiversity Legislation ?

Ans. In India the Biological Diversity Act (also called biodiversity act) was passed by the Central Government in 2002. The purpose of The Biodiversity Act -2002 is to protect India's rich biodiversity (genetic resources) and associated knowledge against their use by foreign individuals, institutions or companies without sharing the benefits arising out of such use, and check bio-piracy.

Q.28. Describe the Structures of Indian Biodiversity Act -2002 ?

Ans. The biodiversity Act provides for establishment of a three tiered structure at the national, state and local level. It includes (i) National Biodiversity Authority (NBA), (ii) State Biodiversity Boards (SBB), and (iii) Biodiversity Management Committees (BMCs).

Q.29. What is the role of National Biodiversity Authority (NBA) ?

Ans. It deals with all matters relating to requests for access by foreign individuals,. institutions or companies, and all matters relating to transfer of results of research to any foreigner.

Q.30. What are the functions of State Biodiversity Boards (SBB) ?

Ans. All matters relating to access by Indians for commercial purposes will be under the purview of the State Biodiversity Boards (SBB). The Indian industry will be required to provide prior intimation to the concerned SBB about the use of biological resource. The State Board will have the power to restrict any such activity , which violates the objectives of conservation, sustainable use and equitable sharing of benefits.

Q.31. What is the role of Biodiversity Management Committees (BMCs) ?

Ans. Institutions of local self government will be required to set up Biodiversity Management Committees in their respective areas for conservation, sustainable use, documentation of biodiversity and chronicling of knowledge relating to biodiversity. NBA and SBBs are required to consult the concerned BMCs on matters related to use of biological resources and associated knowledge within their jurisdiction.

Q.32. Is there need for seeking permission of the National Biodiversity Authority for carrying out research ?

Ans. There is no need for seeking permission of the National Biodiversity Authority for carrying out research, if it is carried out in India by Indians, as well as under collaborative research projects that have been drawn within the overall policy guidelines formulated by the Central Government. The only situations that would require permission of the NBA are: (i) when the results of any research which has made use of the country's biodiversity is sought to be commercialized, (ii) when the results of research are shared with a foreigner or foreign institution, and (iii) when a foreign institution/ individual wants access to the country's biodiversity for undertaking research.

Q.33. What are the Exemptions of Indian Biodiversity Act 2002 ?

Ans. The legislation provides for the following exemptions.

(i) Exemption to local people and community of the area for free access to use biological resources within India.

(ii) Exemptions to growers and cultivators of biodiversity and to vaids and Hakims to use biological resources.

(iii) Exemption through notification of normally traded commodities from the purview of the Act.

(iv) Exemption for collaborative research through government sponsored or government approved institutions subject to overall policy guidelines and approval of the Central Government.

(v) The value added products have been excluded from the definition of biological resources.

(vi) The Indian researchers neither require prior approval nor need to give prior intimation to SBB for obtaining biological resource for conducting research in India.

Q.34. What is the impact of Biodiversity Act on Indian industry ?

Ans. The Indian industry is required to give prior intimation to the concerned SBB about obtaining the biological resources for commercial purposes. The SBB will have the power to prohibit or restrict any such activity , which violates the objectives of conservation, sustainable use and equitable sharing of benefits.

Q.35. What is the impact of Biodiversity Act on Indian Ayurvedic industry ?

Ans. The Act does not aim at banning the use of medicinal plants. It 'provides that for commercial use of resources and related knowledge by Indians only, prior intimation to the State Biodiversity Board is required. Hakims and vaids will continue to have free access to resources and knowledge.

Q.36. Is there overlapping of Biodiversity Bill and Plant Varieties Protection Act?

Ans. There is no overlap between Biodiversity Bill and Plant Varieties Protection (PVP). Act The scope and objectives of these two legislations are different. The PVP legislation, accords intellectual property rights to a person for developing a new plant variety. On the other hand; the biodiversity legislation is primarily aimed at regulating access to biological resources and associated knowledge so as to ensure equitable sharing of benefits arising from their use.

The Act provides that prior approval of NBA is necessary before applying for any kind of IPRs based on any research or information on a biological resource obtained from India. However, in case of patents, permission of the NBA may be obtained after acceptance of the patent but before sealing of the patent. This has been done to protect the priority date of the patent applicant.

Q.37. What are Heritage sites ?

Ans. The Act provides for designating heritage sites. These are areas of biodiversity importance, which harbour rich biodiversity, wild relatives of crops, or areas, which lie outside the protected area network. The purpose is not to cover the already designated protected areas such as national parks and wildlife sanctuaries.

Q.38. What is the provision for Threatened species ?

Ans. The Act (Section 38) provides for notifyjng threatened species and prohibits or regulates their collection. It also provides *for* taking appropriate steps to rehabilitate and preserve those species, thereby ensuring their conservation and management.

Q.39. What is the Check on Bio-piracy by Indian biodiversity Act 2002 ?

Ans. To check bio piracy, the proposed Legislation provides that access to biological resources and associated knowledge is subject to terms and conditions, which secure equitable sharing of benefits. Further, it would be required to obtain the approval of the National Biodiversity Authority before seeking any IPR based on biological material and associated knowledge obtained from India.

Q.40. Who are the Benefit claimers of biological resources ?

Ans. The benefit claimers are conservers of biological resources, creators and holders of knowledge and information relating to the uses of biological resources. While granting approvals, NBA will impose conditions, which secure equitable share in the benefits arising out of the use of biological resources occurring in India or knowledge relating to them. These benefits could include monetary gains, grant of joint ownership of IPRs, transfer of technology, association of Indian Scientists in R&D, setting up of venture capital fund etc.

In cases where specific individuals, or group of individuals are identifiable, the monetary benefits will be paid directly to them. Otherwise, the amount will be deposited in the National Biodiversity Fund.

MODES OF REPRODUCTION AND MATING SYSTEMS

Q. 1. What is reproduction ?

Ans. The process by which living organisms give rise to the offspring of similar kind (species) is called reproduction. In crop plants the mode of reproduction is of following two types.

(i) Asexual reproduction. It refers to multiplication of plants without the fusion of male and female gometes. It is of two types, viz. vegetative reproduction and apomixis.

(ii) Sexual reproduction. It refers to multiplication of plants through fertilized embryos.

Q. 2. What is vegetative reproduction ?

Ans. Multiplication of plants by means of various vegetative plant parts such as stem cutting and root cuttings. It is of two types, viz. *natural* and artificial.

Q. 3. What is natural vegetative reproduction ?

Ans. Multiplication of plants in nature by vegetative means is called natural vegetative reproduction. It may occur by underground stems, subaerial stems, root and bulbs.

(i) Underground stems. These are of four types, viz. Rhizomes, tuber, corm and bulb. Rhizomes are found in turmeric and ginger, tuber in potato, corm in Arvi and Bunda and bulb in garlic and onion.

(ii) Sub-aerial stems. This group includes runner, sucker, stolen etc. and are found in mint, rose, strawberry, banana etc.

(iii) Bulbills. These are modified form of flower and are found in garlic. They develop into plants when fall on the ground.

Q. 4. What is artificial vegetative reproduction ?

Ans. Multiplication of plants by vegetative parts through artificial method is known as artificial vegetative reproduction. It occurs by stem cutting, root cutting, layering, grafting and gootee.

Q. 5. What are advantages of vegetative reproduction ?

Ans. Advantages of vegetative reproduction are given below :

(i) It maintains genetic purity of a genotype generation after generation.
(ii) It is useful in mass multiplication of a genotype.
(iii) In plant breeding, desirable plants can be selected and maintained by vegetative reproduction.
(iv) Bud mutation can be directly released as new varieties.

Q. 6. How reproduction occurs in sugarcane, grapes and roses ?

Ans. In sugarcane, grapes and roses reproduction occurs through stem cuttings.

Q. 7. How reproduction occurs in sweet potato, citrus and lemon ?

Ans. In sweet potato, citrus and lemon, reproduction occurs by stem cuttings.

Q. 8. What is the main drawback of asexual reproduction ?

Ans. The main drawback of asexual reproduction is that it does not allow combining of desirable genes from different sources.

Q. 9. What is amphimixis ?

Ans. The union of male and female gametes is known as amplimixis.

Q. 10. Who coined the term amphimixis ?

Ans. The term amphimixis was first used by Weismann in 1891.

Q. 11. What is the significance of sexual reproduction ?

Ans. The importance of sexual reproduction is given below :

(i) It offers opportunities to combine desirable genes into a genotype from different sources.
(ii) It creates genetic variability by hybridization and crossing over. Crossing over takes place during meiosis and leads to new gene combinations. Crossing over does not take place in asexually propagated species.

Q. 12. What is autogamy ?

Ans. Development of seed by self pollination is known as autogamy; also called self pollination. Main features of autogamy are given below :

(i) It is the closest form of inbreeding.
(ii) It leads to homozygosity.
(iii) Autogamous species develop homozygous balance and do not exhibit significant inbreeding depression.

Q. 13. What are mechanism which promote autogamy ?

Ans. The following mechanisms promote autogamy.

(i) Presence of bisexual or hermaphrodite flowers (bisexuality)
(ii) Maturation of anthers and stigma of a flower at the same time. (Homogamy)
(iii) Pollination and fertilization in unopened flower bud (Cleistogamy)

(iv) Pollination immediately after flowering (Chasmogamy)
(v) Surrounding of stigma by anthers.

Q. 14. What is homogamy ?

Ans. Maturation of anthers and stigma of a flower at the same time in called homogamy. As a rule, homogamy is essential for self pollination.

Q. 15. What is cleistogamy ?

Ans. Cleistogamy refers to pollination and fertilization in unopened flower bud. It ensures self pollination and prevents cross pollination. It has been reported in some varieties of wheat, barley, oats and several other grass species.

Q. 16. What is chasmogamy ?

Ans. In plants, fertilization after opening of flower is called chasmogamy. It promotes self pollination and is found in wheat, barley, rice and oats.

Q. 17. What is bisexuality ?

Ans. Presence of male and female organs in the same flower is known as bisexuality. All self pollinated species have hermaphrodite flowers.

Q. 18. What is semigamy ?

Ans. When male nucleus enters the egg cell but does not fuse with egg nucleus, it is known as semigamy.

Q. 19. Who coined the term semigamy ?

Ans. The term semigamy was coined by Battaglia in 1963.

Q. 20. What is allogamy ?

Ans. Development of seed by cross pollination is called allogamy. Main features of allogamy are given below :

(i) It is the common form of outbreeding.
(ii) It leads to heterozygosity.
(iii) Allogamous species develop heterozygous balance and exhibit significant inbreeding depression on selfing.

Q. 21. What are mechanisms which promote allogamy ?

Ans. The following mechanisms promote allogamy :

(i) *Dictiny.* It refers to unisexual flowers.
(ii) *Dichogamy.* It refers to maturation of anthers and stigma of the same flower at different times.
(iii) *Heterostyly.* It refers to variation in the length of style and filaments in the same flower.
(iv) *Herkogamy.* It refers to hindrance to self pollination due to hyline membrane around the anther, and
(v) Presence of self incompatibility and male sterility.

Q. 22. What is decliny ?

Ans. Presence of unisexual flowers is called decliny. It is of two types as follows :

(i) *Monoecy.* It refers to separate presence of male and female flowers in the same plant. In some crops, male and female flowers are present in the same infloresence such as mango, castor and banana,. In some cases, they are on separate inflorescence such as in maize.

(ii) *Dioecy.* It refers to presence of male and female flowers on different plants. It includes papaya, date palm, spinatch, hemp and asparagus.

Q. 23. What is dichogamy ?

Ans. Maturation of anthers and stigma of the same flower at different times is called dichogamy. It promotes cross pollination even in hermophrodite species. It is of following two types.

(i) *Progogyny.* In this type, pistal matures before anthers such as in pearl millet.

(ii) *Protandry.* In this type anthers mature before pistil such as in maize and sugarbeet.

Q. 24. What is heterostyly ?

Ans. Different lengths of styles and filaments in a flower refer to heterostyly. It promotes cross pollination such as in linseed.

Q. 25. What is herkogamy ?

Ans. Hindrance to self pollination due to some physical barriers such as presence of hyline membrane around anther is called herkogamy. Such membrane does not allow dehiscence of pollen and prevents self pollination as in alfalfa.

Q. 26. What is anthesis ?

Ans. The process of dehiscence of anthers or the period of pollen distribution is called anthesis.

Q. 27. What do you mean by inbreeders ?

Ans. The self pollinated plants are also known as autogamous plants or inbreders. Main features of inbreeders are given below.

(i) They are self pollinated
(ii) They are homozygous and true breeding
(iii) They have homozygous balance and are tolerant to inbreeding
(iv) New gene combination is not possible due to regular self pollination.
(v) Inbreeders are composed of several homozygous lines.
(vi) They have close adaptation.
(vii) They have more of additive gene action.

Q. 28. What are outbreeders ?

Ans. Cross pollinated plants are also known as allogamous plants or outbreders. Main features of outbreeders are given below :

(i) Outbreeders have random mating.
(ii) They are heterozygous and have advantage of heterozygosity.
(iii) They are intolerant to inbreeding.
(iv) They permit new gene combination due to cross pollination.
(v) They have wider adaptability.
(vi) They have more of non-additive gene action.

Q. 29. What are effects of inbreeding on cross pollinated species ?

Ans. Main effects of inbreeding on cross pollinated specris are given below :

(i) The homozygosity is increased due to fixation of genes.
(ii) The heterozygosity is reduced due to fixation of genes.
(iii) The population mean is reduced due to decrease in number of hybrid genotypes.
(iv) The genetic correlation increases between close relatives due to increase in prepotency, which increases with homozygosity.
(v) The hidden variability becomes free.

Q. 30. What are effects of outbreeding ?

Ans. Main effects of outbreeding are given below :

(i) The heterozygosity is increased due to random mating.
(ii) The homozygosity is reduced because outbreeding favours heterozygotes.
(iii) The population mean will increase due to increase in heterozygosity and combining more dominant genes from different lines.
(iv) The genetic correlation will decrease due to decrease in homozygosity.
(v) The variability is increased due to intermating and segregation.

Q. 31. What is mating system ?

Ans. The method by which individuals are paired for crossing is called mating system.

Q. 32. Who coined the term monoecious ?

Ans. The term monoecious was coined by Darwin in 1877.

Q. 33. Who termed the term heterostyly ?

Ans. The term heterostyly was first used by Darwin in 1865.

Q. 34. Who first reported double fertilization ?

Ans. Double fertilization was first reported by Navashin and Guignard in 1899.

Q. 35. Who gave systems of mating ?

Ans. Systems of mating were given by Sewall Wright in 1921.

Q. 36. What are different systems of mating ?

Ans. There are following five systems of mating :

(i) Random mating system
(ii) Genetic assortative mating
(iii) Genetic disassortative mating
(iv) Phenotypic assortative mating
(v) Phenotypic disassortative mating.

Q. 37. What is random mating ?

Ans. A mating system in which each female gamete has equal chance to unite with every male gamete is called random mating. Main features of random mating are given below.

(i) It is a form of outbreeding.
(ii) Without selection, variability, population mean and genetic correlation between relatives remain constant.
(iii) It is useful in development and maintenance of synthetic and composite varieties.

Q. 38. What is genetic assortative mating ?

Ans. It refers to mating between genetically similar or closely related individuals. Its main features are given below :

(i) It is commonly known as inbreeding.
(ii) It leads to increase in homozygosity, genetic correlation and variability among lines (without selection). It leads to reduction in heterozygosity and population mean.
(iii) It is useful for development of inbred lines.

Q. 39. What is genetic disassortative mating ?

Ans. It refers to mating between genetically dissimilar individuals or between less closely related individuals. Its main features are given below :

(i) It is commonly known as outbreeding.
(ii) It leads to increase in heterozygosity, variability and population mean, and decrease in homozygosity and genetic correlation.
(iii) It is useful in population improvement.

Q. 40. What is phenotypic assortative mating ?

Ans. It refers to mating of extreme types, viz. AA × AA and aa × aa. Its main features are given below.

(i) *The variability is divided into two extreme types (AA and aa).*

(ii) *The population mean is divided according to variability.*

(iii) It is useful (with selection) in development of extreme phenotypes.

Q. 41. What is phenotypic disassortative mating ?

Ans. Mating between phenotypically dissimilar individuals (AA × aa) is called phenotypic disassortative mating. Its main features are given below :

(i) *It maintains genetic variability due to production of intermediate types.*

(ii) Heterozygosity and population mean remain constant.

(iii) *Genetic correlation between relatives decreases due to decrease in prepotency, which is decreased due to decrease in homozygosity.*

(iv) This system is used in disruptive mating programs.

Q. 42. What is prepotency ?

Ans. The property of an individual to produce progeny which are identical to each other and to the parent is called prepotency.

Q. 43. What is gamogamy ?

Ans. Gamogamy refers to sexual reproduction.

Q. 44. Who coined the term gamogamy ?

Ans. The term gamogamy was coined by Hartmann in 1904.

Q. 45. Give comparison of random, genetic assortative and genetic disassortative systems of mating.

Ans. Comparison of these systems of mating is presented below in table 4.1

TABLE 4.1. Comparison of random, genetic assortative and genetic disassortative systems of mating

Articulars	Random mating	Genetic assortative mating (inbreeding)	Genetic disassortative Mating (outbreeding)
1	2	3	4
Effect on genetic variability	Variability remains constant	Variability among families increases	Variability increases due to segregation
Homozygosity	Remains constant	Increases due to gene fixation	Decreases
Heterozygosity	Remains constant	Decreases due to fixation ofgenes	Increases

Population Mean	Remains constant	Decreases due to decreases in heterozygosity	Increases
Genetic correlation between . relatives	Remains constant	Increases due to increase in prepotency	Decreases
Uses	Useful in developing synthetics and composites.	Useful for development of inbreds	Useful for intra and interspecific gene transfer.

5

APOMIXIS IN PLANT BREEDING

Q.1. What is apomixis ? Give its main features.

Ans. The term apomixis was first used by Winkler in 1906. Apomixis refers to the development of seed without fertilization. In other words it does not involve sexual fusion for development of seed. It differs from amphimixis which involves union of male and female gametes for development of seed. Main features of apomixis are given below:

1. It is an asexual means of plant reproduction.
2. Apomixis is found in many plant species. It is more frequent in fruit crops than in field crops.
3. In some plant species, reproduction occurs only by apomixis. This type of amomixis is referred to as obligate apomixis.
4. In some plant species, sexual reproduction also occurs in addition to apomixis. This type of apomixis is known as facultative apomixis.
5. Apomixis does not permit segregation. It is useful in maintaining the genetic purity of mother plant.
6. Apomixis does not involve union of male and female gametes hence it does not permit recombination. In other words, it does not lead to generation of variability in the population.
7. It does not permit gene flow. In other words, it does not permit combination of genes from different sources.
8. It helps in rapid development of pure lines. It can be achieved by haploid parthenogenesis.
9. It is useful in conserving superior genotypes as well as hybrid vigour.
10. Apomixis is genetically controlled. In majority of cases it is governed by single dominant gene.
11. Apomixis has been reported in more than 300 plant species of 35 families.

Q.2. What are the differences between apomixis and amphimixis ?

Ans. Apomixis and amphimixis differ from one another in several ways. The important differences between apomixis and amphimixis are presented in Table 5.1

Table. 5.1 Differences between Apomixis and Amphimixis

Particulars	Apomixis	Amphimixis
1. Term first used or coined by	Winkler 1906	Weismann 1891
2. Sexual Fusion or union of male and female gametes	Does not involve	Involved
3. Gene combination or gene flow	Not possible	Possible
4. Segregation	Does not occur	Occurs
5. Rapid development of inbred lines	Possible	Not possible
6. Conservation of heterosis	Possible	Not possible
7. Type of reproduction	Asexual	Sexual
8. Types	Parthenogenesis Apogamy, Apospory Adventive embryony	Autogamy Allogamy
9. Frequency in plants.	Low	High

Q.3. Give classification of apomixis.

Ans. The apomixes is generally classified on the basis of *(i)* cell involved, *(ii)* occurrence, and *(iii)* frequency. These are presented in Table 5.2. Brief description of each type is also given in the following paras.

Table 5.2 Classification of Apomixis

Basis of classification	Types of Apomixis	Brief description
(i) Cell involved	Parthenogenesis	Embryo develops from egg cell.
	Apogamy	Embryo develops from either synergids or antipodal cells.
	Apospory	Embryo develops from the embryosac which has developed from the cell of archisporeum/ nucellus or integument.
	Adventive Embryony	Embryo develops directly from the diploid cell of either nucellus or integument
(ii) Occurrence	Recurrent Apomixis	Embryosac has diploid cells and embryo develops from the diploid cells.

	Non-Recurrent Apomixis	Embryosac consists of haploid cells and embryo develops from haploid cell.
(iii) Frequency	Obligate Apomixis	The reproduction occurs by apomictic means only.
	Facultative Apomixis	The reproduction occurs by both sexual and apomictic means.

CLASSIFICATION BASED ON SITE OF ORIGIN :

Based on the site of origin the apomixis is of four types, viz. *(i)* parthenogenesis, *(ii)* apogamy, *(iii)* apospory and *(iv)* adventive embryony. These are briefly discussed below:

Q.4. What is Parthenogenesis.

Ans. The term parthenogenesis was coined by Owen in 1849. It refers to development of embryo from the egg cell without fertilization. Parthenogenesis has been reported in several crop plants such as grain sorghum, *Hordeum bulbosum,* and cotton. It has also been reported in *Solanum nigram* and grass like *Taraxacum.* Parthenogenesis is of two types, viz. haploid parthenogenesis and diploid parthenogenesis. Thus main points about parthenogenesis are as follows:

(i) The term was coined by Owen in 1849.
(ii) The embroyo develops from egg cell.
(iii) It may be either haploid or diploid.
(iv) It is genetically controlled.
(v) It occurs in nature as well as can be induced artificially.
(vi) It is useful for conserving heterosis (diploid parthenogenesis).
(vii) It is useful in developing purelines (haploid parthenogenesis).

Q.5. What are types of parthenogenesis.

Ans. It is of two types :

(i) Haploid Parthenogenesis

The development of embryo from the haploid egg cell is referred to as haploid parthenogenesis. It has been reported in *Hordeum bulbosum, Solanum nigram* and some other plant species. Haploid parthenogenesis is very useful in developing inbred/purelines. The purelines can be obtained by chromosome doubling of haploid plants.

(ii) Diploid Parthenogenesis.

The development of embryo from the diploid egg cell is known as diploid parthenogenesis. Sometimes embryosac develops without reduction division.

All cells viz. egg cell, antipodals and synergids in such embryosac are idiploid. It gives rise to diploid embryos. Diploid parthenogenesis has been reported in grasses like *Taraxacum*. It is very useful in conserving heterosis.

In some plant species, pollen grains can be induced to develop into embryos. This development of embryos from pollen grains or anthers is called androgenesis. It has been reported in tobacco, rice and some other species.

Q.6. What are Causes of Parthenogenesis.

Ans. There are several causes of parthenogenesis in crop plants. The main causes are presented below:

(i) Inability of the pollen tube to discharge the contents inside the embyosac
(ii) Insufficient attraction between male and female gametes
(iii) Early degeneration of the sperm
(iv) Very long style
(v) Schlerenchymatous style
(vi) Short pollen tube
(vii) Slow rate of pollen tube growth
(viii) Stimulation of pollination in the absence of pollen tube
(ix) Self incompatibility and cross incompatibility.

Q.7. What is Apogamy ?

Ans. The term apogamy was coined by Winkler in 1908. It refers to origin of. embryo either from synergids or antipodal cells of the embryosac. The main points about apogamy are presented below:

(i) The term apogamy was coined by Winkler in 1908.
(ii) In apogamy the embryo originates either from synergids or antipodal cells.
(iii) It may be either haploid or diploid.
(iv) Embryo develops from the cells of normal embryosac.
(v) It has been reported in Allium, Iris and many other plant species-
(vi) It occurs in nature and can also be induced by artificial means.

Q.8. What are types of apogamy ?

Ans. It is of two types :

(i) Haploid Apogamay. The development of embryo from haploid synergids or antipodal cells is referred to as haploid apogamy. It is useful in development of pure lines.

(ii) Diploid Apogamy. The development of embryo from diploid synergids or antipodal cells is called diploid apogamy. Sometimes, embryosac develops without reduction and thus all cells of such embryosac are diploid. This leads to development of diploid apogamy. Diploid apogamy is useful in conserving heterosis. It has been reported in *Allium, Iris* and many other plant species.

Q.9. What is Apospory ?

Ans. In apospory, first diploid cells of ovule lying outside the embryosac develop into another embryosac without reduction. The embryo then develops directly from the diploid egg cell without fertilization. The main points about apospory are presented below:

(i) The term apospory was coined independently in 1886 by De Very and Bower.

(ii) The embryo develops from the diploid egg cell of another embryosac which originates from the cell of ovute lying outside the normal embryosac.

(iii) It gives rise to diploid embryos.

(iv) The another embryosac may develop from the cell of archesporium, nucellus or integument.

(v) It is of two types, viz. *(i)* generative apospory and *(ii)* somatic apospory.

(vi) This is the most frequent type of apomixis.

Q.10. What are types of apospory ?

Ans. It is of two types :

(i) Generative Apospory. The development of embryo from the emryosac that has originated from the cell of archesporium is referred to as generative apospory. It has been reported in Parthenium.

(ii) Somatic Apospory. When the embryo develops from the embryosac that has originated from the cell of either nucellus or integument is known as somatic apospory. It has been reported in *Malus, Rubus, Allium, Crepis, Poa* and *Opuntia.*

Q.11. What is Adventive Embryony ?

Ans. The development of embryo directly from the diploid cells of ovule lying outside the emryosac belonging either to nucellus or integument is referred to as adventive embryony. Main points about adventive embryony are given below:

(i) The term adventive embryony was first used by Strasburger in 1878.

(ii) The embryo develops directly from the diploid cells of either nucellus or integument.

(iii) It always gives rise to diploid embryos.

(iv) In this case another embryosac is not formed.

(v) It is a type of sporophytic budding.

(vi) It has been reported in *Citrus* species.

Q.12. What is recurrent apomixis ?

Ans. Where apomictic lines are obtained from one generation to another, it is known as recurrent apomixis. Main features of recurrent apomixis are given below:

(i) The embryosac develops from diploid cells without reduction in chromosome number.

(ii) The plants produced by this method have diploid *(2n)* chromosome number.

(iii) Such plants are always fertile.

(iv) It is of three types, viz. *(a)* diploid parthenogenesis, *(b)* diploid apogamy and *(c)* apospory.

(v) In this case even the heterozygous plants will breed true.

Q.13. What is non-recurrent apomixis ?

Ans. When apomictic plants are not obtained from one generation to another, it is called non-recurrent apomixis. Main features of non-recurrent apomixis are given below:

(i) The embryosac has usual haploid cells *(n)*.

(ii) Plants produced by this process are haploid and sterile.

(iii) It includes haploid parthenogenesis and haploid apogamy.

(iv) It is useful for development of inbred lines.

Q.14. What is obligate apomixis ?

Ans. In some species, reproduction occurs only by apomixis. Such apomixis is known as obligate apomixis. Such species produce uniform progeny identical to the mother plant. Main features of obligate apomixis are given below:

(i) Reproduction occurs by apomictic means only.

(ii) The progeny is uniform.

(iii) The progeny is identical to mother plant.

Q. 15. What is facultative apomixis ?

Ans. In some species sexual reproduction also occurs in addition to apomixis. Such apomixis is termed as facultative apomixis. Main features of facultative apomixis are as follows :

(i) Reproduction also occurs by sexual process in addition to apomixis.

(ii) The progeny is heterogeneous in such case.

(iii) It has been reported in *Sorghum* and some other plant species.

Q.16. How can you Induce apomixis ?

Ans. Apomixis particularly the parthenogenesis can be artificially induced by various ways as given below:

(i) By the stimulation of widely related pollen or foreign pollen.

(ii) By low temperature treatment to ovule.

(iii) By pollinating by X-ray irradiated pollen.

(iv) By treatment with certain chemicals like belviton .

Q.17. What are advantages of apomixis ?

Ans. Apomixis has several useful applications in plant breeding. Some important applications/ advantages of apomixis are presented below:

1. Development of Purelines

Apomixis is an effective means of rapid production of pure lines. Haploid parthenogenesis and haploid apogamy give rise to haploid plants which after colchicine treatment will produce diploid purelines. Such purelines can be used in breeding programmes for developing high yielding cultivars and hybrids.

2. Maintenance of Purity

The obligate apomixis breeds true for the characteristics of mother plant. Thus it is useful in maintaining the genetic purity from generation to generation. It can maintain a genotype for unlimited number of generations. In other words, apomictic lines always breed true irrespective of homozygosity and heterozygosity.

3. Conservation of Heterosis

The obligate recurrent apomixis is useful in conserving 'heterosis or hybrid vigour for unlimited generations. Apomixis does not permit segregation. Hence heterosis can. be easily conserved.

4. Easy Hybrid Seed Production

Apomixis provides an easy way of hybrid seed production. The hybrid seed is automatically produced by apomictic means. There is no need of crossing. In other words, in apomictic genotypes crossing is not required for hybrid seed production. Once the hybrid is developed using apomictic line as one of the parents, the hybrid seed production will occur automatically. Thus it permits commercial seed production of hybrids.

5. Cheaper Hybrid Seed

Use of apomixis is the cheaper way of hybrid seed production. It does not require crossing. Hence it helps in saving lot of money which is required for engaging labourers for crossing purpose in conventional hybrids for production of hybrid seed.

6. Easy for Handling

The handling of apomictic genotypes is easy. When the hybrid is developed using apomixes; We have to handle one genotype only. When the hybrid is developed using cytoplasmic genic male sterility, the breeder has to handle three type of material *i.e.* A, B and R lines. The purity of apomictic lines/ genotypes is maintained automatically, whereas A, B and R lines require lot of labour and care for maintenance of their genetic purity. In other words, when obligate apomixis is available, there is no need for using male sterility for hybrid development.

7. No Need of Islation

In case of apomixis, there is no need of isolation distance, where as in case of hybrid seed production by conventional method by using male sterility system, the proper isolation distance is required for maintaining the genetic purity of parental lines.

8. No Outcrossing

The apomixis is genetically controlled. It is governed by one or few dominant or recessive genes. Hence no seed setting occurs with cross pollination. Thus it does not permit outcrossing. In other words, contamination of apomictic genotypes through outcrossing is not possible.

9. No Segregation

Apomictic lines always breed true (as stated above) to the characteristics of mother plant. The hybrid seed is produced automatcally. There is no need of making crosses every year to produce hybrid seed. This is the cheaper method of hybrid seed production. The apomictically propagated hybrid can be used as a variety, because it does not permit segregation.

10. Useful to Farmers

Apomictically propagated cultivars and hybrids are useful to the farmers, because farmers can produce their own seed. Thus farmers need not to purchase fresh seed every year from seed companies.

Q.18. What are limitations of apomixis ?

Ans. Apomixis has several advantages as stated above. However, there are some disadvantages or limitations of apomixes. Some limitations of apomixis are briefly presented below:

1. Genetic Diversity

Obligate apomictic lines have low genetic diversity. In other words apomictic lines are highly uniform. As a result such lines have narrow adaptation and narrow genetic base. Such lines are adaptable to a specific environment. The diversity arises through mutation.

2. Lack of Gene Flow

Apomixis particularly obligate type does not permit natural gene flow in the population. As a result recombinations do not occur in such population.

3. Selection in Ineffective

In apomictic (obligate) lines, selection is ineffective due to lack of herltable variation. All the plants of a apomictic line have exactly the same genotype. It may be homozygous or heterozygous.

4. Low Frequency

Apomictic lines have very low frequency. Such lines are available in limited crop plants. However, such lines are more frequent in fruit crops and some grasses.

5. Problems with Facultative Apomixis

It is easy to use obligate apomixis in plant breeding. However, use of facultative apomixis (which reproduces both by apomictic and sexual means) is difficult. It becomes difficult to identify apomictic and sexually reproduced plants. Facultative apomixis is more useful in forage crops.

6. Adverse Effect on Seed Industry

The seed production is very easy in apomictic lines than seed production lines than seed production by sexual means. Farmers can easily produce their own seed and they need not to purchase fresh seed every year from seed companies. This will lead to reduced seed sale resulting in adverse effect on seed Industry.

7. Sexual Reproduction

Sometimes, apomictic lines are influenced by environmental conditions and they reproduce by sexual means. It poses problem in the use of apomictic lines in plant breeding.

Q. 19. What are sources of apomixis ?

Ans. It has been reported that apomixis is genetically controlled. It is governed by oligogenes. In some cases it is governed by dominant gene while in others it is controlled by recessive genes. In few cases, it is governed by epistatic gene. Thus it is governed by one or few genes that can be dominant, recessive or epistatic in expression. Apomictic genes are generally found in wild species. As a result apomictic lines are obtained from interspecific derivatives. Some times, apomictic lines are also obtained from the germplasm of cultivated species.

6

SELF INCOMPATIBILITY AND MALE STERILITY

Q. 1. What is self incompatibility ?

Ans. The inability of a plant with functional pollen to set seeds when self pollinated is called self incompatibility. Self incompatibility is classified into two groups, viz. *heteromorphic* system and *homomorphic* system.

Q. 2. Who coined the term self incompatibility ?

Ans. The term self incompatibility was coined by stout in 1917. Self incompatibility was first reported in *Verbascum phoeniceum* by Koelreuter in the middle of 18th century.

Q. 3. What is heteromorphic self incompatibility ?

Ans. Self incompatibility due to differences in floral biology is known as heteromorphic self incompatibility. It is of two types, viz. *distyly and tristyly.*

Q. 4. What is distyly ?

Ans. When style has two positions, viz. low and high with regard to length, it is called distyly.

Q. 5. What is tristyly ?

Ans. When the length of style has three positions, viz. low, medium and high, it is known as tristyly.

Q. 6. What is Pin ?

Ans. In Primula, flowers with long style and low anthers are referred to as Pin.

Q. 7. What is thrum ?

Ans. In primula, flowers with short style and high anthers are called thrum. In primula, crosses are compatible only between style and stamen of matching length *i.e.* between Pin and Thrum or Thrum and Pin but not between Pin × Pin and Thrum × Thrum.

Q. 8. What is genotype of Pin and Thrum ?

Ans. The Pin is homozygous recessive (*ss*) and thrum is heterozygous (*Ss*). A cross between Pin and Thrum produces Pin and Thrum in 1 : 1 ratio.

Q. 9. What is homomorphic SI System ?

Ans. Presence of self incompatibility due to physiological or genetic causes rather than differences in floral morphology is called homomorphic self incompatibility. It is of two types, viz. gametophytic self incompatibility and sporophytic self incompatibility.

Q. 10. What is gametophytic self incompatibility ?

Ans. Self incompatibility which is controlled by the genetic constitution of gametes is known as gametophytic self incompatibility.

Q. 11. Who first reported gametophytic SI ?

Ans. Gemetophytic self incompatibility was first reported by East and Mangelsdorf in 1925 in *Nicotiana sanderae.*

Q. 12. What are main features of gametophytic system ?

Ans. Main features of gametophytic SI system are given below :

(i) SI is controlled by genetic constitution of pollen.
(ii) SI is governed by a single gene with multiple alleles.
(iii) Alleles show individual action in the style without interaction.
(iv) The pollen tube growth is usually inhibited in the style or ovary.
(v) It does not exhibit reciprocal differences.
(vi) It permits recovery of male parent only.
(vii) It does not permit production of homozygotes.
(viii) Crosses are of there types, viz. fertile, partially fertile and sterile.

Q. 13. What are crops in which gometophytic SI in found ?

Ans. Gometophytic self incompatibility system has been reported in Red Clover, White Clover, Rye, Potato, Tomato, etc.

Q. 14. What is sporophytic self incompatibility ?

Ans. Self incompatibility that is governed by the genotype of pollen producing plant *i.e.* sporophyte is called sporophytic self incompatibility.

Q. 15. Who first discovered sporophytic SI system ?

Ans. The sporophytic self incompatibility system was first discovered by Hughes and Babcock (1950) in *Crepis foetida* and Gerstel (1950) in Parthenium argentatum (Guayule).

Q. 16. What are main features of sporophytic SI system ?

Ans. Main features of sporophytic self incompatibility system are given below.

(i) SI is controlled by the genotype of pollen producing plant.
(ii) SI is governed by single gene with multiple allele.
(iii) Alleles may show dominance, individual action or interaction either in pollen or style.

(iv) Pollen germination or pollen tube growth is inhibited in the stigma.
(v) Reciprocal differences are observed.
(vi) Recovery of both male and female parent is possible.
(vii) It permits production of some homozygotes
(viii) Crosses are either fully fertile or fully sterile.

Q. 17. What are crops in which sporophytic SI system is found ?

Ans. The sporophytic self incompatibility system has been reported in Radish, Cabbage, Cauliflower, etc.

Q. 18. What is monoallelic self incompatibility ?

Ans. Self incompatibility which is governed by single gene is called monoallelic self incompatibility. It is found in same species of the family *leguminasae, solanaceae* and *cruciferae.*

Q. 19. What is diallelie self incompatibility ?

Ans. The SI which is controlled by two genes is called diallelic self incompatibility. It is found in the family *Graminae.*

Q. 20. What is polyallelic self incompatibility ?

Ans. SI which is governed by several genes is known as polyallelic self incompatibility. It is found in sugarbeet.

Q. 21. What are types of SI based on site of gene expression ?

Ans. Based on the site of expresson of gene, self incompatibility is of three types as follows :

(i) *Stigmatic inhibition.* It has been reported in radish, cabbage, cauliflower, sunflower etc.
(ii) *Stylar inhibition.* It is found in sugarbeet, helianthus, Bougainvillea and Fagopyrum.
(iii) *Ovarian inhibition.* It is found in Narcissus, Lilium, Gasteria, Ribes, Anona etc.

Q. 22. What are the hypothesis to explain mechanism of self incompatibility ?

Ans. The following two hypotheses have been proposed to explain the mechanism of self incompatibility.

(i) *Complementary hypothesis.* According to this hypothesis, self incompatibility results due to absence of substances in the pistal or poller which are essential for pollen tube penetration.
(ii) *Oppositional hypothesis.* According to this hypothesis, interaction between like allele and (S_1 S_2 × S_1 S_2) leads to production of a substance which inhibits the growth of pollen tube in the pistal.

Q. 23. What are ways and means to overcome self incompatibility ?

Ans. In crop plants, following ways and means (measures) are used to overcome self incompatibility.

(i) Pollination of immature buds with mature pollen.
(ii) Pollination of aged pistils with incompatible pollen.
(iii) Self pollination at the end of the flowering season.
(iv) Irradiation of style with X-rays before selfing.
(v) Treatment of style with temperature ranging from 30°C to 60°C.
(vi) Removal of stigma in species having stigmatic self incompatibility.
(vii) Placing of pollen in direct contact of ovules.

Q. 24. What is significance of self incompatibility ?

Ans. Self incompatibility has great significance in plant breeding. It prevents autogamy and promotes allogomy. It is used for hybrid seed production in *Brassica* and sunflower. Two self incompatible lines are planted in alternate rows for hybrid seed production. The harvest from both lines would be hybrid seed.

Q. 25. What are types of SI based on flower morphology ?

Ans. On the basis of flower morphology, self incompatibility is classified into two types, viz. *heteromorphic* and *homomorphic*. Heteromorphic system includes distyly and tristyly and homomorphic system again is of two types, viz. gametophytric system and sporophytic system.

Q. 26. What are types of SI based on genes involved ?

Ans. On the basis of gene involved, self incompatibility is of three types, viz. *monoallelic, diallelic* and *polyallelic.*

Q. 27. How will you classify SI on the basis of site of expression?

Ans. On the basis of site of gene expression, self incompatibility is of three types, viz. stigmatic, stylar and ovarian.

Q. 28. How will you classify SI on the basis of cytology ?

Ans. On the basis of pollen cytology, self incompatibility is of two types as follows.

(i) *Binucleate.* Pollen grains have two nuclei.
(ii) *Trinucleate.* Pallen grains have three nuclei.

Q. 29. Who proposed complementary hypothesis of self incompatibility ?

Ans. The complementary hypothesis of self incompatibility was proposed by Bateman in 1952.

Q. 30. Who first reported diallelic self incompatibility ?

Ans. The diallelic gametrophytic self incompatibility was first reported by Lundquist in 1956 in Rye.

Q. 31. How will you identify self incompatibility ?

Ans. The self incompatibility can be identified as follows :

(i) If functional pollen grains are unable to fertilize same plant or by absence of seed set by functional pollen on selfing.

(ii) By difference in flower morphology.

Q. 32. In gametophytic self incompatibility what will be the outcome of cross S_1S_2 x S_1 S_2?

Ans. This cross will be fully incompatible because both the alleles are same in female and male parents.

Q. 33. What will be the result of cross $S_1S_2 \times S_1S_3$ in gametophytic self incompatibility ?

Ans. This cross will be partially fertile because the male parent differs in one allele.

Q. 34. What will be the result when both parents have different alleles (S_1S_2 $\times S_3$ S_4) in the gametophytic self incompatibility ?

Ans. Such cross is fully compatible and produces four types of progeny, viz S_1 S_3, S_1S_4, S_2S_3 and S_2S_4.

Q. 35. What is the result of the cross $S_1S_2 \times S_1S_3$ in sporophytic self incompatibility ?

Ans. This cross will be fully incompatible because in male parent S_3 will also behave as S_1 due to dominance.

Q. 36. What will be outcome of the cross $S_2S_3 \times S_1S_2$ in sporophytic self incompatibility ?

Ans. The cross will be fully compatible because S_2 also behaves as S_1 in male parent due to dominance. This cross will produce four types of progemy viz. S_1S_2, S_2S_2, S_1S_3 and S_2S_3. Thus one homozygote is also obtained.

Q. 37. What will be outcome of the cross $S_1S_3 \times S_2S_3$ in sporophytic self incompatibility ?

Ans. The cross is fully compatible because in male parent S_3 will also behave as S_2 due to dominance. This cross will produce four types of progeny, viz. S_1S_2, S_1S_3, S_2S_3 and S_3. This cross permits recovery of both parents S_1S_3 and S_2S_3 and also produces one homocygote *i.e.* S_3S_3 which is not possible in gametophytic system.

Q. 38. What is male sterility ?

Ans. Male sterility refers to condition in which either pallen is absent or non-functional in flowering plants. Male sterility is of three types, viz. genetic male sterility, cytoplasmic male sterility and cytoplasmic genic male sterility.

Q. 39. Who first reported male sterility in flowering plants ?

Ans. In flowering plants, male sterility was first reported by Koelreuter in 1763.

Q. 40. What is genetic male sterility ?

Ans. Pollen sterility which is caused by nuclear genes is called genetic male sterility. Main features of genetic male sterility are given below.

(i) It is usually governed by recessive genes and rarely by dominant genes.
(ii) In majority of cases, it is governed by single recessive genes. However, some times it is governed by two recessive genes.
(iii) The male sterile and fertile plants are observed in 50 : 50 ratio. Thus 50% plants are to be rogued out.
(iv) The male sterility is influenced by environmental factors such as low or high temperature.

Q. 41. What is cytoplasmic male sterility ?

Ans. Pollen sterility which is governed by cytoplasmic (plasma genes) is called cytoplasmic male sterility. Main features of CMS are given below :

(i) It is governed by plasma genes.
(ii) It consists of male sterile (A) and male fertile (B) lines.
(iii) It can be used for hybrid seed production in vegetatively propagated crops where grain is not the economic product.
(iv) CMS is not influenced by environmental factors such as low or high temperature.

Q. 42. Who first reported CMS in crop plants ?

Ans. Cytoplasmic male sterility was first reported by Welch and Grimball in 1947.

Q. 43. What is cytoplasmic genic male sterility ?

Ans. Pollen sterility which is controlled by both cytoplasmic and nuclear genes is called cytoplasmic genic male sterility. Main features of CGMS are given below :

(i) It is governed by both plasma genes and nuclear genes.
(ii) It consists of three lines i.e. A, B and R lines.
(iii) Male sterility is maintained by crossing it with maintainer line.

Q. 44. Who first reported CGMS in crop plants ?

Ans. The cytoplasmic genic male sterility was firrt reported by Jones and Davis in onion.

Q. 45. What are demerits of genetic male sterility ?

Ans. There are two drawbacks of genetic male sterility as given below :

(i) It is influenced by environmental factors. For example, in rice sterile plants become fertile at temperature below 23°C.
(ii) It segregates in 50 : 50 fertile and sterile plants in each generation. Hence 50% fertile plants have to be removed.

Q. 46. What are merits of genetic male sterility ?

Ans. There are two main advantages of genetic male sterility as follows.

(i) It can be used for hybrid seed production in both seed propagated and vegetatively propagated species.

(ii) It generally does not have undesirable effects.

Q. 47. What are merits of cytoplasmic male sterility ?

Ans. Main advantages of CMS are given below :

(i) It is highly stable. It is not influenced by low or high temperature.

(ii) It has only A and B lines and hence requires less area than CGMS system.

Q. 48. What are demerits of cytoplasmic male sterility ?

Ans. Main demerits of CMS are given below :

(i) It can be used for development of hybrids only in vegetatively propagated species.

(ii) Sometimes CMS lines have inferior agronomic performance.

Q. 49. What is A line ?

Ans. The male sterile line is called A line.

Q. 50. Define B line.

Ans. The male fertile line which is similar to A line except for male fertility is called B line. In other words, it is isogenetic line of A line.

Q. 51. What is R line ?

Ans. A line which restores fertility when crossed with cytoplasmic genic male sterile lines is referred to as R line.

Q. 52. What is gynodioecy ?

Ans. The co-existence of female (male sterile) and hermaphrodite individuals in a population is known as gynodioecy.

Q. 53. What are male gametocides ?

Ans. Chemicals which are used for large scale induction of male sterility in crop plants are called male gametocides or chemical hybridization agents. Such agents include Sodium methyl arsenate, Zinc methyl arsenate, NAA, Gibbrellins, FW 450, Ethrel etc.

Q. 54. What are main drawbacks of male gametocides ?

Ans. Main drawbacks of male gametocides are given below :

(i) Pollen abortion is incomplete and erratic.

(ii) The effect is short lived hence repeated application is required.

(iii) Ovule fertility is also adversely affected resulting in low seed setting.

(iv) Male gametocides are expensive.

(v) Male gametocides have adverse effects on plant growth.

Q. 55. What is difference between mutagen induced and gametocide induced male sterility ?

Ans. The main difference between mutagen induced male sterility and gametocide induced male sterility is that the former is heritable and the latter is non-heritable.

Q. 56. What are advantages of gametocide induced male sterility?

Ans. Main advantages of gametocide induced sterility are given below :

(i) The sterility can be induced in the same season. It takes 4-5 years to transfer male sterility from one genotype to another by backcross method.

(ii) It is less labourous than backcross method.

(iii) There is no need of maintaining A, B and R lines in this method.

Q. 57. What is transgenic male sterility ?

Ans. The male sterility which is induced/developed by the technique of genetic engineering is called transgenic male sterility. Such sterility has been induced in tobacco and rapeseed by transferring a gene from *Bacillus amyloliquefaciens.*

Q. 58. What are sources of male sterility ?

Ans. There are four important sources of male sterility as given below :

(i) Spontaneous mutations

(ii) Induced mutations

(iii) Interspecific crosses, and

(iv) Biotechnology.

Q. 59. How will you identify male sterility ?

Ans. Main features of male sterile plants are given below :

(i) Anthers are shrivelled.

(ii) Anthers are smaller than normal.

(iii) Anthers do not contain pollen grains.

(iv) Anthers may contain non-functional (aborted) pollen grains.

Q. 60. Compare self incompatibility and male sterility.

Ans. Comparison of self incompatibility and male sterility is presented in Table 6.1

TABLE 6.1 Comparison of self incompatibility and male sterility

Self incompatibility	Male sterility
Similarities :	
1. It is an important outbreeding mechanism.	It is also an important outbreeding mechanism.

2. It prevents autogamy and promotes allogamy.	Same as incompatibility.
3. It is found in nature.	Also found in nature.
4. It is used for hybrid seed production.	Also used for development of hybrids.
Dissimilarities :	
1. Pollen is functional.	Pollen is absent or nonfunctional.
2. It results, due to morphological, genetic, physiological and biochemical causes.	It results due to genetic cytoplasmic or both causes.
3. It may be heteromorphic, homomorphic, gametophytic, or sporophytic.	It may be genetic, cytoplasmic or cytoplasmic genetic.
4. Artificial induction is difficult.	Artificial induction is easy.

Q. 61. Compare gametophytic and sporophytic systems of self incompatibility.

Ans. Comparison of gametophytic and sporophytic system of self incompatibility is presented below in Table 6.2.

TABLE 6.2 Comparison of gametophytic and sporophytic system of self incompatibility.

Gametophytic system	Sporophytic system
1. Self incompatibility is controlled by the genetic constitution of pollen.	Self incompatibility is controlled by the genotype of pollen producing plant (sporophyte).
2. Incompatibility is governed by a single S gene with multiple alleles.	Self incompatibility is also governed by a single S gene with multiple alleles.
3. Alleles have individual action in the style without interaction.	Alleles may show dominance, individual action or interaction in either pollens or style.
4. The pollen tube growth is usually inhibited in the style or ovary.	Pollen germination or pollen tube growth is inhibited on the stigma.
5. Plant species belonging to this sytem have binucleate pollen grains.	Plant species belonging to this system have trinucleate pollen grains.
6. Reciprocal differences are not observed.	Reciprocal differences are observed.
7. Recovery of only male parent is possible from crosses.	Recovery of both male and female parents is possible from crosses.

8. Does not permit production of homozygotes.	Permits production of some homozygotes.
9. Crosses may be sterile, partially fertile or fully fertile.	Crosses would be either fully sterile or fully fertile.
10. Biochemical substance associated with incompatibility is formed during pollen formation.	Biochemical substance associated with incompatibility is formed before pollen formation.
11. Such incompatibility can be overcome by polyploidy.	This cannot be overcome by polyploidy.
12. Examples of this system include Red clover, White clover, Rye, Potato, Tomato, etc.	Examples of this system include Radish, Cabbage, Cauliflower, etc.

Q. 62. Compare genetic male sterility (GMS) and cytoplasmic genic male sterility (CGMS).

Ans. Comparison of GMS and CGMS is presented below in Table 6.3.

TABLE 6.3 Comparison of GMS and CGMS

Genetic male sterility	Cytoplasmic genetic male sterility
1. It is controlled by nuclear genes.	Controlled by both nuclear and cytoplasmic genes.
2. Consists of *A* and *B* lines.	Consists of *A*, *B* & *R* lines.
3. May become fertile at low temperature.	No effect of temperature.
4. Maintained by mating sterile plants with heterozygous male fertile plants.	Maintained by crossing of *A* line with *B* line.
5. Used for production of hybrids in both seed and vegetatively propagated crops.	Can be used in both seed propagated and vegetatively propagated species.
6. The male sterile and fertile plants are observed in 1 : 1 ratio. The fertile plants have to be removed in hybrid seed production plot.	It requires handling of three types of material, *i.e. A, B* and *R* lines.
7. Barley, wheat, lucerne, water-melon, muskmelon, pumpkin, rice, cotton, tomato, tobacco, sunflower etc. are examples.	Pearlmillet, sorghum, cotton, wheat, maize, rice, sugarbeet, sunflower, tobacco, tomato etc. are examples.

7

GENE FLOW IN CROP PLANTS

Q.1. What is gene flow ? Give main points related to gene flow.

Ans. Gene flow refers to movement of various alleles from exotic or unadapted or wild germplasm to the cultivated genotypes. In other words, the movement of alleles from one genotype to another due to open pollination/cross pollination is referred to as gene flow. The term gene flow was coined by Birds sell in 1950. Main points related to gene flow are given below :

(i) It involves transfer of alleles from one genotype to another.

(ii) Generally gene transfer is from exotic or unadapted germplasm to cultivated genotypes.

(iii) It also includes movement of alleles from wild relatives or wild species to the cultivated genotypes.

(iv) Open pollination or cross pollination is essential for gene flow.

(v) The rate of gene flow is higher in cross-pollinated species than in often cross pollinated species and self pollinated species.

(vi) The gene flow is influenced by several factors which will be discussed later.

(vii) Gene flow has several useful consequences (see later).

(viii) The rate of gene flow can be enhanced through artificial hybridization.

(ix) Some natural devices such as open pollination, male sterility, self incompatibility promote gene flow among the population.

(x) The gene flow may take place among the genotypes of the same species or between distantly related genotypes belonging to two different species of the same genus.

Q.2. Explain types of gene flow.

Ans. The gene flow occurs in nature through open pollination or cross pollination. Hence gene flow is higher in cross pollinated species than often cross pollinated species and self pollinated species. The gene flow can also be achieved through artificial hybridization. The gene flow takes place between various types of plant populations. Based on mode, and species involved, gene flow can be of different types as follows :

Table 7.1 Types of Gene Flow in crop plants

Bases and Type	Brief description
I. Mode of Allele Transfer	
(a) Natural Gene Flow	Movement of alleles from one genotype to another through open pollination or cross pollination.
(b) Artificial Gene Flow	Movement of alleles from unadapted genotype to cultivated genotype using artificial hybridization.
2. Genotypes Involved	
(a) Intraspecific Gene Flow	Movement of alleles (natural or artificial) among the genotypes of the same species *i.e.* from germplasm to cultivated genotypes.
(b) Interspecific Gene Flow	Movement of alleles (natural or artificial) between two species of the same genus or from wild species to the cultivated species.
(c) Intergeneric Gene Flow	Movement of alleles by artificial means between two genera of the same family.

The gene flow is very common among the genotypes of the same species especially in open pollinated crops. The interspecific and intergeneric gene flow is achieved by artificial methods.

Q.3. What are modes of gene flow ?

Ans. In plant breeding populations, the gene flow takes place in two ways, viz. *(i)* natural, and *(ii)* artificial. These are briefly discussed below:

(I) Natural Gene Flow. The movement of alleles from one genotype to another in nature by open pollination is called natural gene flow. It occurs in cross pollinated and often cross pollinated species. All those mechanisms that promote cross pollination also promote natural gene flow.

***(ii)* Artificial Gene Flow.** The movement of alleles from one genotype to another through man made hybridization is referred to as artificial or man made gene flow. This type of gene flow is possible in all type of crops viz. self pollinated, cross pollinated and often cross pollinated species.

Q.4. What are the factors affecting gene flow ?

Ans. The gene flow is a natural process. However, gene flow is promoted by human through artificial crossing between different species. In nature, the gene flow is influenced by various factors. In the plant populations, the important factors affecting gene flow include *(i)* mode of pollination, *(ii)* cross compatibility, *(iii)* pollen mobility and *(iv)* hybrid viability. These are briefly discussed below:

1. Mode of Pollination

The rate of gene flow varies depending upon the mode of pollination of a plant species. The gene flow is high in cross pollinated species such as corn, pearlmillet, sunflower, rapeseed etc. and low in self pollinated species such as wheat, barley, rice, urdbean, mungbean, soybean etc. Thus those factors which promote cross pollination such as dicliny dichogamy, heterostyly, herkogamy, self incompatibility and male sterility also promote gene flow. Similarly those factors which prevent cross pollination such as bisexuality, homogamy cleistogamy, chasmogamy and cross incompatibility etc. also prevent gene flow.

2. Cross Compatibility

The another important factor that affects the rate of gene flow is cross compatibility. The gene flow can occur between those cultivated and wild species which are sexually compatible. In other words, gene flow will take place when the cross between two individuals or genotypes or species is compatible. The gene flow will not occur between those species having cross incompatibility. The gene flow can occur between genetically engineered and non-genetically engineered genotypes if there is cross compatibility. The cross incompatibility is observed in distantly related species or between species having ploidy differences both in cross and self pollinated species.

3. Pollen Mobility

The gene flow is governed by pollen mobility. The pollen mobility depends on pollen weight, pollen longivity and pollen production. The pollen grains with light weight, abundance production and more longivity are expected to travel long distance and effect cross pollination. This promotes gene flow, because such pollen grains can blow with wind to long distance. Heavy pollen grains with low longivity will restrict or prevent gene flow.

4. Hybrid Viability

The hybrid viability also governs the rate of gene flow. The cross between two species say cultivated and wild may be compatibly but the hybrid may not be viable. This prevents gene flow between species. If the hybrid between two species is viable and fertile it will enhance the rate of gene flow. The interspecific hybrids are fertile when there is chromosomal homology between two species. When there is lack of chromosomal homology between two species, the hybrid would be sterile.

Q.5. What are methods of gene flow ?

Ans. The gene flow occurs in nature through open pollination. It can also be promoted by various breeding approach. There are two broad methods of man made gene flow, viz.(*i*) conventional methods, and (*ii*) modern methods. These are briefly discussed below :

1. Conventional Methods

The conventional methods of gene flow among the plant breeding populations include backcross method, bulk method, pedigree method, single seed descent method. There are four populations improvement approaches, viz. recurrent selection, distruptive selection, biparental mating and diallel selective mating which are also used for man made gene flow. All these methods involve crossing and hence are referred to as hybridization methods. These methods permit combining of characters or genes from different sources into one genotype and hence are also known as combination breeding approaches. Backcross methods is used for transfer of one or two characters from donor parent to the good agronomic base. Pedigree, bulk an single seed descent methods are used for combining several characters of two genotypes together. Population improvement procedures are used for creating genetic variability, breaking undesirable linkages and broadening genetic base of the population.

2. Modern Methods

Modern methods include (*i*) direct gene transfer using biotechnological tools and (*ii*) somatic hybridization. There are four methods of direct gene transfer viz. Agrobacterium mediated gene transfer, particle gun method, electroporation method and micro-injection. These techniques are used for direct gene transfer bypassing sexual process. The gene can be transferred from one species to another in single step. In somatic hybridization, the gene transfer is achieved through protoplast fusion. These techniques are used for gene flow across the species and genera. It permits gene transfer even from unrelated species or organisms.

Q.6. What are advantages of gene flow ?

Ans. Gene flow has several desirable consequences such as (*i*) creation of genetic diversity, (*ii*) combining desirable genes, (*iii*) formation of new species, (*iv*) development of hybrids, (*iv*) improvement in adaptation, (*v*) broadening genetic base etc. These are briefly discussed below:

1. Genetic Diversity

Genetic diversity refers to variety of genes and genotypes found in a particular crop species. It provides protection from biotic and abiotic stresses resulting in insurance against crop losses. The movement of alleles from one genotype to an other enhances genetic diversity. The new variability is also created through combination and recombination of genes between two genetically different genotypes. Genetic diversity in a population provides opportunities for selection of superior genotypes. In other words, selection is effective when there is genetic diversity in a population. Genetic diversity also provides broad genetic base and wide adaptation to a population. In other words, it provides buffering capacity to a population.

2. Combining Desirable Genes

The gene flow permits transfer of desirable alleles from different sources into one genotype. For example, alleles for yield, quality, resistance to biotic and abiotic stresses, earliness, suitable plant type can be combined together through man made gene flow.

3. Formation of New Species

Gene flow permits formation of new crop species. The natural gene flow has resulted in formation of new species in various field crops such as wheat, tobacco, Brassica, cotton, oats etc. New species have also been developed through man made gene flow. Such species include Raphanobrassica (Radish x Cabbage), Triticale (Wheat x Rye) and Nicotiana digluta (N. tabacum x N. glutinosa). Several other such examples can be cited.

4. Development of Hybrids

Gene flow permits exploitation of heterosis or hybrid vigour in crop plants. The heterosis can be fully exploited by developing hybrids and partially by developing synthetic and composite varieties. Hybrids for commercial, cultivation have been developed in several crops such as maize, sorghum, pearlmillet, cotton, caster, cucumber, sunflower, pigeonpea, tomato, rice, brinjal, jute etc. The composite and synthetic varieties have been developed in maize.

5. Improvement in Adaptation

Adaptation refers to fitness of a genotype to particular environment and adaptability refers to capacity of a genotype to adjust to environmental changes. The varietal adaptability helps in stabilizing crop production over regions and seasons. Gene flow among plant breeding population leads to improvement in varietal adaptation and adaptability due to gene combination from different sources. In other words, gene flow improves buffering capacity of plant breeding populations.

6. Improvement in Genetic Base

The gene flow increases genetic diversity in a plant breeding population which leads to broadening of genetic base of a population. Broad genetic base provides protection against biotic and abiotic stresses to crop plants.

Man made gene flow sometimes leads to narrow genetic base as in cultivars developed by backcross, pedigree and SSD methods.

Q. 7. What are practical achievements of gene flow ?

Ans. Remarkable practical achievements have been made in different crop plants through natural and man made gene flow. Some important achievements are given below :

1. Several desirable alleles have been combined with cultivated genotypes from unadapted germplasm and wild species.
2. Sources of male sterility have been identified through interspecific crosses between cultivated and wild species.
3. Large number of hybrids have been evolved through intraspecific and interspecific hybridization *i.e.* by man made gene flow.
4. In some crop plants, new species have been developed both by natural gene flow and man made gene flow. Examples of development of new species are wheat, tobacco, Brassica, cotton and oats. The man made gene flow resulted in development of three new species viz Raphanobrassica, Triticate and *Nicotiana digluta.*
5. Man made gene flow has resulted in development of thousands of cultivars in field crops, vegetable crops and fruit crops.
6. Gene flow has played key role in improving various agronomic and quality traits in different field crops.
7. Gene flow has helped in improving varietal adaptation, broadening genetic base and breaking undesirable linkages in plant breeding populations.

Q.8. What are undesirable effects of gene flow ?

Ans. There are some undesirable effects of gene flow in nature which are briefly discussed below:

1. The gene flow from herbicide resistant genetically engineered (GE) plants to non-GE wild relatives may lead to development of herbicide resistant weeds what is known as super weed.
2. The gene flow from GE crop to non GE organic crop will lead to contamination of organic crop resulting in financial loss to farmers in terms of price.
3. The gene flow can make distant populations genetically similar to one another and hence reduce the chance of speciation. Lesser the gene flow between two populations more will be chances of evolving two species from such populations.

8

GENETIC BASE IN BREEDING POPULATIONS

Q.1. Define genetic base and explain it briefly.

Ans. The amount or quantum of genetic diversity available in a breeding population is refered to as genetic base of that population. In other words, genetic base represents spectrum of genetic variability in a plant breeding population. Main points related to genetic base are listed below:

(i) The genetic base of a population is determined by the number of constituent genotypes.

(ii) The genetic base has positive association with the amount of genetic diversity present in a population.

(iii) The genetic base has also positive association with the buffering capacity of a population to environmental changes.

(iv) Genetic base of a population is influenced by various factors (see later).

(v) Genetic base can be broaden by various breeding approaches.

(vi) The genetic base is of two types, viz. (a) broad genetic base and (b) narrow genetic base.

(vii) Broad genetic base provides protection to crop plants from biotic and abiotic stresses.

Q.2 What are types of genetic base ?

Ans. Depending upon the amount of genetic diversity present in a plant breeding population, genetic base is of two types, viz. *(i)* broad genetic base and *(ii)* narrow genetic base. These are discussed below:

(I) Broad Genetic Base. Plant populations that are composed of several pure lines, inbred lines or heterozygotes and homozygotes are said to have broad genetic base. Examples of such populations are mass selected varieties, multiline varieties, synthetics, composites and land races. Such genotypes have wider adaptation to environmental changes.

(ii) Narrow Genetic Base. Those plant populations that have been developed from single homozygote or heterozygote are-said to have narrow genetic base. Examples of such populations are pure line variety, varieties developed by backcross, pedigree method bulk method, and single seed decent method, clonal variety and hybrids between two inbred lines or pure lines. Such populations have danger of uniformity.

Q.3. What are Homogeneous Populations ?

Ans. Genetically similar plants constitute homogeneous populations. Such populations may be homozygous or heterozygous. Examples of homogeneous populations are pure lines, inbred lines, F_1 between two inbred lines or pure lines and progeny of a clone. These populations generally have narrow genetic base. The genetic base of F_1 hybrid depends on the genetic base of parental lines.

Q.4. What are Heterogeneous Populations ?

Ans. Genetically dissimilar plants constitute heterogeneous populations. Such populations may be homozygous such as landraces of self pollinated species, mass selected varieties and multiline varieties; and also heterozygous such as synthetics and composites. Heterogeneous populations have wise stability and stable performance under different environments.

Q.5. Explain Homozygous Populations ?

Ans. Non segregating genotypes constitute homozygous populations. Examples of homozygous populations are pure lines, inbred lines, mass selected varieties of self pollinated crops and multiline varieties. Thus pure lines and inbred lines are homozygous and homogeneous and multilines and mass selected varieties are homozygous and heterogeneous, because they are mixture of several purelines.

Q.6. What are Heterozygous Population ?

Ans. Genotypes that segregate into various types on selfing constitute heterozygous populations. Examples of heterozygous populations are F_1 hybrids, composites and synthetics. F_1 hybrids are heterozygous but homogeneous. Composite and synthetics are heterozygous and heterogeneous and have greater buffering capacity to environmental fluctuations.

Q.7. What are the factors affecting genetic base of a population ?

Ans. There are various factors that affect the genetic base of a population. Such factors include *(i)* mode of pollination, *(ii)* genetic diversity, *(iii)* breeding method, *(iv)* genetic polymorphism, and *(v)* genetic material. These are briefly discussed below:

(i) Mode of Pollination. Cross pollinated species have more genetic diversity (heterogeneity and heterozygosity) than self pollinated species. As a result cross pollinated species have broad genetic base which provides them more buffering capacity to environmental fluctuations.

(ii) Genetic Diversity. Genetic diversity refers to presence of heritable variation in a population. Generally heterogeneous populations give more stable yield over several environments than pure lines. Such populations have broad genetic base which provides them higher degree of buffering capacity to environmental changes. Thus genotypes or population with high genetic diversity have broad genetic base and those with low genetic diversity have narrow genetic base.

(iii) Breeding Method. The genetic base of a population is also affected by breeding method. For example, in self pollinated species, varieties developed by mass selection and multiline breeding have broad genetic base. While varieties developed by backcross, pedigree, bulk and single seed descent methods have narrow genetic base.

(iv) Genetic Polymorphism. Genetic polymorphism refers to regular occurrence of several phenotypes in a population. It is maintained due to superiority of heterozygotes over homozygotes. It enhances diversity of genotypes in a population and thus provides more buffering capacity to a population. Thus polymorphic populations have broad genetic base.

(v) Genetic Material. The genetic base is also influenced by genetic material. For example land races, multilines, synthetics and composites have broad genetic base, where as pure lines and inbred lines have narrow genetic base. Similarly F2 populations and double cross hybrids are more stable than F_1 population, because F_2 population has more genetic diversity and broad genetic base.

Q.8. What are the advantages of broad genetic base ?

Ans. In plant breeding populations, broad genetic base has several advantages. It provides protection from *(i)* biotic stresses, *(ii)* abiotic stresses, *(iii)* wide adaptation, *(iv)* yield stability, and *(v)* high heterosis. These are briefly discussed below :

(i) Protection from biotic stresses. Populations with broad genetic base are composed of several genotypes which provide them protection from both biotic (insects, diseases) and abiotic stresses (drought, soil salinity, frost, cold, heat etc.). The broad genetic base provides high buffering capacity which provides protection form biotic stresses because all the component lines of the population will never become susceptible to a particular disease or insect. Thus it provides insurance against yield losses caused by biotic stresses.

(ii) Protection from Abiotic Stresses. Abiotic stress refers to adverse effects of environmental factors such as moisture stress, salinity, frost, heat etc. on growth and reproduction of crop plants. Populations with broad genetic base have several component lines in their genetic composition which provide them high buffering capacity to environmental changes.

(iii) Wide Adaptation. Populations with broad genetic base have wider adaptation due to presence of more genetic diversity (heterogeneity and heterozygosity). Broad genetic base provides high buffering capacity against environmental changes, resulting in wider adaptability .

(iv) Yield stability. Populations with broad genetic base provide stable yield over regions and seasons due to their wide adaptation or high buffering capacity to environmental changes.

(v) High heterosis. It has been observed that genotypes with broad genetic base result in manifestation of high level of heterosis when used as a parent in development of hybrid.

Q.9. What are demerits of broad genetic base ?

Ans. There are two main disadvantages of populations with broad genetic base, viz. *(i)* less uniform, and *(ii)* low yield. These are discussed below:

(i) Less Uniform. The produce of heterogeneous population that have broad genetic base is lesser uniform and lesser attractive than those of pure lines, inbred lines and F_1 hybrids which are homogeneous and have narrow genetic base.

(ii) Low yield. The yield of heterogeneous populations is always lesser than that of most productive pure line cultivar under normal conditions. However, under adverse conditions the yield of heterogeneous population is higher than uniform pureline cultivars.

Thus populations with narrow genetic base have two main advantages, viz. *(i)* uniform produce, and *(ii)* higher yield. For example the produce of pure lines, inbred lines and F_1 hybrid between two inbred lines/pure lines is very attractive and uniform. However, such populations have two main disadvantages, viz. (i).danger of uniformity and *(ii)* poor adaptation. These are briefly discussed below:

(i) Danger of Uniformity. Populations with narrow genetic base have danger of uniformity. Such populations are derived from single homozygote (such as pure lines and inbred lines) or single heterozygote (such as F_1 hybrid). If such population is attacked by a disease or insect it may lead to crop failure due to lack of heterogeneity.

(ii) Poor Adaptation. Uniform populations such as pure lines and inbred lines have poor adaptation due to lack of heterogeneity. Such population can not withstand adverse effects of environment due to poor buffering capacity.

Q.10. Explain briefly various ways of broadening the genetic base of a population.

Ans. The genetic base of plant breeding populations can be broaden in various ways as suggested/indicated below:

1. In self pollinated species, mass selected varieties, multiline varieties and varietal blends can be used for broadening the base of populations.

2. In cross pollinated species, mass selected varieties, synthetics and composites can be used to develop populations with broad genetic base. Three-way, double and composite or multiple crosses can be made for broadening the genetic base of the populations.
3. Population improvement approaches, viz. recurrent selection, disruptive selection, biparental mating and diallel selective mating should be used for developing populations with broad genetic base. All these approaches can be used for broadening of genetic base in self pollinated species. In cross pollinated species, the first three approaches are used.
4. The biotechnology can be used by creating genetic diversity in plant breeding populations and thereby in broadening the genetic base.
5. The introgressive hybridization is a very potent tool for increasing genetic diversity in breeding populations. It can be used for broadening the genetic base of breeding populations.
6. The gene pyramiding technique can be used for broadening the genetic base against biotic and abiotic stresses.

GENETIC ENHANCEMENT OF CROP PLANTS

Q.1. Define genetic enhancement and give its main features.

Ans. The term genetic enhancement was first used by Jones in 1983. It refers to transfer of useful genes from exotic or wild types into agronomically acceptable background. In 1984, Rick used the term pre-breeding to describe the same activity. Now terms genetic enhancement and pre-breeding are used as synonyms and interchangeable. However, the term genetic enhance is in more usage. Main points related to pre-breeding/genetic enhancement are briefly discussed below.

1. **Main Purpose.** Genetic enhancement is carried out for broadening the genetic base of the population. It increases genetic diversity in the population. In other words, it helps in developing population with more genetic diversity.
2. **Main Steps.** GE consists of three main steps, viz. *(i)* identifying a useful character *(ii)* Capturing its genetic diversity, and *(ii)* putting those genes into usable form.
3. **Base Material.** For launching genetic enhancement programme, three types of material viz. agronomic base, exotic germplasm and wild species are required. The former is used as the recepient parent and last two as donor parents.
4. **Breeding Methods used.** The backcross method is widely used for genetic enhancement, which helps in transferring useful genes from exotic germplasm and wild species into acceptable agronomic base. Now biotechnological approaches are also used for genetic enhancement.
5. **Impact.** The major impact of genetic enhancement is value addition in the germplasm. In other words it leads to genetic improvement of germplasm for various economic characters. It also leads to creation of new variability .It improves adaptation of the population.
6. **End product.** The end products of genetic enhancement programme are improved germplasm lines which can be used as parents for developing productive cultivars/hybrids in traditional breeding.

7. **Time Frame.** The genetic enhancement is a long term programme as compared to traditional plant breeding or sustainable plant breeding.
8. **Undertaken by.** The genetic enhancement work is undertaken by public sector plant breeding organizations. It is seldom undertaken by private sector plant breeding organizations.
9. **Phase of Breeding.** Pre-breeding is the early component of sustainable plant breeding which refers to the transfer or introgression of useful genes and gene combinations from non-adapted sources into breeding material (FAO, 1996). The non-adapted material mainly consists of exotic germplasm and wild species/relatives of the crop plants. It is considered as prior step of sustainable plant breeding.
10. **Adaptation.** Pre-breeding helps in improving the genetic adaptation of the population in two ways. Firstly by creating genetic diversity and secondly by broadening the genetic base of the population.

Q.2. Give a brief comparison of pre-breeding and traditional breeding.

Ans. The strategy of pre–breeding does not differ significantly from the general approach of traditional plant breeding. The pre-breeding does not include cultivar development and is confined only to improvement of germplasm. The enhanced germplasm can more readily be used in breeding programmes for development of cultivars/hybrids. A brief comparison of pre-breeding and traditional breeding is presented below:

Table 9.1 Comparison of pre-breeding and traditional breeding

Pre Breeding	Traditional Breeding
1. It is also known as genetic enhancement.	It is also known as sustainable plant breeding.
2. It leads to genetic enhancement of germplasm.	It leads to development of productive cultivars/hybrids :
3. It leads to value addition.	It does not lead to value addition.
4. It leads to broadening the genetic base of the population.	In improved cultivars the genetic base becomes narrow.
5. The chief breeding method is backcross method.	All breeding methods such as introduction, selection, hybridization and mutation are used.
6. The end products are improved germplasm lines.	The end product is cultivar or hybrid.
7. The end product is used as parent for developing improved cultivar, hybrid.	The end product is used for commercial cultivation.

8. It involves adapted and non-adapted genotypes in crossing programme.	It includes only adapted genotypes.
9. It is a long term breeding programme.	It is a short or medium term breeding programme.
10. It is taken up by public sector plant breeding organizations.	It is taken up by both public and private sector organizations.

Q.3. What are base materials used for genetic enhancement ?

Ans. The breeding material that is required for launching genetic enhancement programme consists of an adapted cultivar, exotic germplasm, land races and wild species.

Q.4. Explain briefly main features of adapted cultivars.

Ans. Commercially cultivated varieties are referred to as adapted cultivars. Such cultivars are also known as popular varieties or good agronomic bases. Adapted cultivars are of two types, viz. obsolete varieties and modern cultivars. Improved varieties of recent past are known as absolute cultivars and currently cultivated high yielding varieties are known as modern cultivars. The main features of adapted cultivars are given below:

(i) They are successful under specific environmental conditions.
(ii) They are also known as popular varieties or good agronomic bases.
(iii) They are high yielding and uniform.
(iv) They have narrow genetic base due to less genetic diversity.
(v) They have local or narrow adaptation.
(vi) They have danger of uniformity due to narrow genetic base.
(vii) They possess most of the desirable traits and are deficient in one or two characters.
(viii) They are generally used as female parents or recurrent parents in pre-breeding programmes.

Q.5. Give main points related to exotic germplasm.

Ans. The germplasm which is collected from other countries is known as exotic germplasm. Such germplasms do not have immediate use without selection for adaptation in a given environment. Exotic germplasm has to undergo conversion before use in plant breeding. Main points about exotic germplasm are given below:

(i) Exotic germplasm is also known as un-adapted material which possesses several inferior alleles in addition to some desirable alleles.
(ii) Use of such material may lead to introduction of inferior alleles into adapted genotypes. Several backcrosses will be required to eliminate undesirable alleles.

(iii) Sometimes, there is linkage between desirable and undesirable alleles which restricts the use of exotic germplasm in breeding programmes.

(iv) It may negatively affect adaptedness when introduced into locally adapted genetic.

(v) It may lead to disruption of co-adapted gene complexes in the elite breeding material.

(vi) Exotic germplasm is less productive and has poor adaptation.

(vii) It cannot be used as such in traditional breeding. It requires some genetic modification or enhancement before use in traditional breeding.

(viii) In Pre-breeding exotic germplasm is generally used as male parents or donor parent.

Q.6. What do you mean by landraces ? Give their main features.

Ans. Land races refer to primitive crop cultivars which were selected and cultivated by the farmers for many generators. Main features of land races are given below:

(i) Land races evolved under subsistence agriculture. They were not deliberately bred like modern cultivars.

(ii) Land races have high level of genetic diversity which provides them high degree of resistance to biotic and abiotic stresses.

(iii) Land races have broad genetic base which provides them wider adaptability.

(iv) Main drawbacks of land races are that they are less uniform and low yielders.

Q.7. Explain main features of wild species.

Ans. Wild species refer to wild forms of cultivated crop plants. Wild species have high degree of resistance to biotic and abiotic stresses. Main features of will given below:

(i) They are wild relatives of cultivated plants.

(ii) They have common ancestry with crop plants.

(iii) They can cross with crop plants.

(iv) They are perenials, less adapted and poor yielders,

(v) When crossed with cultivated species, three types of problems viz. hybrid sterility, hybrid inviability and transfer of undesirable genes; are faced. Because of these problems, wild species are being used to lesser extent in pre-breeding.

Q.8. What are methods of genetic enhancement ?

Ans. Breeding methods used for genetic enhancement are slightly different from those which are used in traditional breeding. The breeding methods/procedures which are commonly used for genetic enhancement include back cross, convergent improvement and marker assisted backcross.

Q.9. Explain role of backcross method in genetic enhancement.

Ans. Backcross refers to crossing of F_1 with either of its parents. Backcross is widely used for introgression and incorporation of desirable genes from exotic germplasm and wild species into well adapted cultivars. Introgression refers to transfer of one or a few alleles from exotic stocks to adapted breeding populations, Incorporation refers to transfer of alleles from exotic germplasm to develop material with broad genetic base and adaptation to local conditions. Main points about backcross method are given below:

(i) It is used for genetic enhancement of both self and cross pollinated species.

(ii) It is used for transfer of both oligogenic and polygenic traits. However, it is more effective with oligogenic traits than polygenic characters.

(iii) It is more successful when the character under transfer has high heritability.

(iv) In this method, cultivated variety or adapted variety is used as maternal parent and the exotic germplasm or un-adapted line as donor parent. The recipient parent is also known as recurrent parent because repeated backcrossing is done with recipient parent.

(v) Generally 5 to 6 backcrosses are sufficient for the transfer of a character.

(vi) Continuous backcrosses are made when the character is governed by dominant gene. Each backcross is followed by one selling when the character is governed by recessive gene.

(vii) The new line differs from the parental line only for the character under transfer. In other words, the end product resembles the parent variety in all the characters except for the character under transfer.

(viii) Various modifications of backcross method have been suggested by different workers to achieve specific objectives. To get intermediate expressions alternate backcross can be made with recurrent and donor parents. To obtain several populations of common genetic backgrounds, 2-3 backcrosses are made and then selfing is done for several generations.

Q.10. How convergent Improvement is useful in genetic enhancement.

Ans. In this method two genotypes say A and B are selected for crossing. The crossing is done between these genotypes and the F_1 is backcrossed to both A and B parents. The main objective of this scheme is to improve both the lines simultaneously. The desirable character of A is transferred to B and vice versa. This method is used when each of the two parent is deficient in one character which is present in the other parent. Main features of this method are given below :

(i) Each parent is deficient in a character which is present in the other parent. In other words, deficiency of one parent is rectified from the allele present in the other parent.

(ii) It simultaneously improves two populations.

(iii) The progress is rapid when the heritability of both the characters is high.

(iv) It can be used for both self and cross pollinated species.

Q.11. Describe role of Marker Assisted backcrossing in genetic enhancement.

Ans. Indirect selection for a desired plant phenotype on the basis of banding patterns of linked DNA (molecular) markers is known as marker assisted selection and improvement of crop plants using such markers is called molecular breeding. Main features of marker assisted selection (MAS) are briefly presented below:

(v) MAS is applicable to both self and cross pollinated species.

(vi) Molecular markers that are commonly used in plant breeding include RFLP, AFLP , RAPD, SSR etc.

(vii) Molecular markers have very high accuracy.

(viii) MAS permits identification of recessive alleles even in heterozygous condition and thus speeds up the progress of crop improvement work.

(ix) DNA markers are free from environmental effects.

(x) DNA markers permit tagging of quantitative trait loci (QTL).

(xi) MAS requires well trained manpower for handling of equipments, isolation of DNA molecule and study of DNA markers.

(xii) It requires sophisticated and well equipped laboratory to initiate work on marker assisted selection.

(xiii) This is very expensive method of genetic enhancement.

In backcross breeding method, use of DNA markers helps in identification of genotypes having desired allele. Thus it helps in speeding up the progress of gene transfer through backcross method. The DNA markers that are linked to the desired phenotypic trait are used for selection.

Q.12. What are types of gene pool used for genetic enhancement ?

Ans. There are various types of genepool that are used for genetic enhancement. Each type of genepool has its own significance and limitations. Four types of genepool have been identified so far. Their brief description is presented in Table 9.2.

Table 9.2 Types of genepool and their description

Type of Genepool	Designation	Definition/brief description
1. Primary Genepool	GP1	Intermating is easy and leads to production of fertile hybrids. It refers to genotypes of the same species. It is of prime breeding importance.
2. Secondary Genepool	GP2	It leads to partial fertility on crossing with primary genepool. It includes related species.
3. Tertiary Genepool	GP3	It leads to production of sterile hybrids on crossing with primary genepool. It includes unrelated species.
4. Quaternary Genepool	GP4	It refers to crop cultivars that have been developed through biotechnological approaches. It includes transgenic varieties and other value added genotypes.

The first three types of genepool are found in the nature while the fourth type (GP4) is created by biotechnological approaches. The primary genepool is used for transfer of desirable alleles and broadening the genetic base of breeding populations. The secondary and tertiary genepools are used for creating vast genetic variablity for various oligogenic and polygenic traits.

Q.13. What are advantages of genetic enhancement ?

Ans. In the past, more emphasis has been laid in developing productive cultivars and hybrids. This has resulted in narrowing down of genetic base in some crop plants. The improved cultivars are productive and uniform but have narrow genetic base and narrow adaptation. The land races, on the other hand are less productive and less uniform but have broad genetic base and wide adaptation. The exotic germplasm is not of much use without selection for local adaptation and enhanced yield potential. The genetic enhancement is essential because of following reasons.

1. It helps in broadening the genetic base of the population which is essential for achieving stability in yield over regions and seasons.
2. The genetic diversity has depleted in the improved cultivars which invites danger of uniformity .Genetic enhancement is useful in restoring genetic diversity in such cultivars.
3. It helps in combining useful genes or gene combination from landraces, perennials and wild species into the cultivated or well adapted

genotypes. Such traits include, resistance to biotic and abiotic stresses, earliness and improvement in quality parameters.

4. It helps in developing plant types which are suitable for machine harvesting.
5. It also leads to creation of new genetic variability in various economic traits. Thus it leads to value addition in the germplasm.
6. The germplasm lines developed through genetic enhancement become usable in traditional breeding programmes for development of productive cultivars/hybrids.

Q.14. What practical achievements of genetic advancement ?

Ans. The genetic enhancement or pre-breeding work has been carried out in almost all the important field crops and remarkable results have been achieved. Some examples are cited below:

1. **Maize.** In maize new inbred lines have been developed through pre-breeding which are used as parents for development of productive single cross hybrids. These single cross hybrids are superior to double cross hybrids. The heterotic pool has been developed.
2. **Sorghum.** In sorghum, desirable traits have been introgressed from Ethiopian and Sudanense land races into adapted Indian cultivars.
3. **Sugarcane.** In sugarcane, pre-breeding has resulted in development of interspecific genetic stocks of S. spontaneum, S. sinense and S. offcinarum.
4. **Barley.** In barley, pre-breeding has resulted in creating vast genetic diversity and broadening the genetic base of breeding populations.
5. **Potato.** In potato, day long tretraploid populations have been developed through introgressive hybridization. The new germplasm is being used widely in potato Improvement. In potato, genetic enhancement has been achieved mainly through polyploidy breeding.

   The genetic enhancement has also been achieved in other crops such as oil seeds, lentil, tomato, chick pea, ground nut, cassava etc. This work is in progress at various International and National Crop breeding centres. Various plant characters such as yield, quality parameters, resistance to biotic and abiotic stresses, adaptation, genetic base, crop maturity, plant type, photo and thermo insensitivity, etc. are kept in mind during pre-breeding programmes.
6. **Cotton.** In cotton, fibre quality parameters and resistance to biotic and abiotic stresses have been introgressed from wild species.

Q.15. What are limitations of genetic enhancement ?

Ans. There are several problems that are associated with genetic enhancement programmes particularly when genes are introgressed from wild species. Some problems are listed below:

1. In the Introgressive hybridization using wild species, there are problems of cross "incompatibility, hybrid inviability and hybrid sterility.
2. Linkage between desirable and undesirable alleles poses problems in utilization of desirable alleles.
3. Generally, the genetic recombinations are restricted in introgressive breeding.
4. Small populations are available due to poor seed setting in interspecific crosses.

10

GENETIC POLYMORPHISM

Q. 1. Define genetic polymorphism ?

Ans. The term genetic polymorphism was first used by Ford in 1940. Genetic polymorphism refers to the regular occurrence of several phenotypes in the genetic population. Main points related to genetic polymorphism are given below:

1. The term genetic polymorphism was coined by Ford in 1940.
2. It has been reported that two third of the loci in a population exhibit polymorphism.
3. The genetic polymorphism is usually maintained due to superiority of heterozygotes over both the homozygotes.
4. When polymorphism is maintained as a result of heterozygote advantage, it is known as balanced polymorphism.
5. Polymorphism can be detected on the basis of morphological, biochemical and molecular traits or markers.
6. Genetic polymorphism increases the buffering capacity of a population by providing increased diversity of genotypes in a population.
7. Genetic polymorphism broadens the genetic base of a population and thus enhances the adaptability of the population.

Q. 2. What are types of genetic polymorphism ?

Ans. There are six types of genetic polymorphism, viz. *(i)* balanced polymorphism, *(ii)* transient polymorphism, *(iii)* neutral polymorphism, *(iv)* Regional polymorphism, *(v)* unisexual Polymorphism, and *(vi)* cryptic polymorphism. Table 10.1

Table .10.1 Classification of Genetic Polymorphism

Basis of Classification	Types of Polymorphism	Brief description
1. Nature of Polymorphism	*(i)* Balanced Polymorphism	It is due to superiority of heterozygote over homozygotes.
	(ii) Transient Polymorphism	It is limited to a particular period.

	(iii) Regional Polymorphism *(iv)* Neutral Polymorphism *(v)* Unisexual Polymorphism *(vi)* Cryptic Polymorphism	It is found in different regions of the habitat. It is neutral in the surviving genotype. It is confined to one sex only. It refers to chromosomal Polymorphism.
2. Detection of Polymorphism	*(i)* Phenotypic Polymorphism *(ii)* Biochemical polymorphism (iii)Molecular Polymorphism	It is detected on the basis of morphological markers. It is detected on the basis of protein markers. It is detected on the basis of variation in the DNA markers.

Q. 3. What is Balanced Polymorphism ?

Ans. The genetic polymorphism which is maintained due to superiority of heterozygote over both the homozygotes is referred to as balanced polymorphism. This leads to regular occurrence of several phenotypes in a population. The term balanced polymorphism was first used by Ford in 1940. Main features of balanced polymorphism are given below:

(a) The polymorphism is maintained due to heterozygote advantage.

(b) This was first reported by Ford in 1940.

(c) This is the most common type of polymorphism observed in plant breeding populations.

Q. 4. What is Transient Polymorphism ?

Ans. A genetic polymorphism that is limited to a particular `period is called transient polymorphism. Sometimes one allele undergoes replacement by a superior allele. The genetic polymorphism during such period is known as transient polymorphism. It is not a regular phenomenon like balanced polymorphism. Thus there are three main features of transient polymorphism.

(a) It is for a limited period.

(b) It is not a regular feature

(c) It was also reported by Ford in 1940.

Q. 5. What is Neutral Polymorphism ?

Ans. It refers to the genetic polymorphism that is dependent on a gene action which is almost neutral in its effect on the survival of the genotype in which it is contained. In other words, the effect on the career genotype is neutral. It results due to neutral mutations. The main features of neutral polymorphism are given below :

(a) It was coined by Ford in 1940.
(b) It results due to neutral mutation.
(c) Neutral mutations take long time in contributing to polymorphism.

Q. 6. What is Regional polymorphism ?

Ans. This refers to occurrence of two or more phenotypes in a population in different regions of the habitat. It results due to adaptation of different individuals in different environment. Main features of regional polymorphism are given below :

(a) It results due to adaptive variation of alleles.
(b) It is not related to superiority of heterozygotes.
(c) It is also known as geographical polymorphism.

Q. 7. Explain unisexual polymorphism ?

Ans. It refers to the genetic polymorphism that is confined to one sex only. It results due to sex limited manifestation of genes. However, such gene can recombine in both sexes.

Q. 8. What is cryptic polymorphism ?

Ans. It refers to genetic polymorphism in which the genetically different alleles can not be identified on the basis of their phenotype. It may include chromosal polymorphism.

Q. 9. What are causes of polymorphism ?

Ans. There are various causes of polymorphism. The possible causes of polymorphism include *(i)* heterozygote advantage, *(ii)* frequency dependent selection, *(iii)* heterogeneous environment, *(iv)* transition and *(v)* neutral mutation. These are briefly discussed below :

(i) Heterozygote Advantage

The natural selection usually favours heterozygotes than homozygotes because heterozygotes are more adaptable than homozygotes. In other words, heterozygotes have more buffering capacity to environmental changes than homozygotes. The heterozygotes maintain genetic polymorphism in a population.

(ii) Frequency Dependent Selection

The frequency dependent selection also leads to maintenance of polymorphism in a population. Generally selection favours those alleles that have low frequency but produce rare phenotype. The selection goes against the alleles that have high frequency. This type of frequency dependent selection maintains balanced polymorphism in a population.

(iii) Heterogeneous Environment

The environment differs from region to region and season to season. The balanced polymorphism is maintained when one allele is advantageous in one environment and another in different environment. In such situation stable polymorphism can be maintained even without heterozygote advantage.

(iv) Transition

In the evolutionary process, sometimes one allele is replaced by another which is more advantageous for adaptation. This may lead to polymorphism in a population. However, such polymorphism is for a limited period and hence is called as transitional polymorphism.

(v) Neutral Mutation

In a population, mutations do arise. However, the majority of mutants are harmful and deleterious. Such mutants are lost only few mutants will survive and replace the original allele. The changes in gene frequency depend on chance. Thus spread of a mutant through the population is erratic. The frequency of a mutant is sometimes increasing and sometimes decreasing. Only those mutants that have selective advantage will survive and contribute to polymorphism in a population. The surviving few mutants take a very long time to spread in the population but contribute to the polymorphism.

Q. 10. What are the methods of detecting genetic polymorphism ?

Ans. In a plant breeding population, the genetic polymorphism can be detected in three main ways, viz. *(i)* on the basis of phenotype *(ii)* biochemical markers and *(iii)* molecular markers. There are briefly discussed below:

(i) On the basis of Phenotype

The best way of detection of genetic polymorphism is the average heterozygosity at various loci. The higher the heterozygosity, the higher the polymorphism. The regular occurrence of several phenotypes in a population is an indication of genetic polymorphism. The polymorphism can be detected on the basis of plant characters such as shape, colour, surface, size of various plant characters, The polymorphism is observed for both oligogenic and polygenic traits.

(ii) Biochemical Markers

Sometimes it is difficult or impossible to identify the polymorphic alleles by visual observations. In such situation the best way of detecting the polymorphic alleles is the isozyme studies or gel electrophoretic studies. Sometimes mutations give rise to protein polymorphism and the variant forms of protein differ only at few amino acid sites. This type of polymorphism can be easily detected by gel electrophoretic studies which throw light on amino acid banding pattern.

(iii) Molecular Markers

Sometimes polymorphic differences are at molecular or DNA level. In other words, the differences are in nucleotide sequences in the DNA. These can be observed by restriction fragment length polymorphisms (RFLPs), amplified fragment length polymorphisms (AFLPs), random amplified polymorphic DNA (RAPD), single sequence repeat (SSR) etc. The molecular or DNA markers are very accurate in detecting the level of polymorphism in a population.

Q. 11. What are the theories of polymorphism ?

Ans. Two theories have been putforth to explain the wide spread existence of polymorphic variation. These are selectionist theory and neutralist theory. A brief account of these theories is presented below :

(i) Selecton Theory

This theory states that polymorphism is balanced or stable and the stable equilibrium is maintained by selective forces. For example, the balanced polymorphism is the selection in favour of heterozygotes.

(ii) Neutral Mutation Theory

This theory was proposed by Kimura (1983) and further elaborated by Crow (1986). According to this theory some polymorphisms are due to presence of mutant alleles that are nearly neutral with regard to fitness. Such alleles were mutated in distant past and are still present in the population contributing to polymorphism. This theory is widely accepted,

Q. 12. What are advantages of genetic polymorphism ?

Ans. There are several advantages of genetic polymorphism which are briefly presented below :

(i) Genetic Diversity

Polymorphic population has greater genetic diversity than pure lines and inbred lines. The genetic diversity avoids danger of uniformity and provides protection from biotic and abiotic stresses to the population.

(ii) Broad Genetic Base

Polymorphic population has broad genetic base due to presence of several phenotypes. Such population has greater buffering capacity to environmental changes.

(iii) Adaptation

Genetic polymorphism enhances the adaptive value of a population by providing increased diversity of genotypes in a population. It also enhances adaptability of a population, because heterozygotes are more adaptable than homozygotes. Genetic polymorphism gives rise to variation of quantitative characters.

Q. 13. What are disadvantages of genetic polymorphism ?

Ans. There are some demerits or disadvantages of genetic polymorphism which are briefly discussed below:

(i) Difficult to get Purelines

It is difficult to get purelines from a polymorphic population. Inbreeding does not have much effect in polymorphic population. It is difficult to control the number of loci that have to be kept in polymorphic state.

(ii) Less Uniform

The polymorphic populations are less uniform due to presence of genetic diversity. The produce of such population is also less uniform and less attractive.

(iii) Low Yield

The yield of polymorphic population is poorer than the best genotype present in the polymorphic population.

ADAPTATION IN CROP PLANTS

Q. 1. Define stability and explain its main features.

Ans. Varietal adaptability to environmental fluctuations is important for the stabilization of crop production both over regions and years. Adaptability is the ability of a genotype to exhibit relatively stable performance in different environments. Adaptability is measured in terms of phenotypic stability of a genotype over several environments. Stability refers to the suitability of a variety for general cultivation over a wide range of environments. The main points of stability analysis are briefly presented below:

1. Study of varietal stability requires multilocation data or data of several years of the same genotypes. In other words, analysis of stability is based on multilocational or multi seasonal data.
2. Stability analysis can be performed with both parental as well as segregating populations. Usually, stability analysis is performed with varieties or strains.
3. There are different models (see later) which are used for stability analysis. Three models, *viz.* Eberhart and Russell model, Freeman and Perkins model, and Perkins and Jinks model are commonly used for stability analysis in plant breeding programmes.
4. Stability analysis helps in the identification of adaptable genotypes which can be grown over wide range of environmental conditions.
5. Stability analysis is useful in predicting varietal response under different environments.
6. The estimates of stability are based on first order statistics and, therefore, are statistically more robust and reliable.

Q. 2. What do you mean by adaptation ?

Ans. Adaptation refers to those changes in structure or function of an individual or population which lead to better survival or greater fitness in a given environment. The main features of adaptation are given below:

1. Adaptation is the process of adjustment of living organisms to the changes occurring in the environment.
2. Adaptation favours those characters which are advantageous for survival and through which an individual acquires adaptive value or fitness in a given environment.

3. In the process of adaptation survival is the main concern.
4. Natural selection plays an important role in the process of adaptation.

Q. 3. What is adaptability ?

Ans. Adaptability refers to the capacity of a genotype or population for genetic changes in adaptation. The main features of adaptability are presented below:

1. Adaptable genotypes produce narrow range of phenotypes in different environments.
2. Adaptability leads to stable performance of a genotype over a wide range of environments.
3. General genotypic and general population adaptations are the examples of wide adaptability .
4. Varietal adaptability is the result of genetic and physiological homeostasis.
5. Productivity is the major concern in varietal adaptability. Human selection plays an important role in combining adaptability and productivity.

Q. 4. What are different types of adaptation ?

Ans. There are four types of adaptation, *viz.* (1) specific genotypic adaptation, (2) general genotypic adaptation, (3) specific population adaptation, and (4) general population adaptation.

Q. 5. What is specific genotypic adaptation ?

Ans. It is the close adaptation of a genotype to a limited environment. For the production of rice in a deep water area, a variety's capacity for rapid inter-node elongation is an essential feature of its specific adaptation.

Q. 6. What is general genotypic adaptation.

Ans. It refers to the capacity of a genotype to produce a wide range of phenotypes compatible with a range of environments. Semi-dwarf varieties of wheat and rice which can be grown over a wide range of environmental conditions are examples of this type of adaptation (*i.e.,* adaptability).

Q. 7. What is specific population adaptation ?

Ans. It refers to the capacity of a heterogeneous population to adapt to specific environment. A composite or a varietal mixture giving stable production is an example of this category. Here the competition is between the components of variety or mixture rather than adaptation of components themselves.

Q. 8. What is general population adaptation ?

Ans. It is the capacity of heterogeneous populations to adapt to a variety of environments. Synthetic varieties of forage crops are example of this type of adaptation. This property of adaptation is specific to an individual genotype or a group of genotypes and is termed as homeostasis.

Q. 9. What are causes of adaptation ?

Ans. Varietal adaptability or stability is the result of homeostasis, which refers to the buffering capacity of a genotype to environmental fluctuations. The homeostasis is of two types, *viz.* (I) genetic, and (2) physiological.

Q. 10. What is genetic homeostasis ?

Ans. It refers to the genetic buffering capacity of a genotype to environmental fluctuations. In other words, it is the ability of a genotype to withstand environmental fluctuations (Lerner, 1954). High buffering ability indicates consistent performance of a genotype or population over a wide range of environments. A more homeostatic genotype is one which shows less variation in phenotypic values in different environments. Thus variability in performance over a wide range of environments can be used as a criterion for measure of phenotypic stability (Gupta *et al.* 1977).

Q. 11. What is physiological homeostasis ?

Ans. It refers to physiological or developmental buffering capacity of a genotype to the environmental fluctuations. The internal self regulatory mechanisms enable the individual to adjust to the fluctuating environments by resisting such changes. Physiological homeostasis is generally higher in heterozygous genotypes than in homozygous ones.

Q. 12. Describe the factors affecting adaptability.

Ans. There are four important genetic factors, *viz.* heterogeneity, heterozygosity , genetic polymorphism and mode of pollination, which affect the adaptability of a genotype over a series of environments (Simmonds, 1979). These are briefly discussed below:

1. Heterogeneity

The heterogeneous populations have broad genetic base. Such populations have greater capacity to stabilize productivity over a wide range of changing environments because such heterogeneity provides a higher degree of population buffering. Generally diverse (heterogeneous) populations give more stable yields over several environments than pure lines.

2. Heterozygosity

It has been observed that heterozygous individuals such as F_1 hybrids, are more stable than their homozygous parents to environmental variation. The stability of heterozygous individuals seems to be related to their ability to perform better under stress conditions than homozygous plants. Such property constitutes physiological homeostasis. In barley, F2 populations were more stable than their parents. Thus heterozygosity promotes the homeostasis. In maize, double cross hybrids are more stable than single crosses. This is probably due to more heterogeneity in double crosses than single crosses.

3. Genetic Polymorphism

The regular occurrence of several phenotypes in a genetic population is known as genetic polymorphism. The genetic polymorphism is usually maintained due to superiority of heterozygotes over both the homozygotes. When polymorphism is maintained as a result of heterozygote advantage,' it is known as balanced polymorphism. Sometimes, it is difficult to identify the polymorphic allelic forms by visual observations. The best way of detecting the polymorphic alleles is the isozyme studies or gel electrophoretic studies. It has been reported that two third of the loci in a population exhibit polymorphism.

Genetic polymorphism increases the adaptive value or buffering capacity of a population by providing increased diversity of genotypes in a population. Thus genetic polymorphism enhances the adaptability of a population, because heterozygotes are more adaptable than homozygotes.

4. Mode of pollination

The cross-pollinated species have better buffering capacity than self-pollinated because of more heterozygosity in the former than latter.

Q. 13. What are types of environments ?

Ans. The environment refers to the external conditions that affect expression of genes of an individual or genotype. Environment has several component factors. These factors include topography, soil type and climatic conditions such as day length, temperatures, relative humidity, etc. Each of these factors exerts its influence on the performance of a genotype. The environments can be classified into two groups : I. Micro-environment, and 2. Macro-environment.

Q. 14. What is micro-environment ?

Ans. The environment of a single organism or genotype as opposed to that of another growing at the same time in almost the same place is referred to as micro-environment. The main features of micro-environment are presented below:

1. The environmental effects are not easily recognizable such as difference in humidity, temperature, etc. at the same place.
2. Micro-environment is governed by unpredictable factors like rain fall, temperature and relative humidity which fluctuate inconsistently.
3. Micro-environment is not under human control.

Q. 15. What is macro environment ?

Ans. Macro-environment refers to the environment of a population in relation to another population growing at the same time in almost in the same place. The main features of macro-environment are given below:

1. The environmental effects are easily detectable such as fertilizer doses, sowing dates, spacings, irrigation schedules etc.
2. Macro-evironment is controlled by predictable factors such as soil types, planting dates, row spacing etc.
3. Macro-environment is under human control.

Q. 16. What are types of environmental variables ?

Ans. Environmental variable can be divided into two groups, *viz.,* (I) predictable, and (2) unpredictable factors. Predictable factors include permanent features of environment which are under human control, such as soil type, planting date, row spacing, plant population, and doses of nutrient application. Unpredictable factors are those which fluctuate inconsistently like rainfall, temperature and relative humidity and are not under human control.

Q. 17. What are the models used for stability analysis in plant breeding ?

Ans. In plant breeding programmes three stability approaches, *viz.* Eberhart and Russell, Perkins and Jinks, and Freeman and Perkins models are used. The ranking of varieties/genotypes for stability remains the same in all these models. However, the model of Eberhart and Russell is in common use as it is comparatively simple one. A brief comparison of three models of stability analysis is presented in Table 11.1

Q. 18. Give Comparison of three models of stability analysis ?

Table 11.1 Comparison of Various Stability Models

Eberhart and Russell Model	Perkins and Jinks Model	Freeman and Perkins Model
1. Involves 3 parameters *i.e.* mean yield, regression coefficient and deviation from regression.	Same as in first model.	Same as in first model.
2. Variation is divided into two fractions, *i.e.,* genotype and E + G×E.	Variation is divided into 3 fractions *i.e.* genotypes, environments and G×E.	Variation is divided into 3 fractions, *i.e.* genotypes, environment and G×E.

3. The E+G ×E is subdivided into 3 parts, *i.e.* Env. (linear) G×E linear and pooled deviations.	The G×E is subdivided into heterogeneity due to regression and ss due to remainder.	The Env. Is divided into combined regression and residual 1. The G×E is subdivided into heterogeneity of regression and residual 2.
4. Calculation is simple.	Calculation is difficult.	Calculation is very difficult.
5. Does not provide independent estimation for mean performance and environmental index.	Same as in first model.	It provides independent estimation of mean performance and environmental index.
6. Less expensive.	Same as in first model.	More expensive than first and second model.
7. The degrees of freedom for environment is 1.	The degrees of freedom for environment is e-2.	The degrees of freedom for environment is e-2

Q. 19. What are applications of stability analysis in crop improvement ?

Ans.

1. Stability analysis helps in understanding the adaptability of crop varieties over a wide range of environmental conditions and in the identification of adaptable genotypes.
2. The use of adaptable genotypes for general cultivation over wide range of environmental conditions helps in achieving stabilization in crop production over locations and years.
3. Use of stable genotypes in the hybridization programme will lead to development of phenotypically stable high potential cultivars of crop species.
4. Yield stability is genetically controlled, as has been shown for maize (Scott, 1967) and hence selection for yield stability can be effective.
5. Stability analysis is an important tool for plant breeders in predicting response of various genotypes over changing environments.

12

COMPONENTS OF GENETIC VARIANCE

Q. 1. What is genetic variance ?

Ans. The heritable portion of phenotypic (total) variance is called genetic variance. It is of three types, viz. *additive, dominance* and *epistatic* variances.

Q. 2. What is additive variance ?

Ans. Average effects of genes on all segregating loci refers to additive variance. It is the component which arises from differences between two homozygotes of a gene *i.e.* AA and *aa*.

Q. 3. What are main features of additive variance ?

Ans. Main features of additive genetic variance are given below :

(i) Additive genes show lack of dominance i.e. intermediate expression.

(ii) Additive variance is fixable.

(iii) Transgressive segregation is the result of additive gene action.

(iv) The gca is a measure of additive variance.

(v) Additive variance responds to selection.

Q. 4. What is dominance variance ?

Ans. The deviation from the additive scheme resulting from intra-allelic interaction is called dominance variance. It is due to deviation of heterozygote (Aa) from the mean of two homozygotes (AA and aa).

Q. 5. What are main features of dominance variance ?

Ans. The main features of dominance variance are given below :

(i) The heterozygote resembles to one of the two parents.

(ii) Dominance variance is not fixable.

(iii) It is associated with heterozygosity.

(iv) The sca is a measure of dominance variance.

(v) Selection is not effective with dominance variance.

(vi) It is the chief cause of heterosis.

Q. 6. What is epistatic variance ?

Ans. Deviation from the additive scheme as a consequence of interallelic interaction. The main features of epistatic variance are given below :

(i) It includes both additive and non-additive components.

(ii) It is of three types, viz. additive x additive, additive x dominance, and dominance x dominance.

(iii) Additive x additive type of epistasis is fixable and rest are non-fixable.

Q. 7. What is additive × additive epistasis ?

Ans. It refers to interaction between two or more loci, each exhibiting lack of dominance individually. It is denoted as A×A and is fixable. It comes under additive gene action. Selection is effective for traits governed by this type of epistatic variance.

Q. 8. What is additive × dominance epistasis ?

Ans. It refers to interaction between two or more loci, one exhibiting lack of dominance and the other dominance individually. It is denoted as A×D and is non-fixable.

Q. 9. What is dominance × dominance epistasis ?

Ans. It refers to interaction of two or more loci, each exhibiting dominance individually. It is represented as D × D and is non-fixable.

Q. 10. What is gene action ?

Ans. The behaviour or mode of expression of genes in a genetic population is called gene action. Gene action is of two types, viz. additive gene action and non-additive gene action.

Q. 11. What is additive gene acton ?

Ans. Joint effects of additive variance and additive x additive type of epistasis refer to additive gene action.

Q. 12. What is non-additive gene action ?

Ans. It includes dominance variance and additive x dominance, and dominance x dominance types of epistasis.

Q. 13. What is fixable variance ?

Ans. It refers to gene action/genetic variance which can be fixed as a true breeding line. It includes additive variance and additive x additive type of epistasis.

Q. 14. What is non-fixable variance ?

Ans. The genetic variance which can not be fixed as a true breeding line is called non-fixable variance. It included dominance variance and additive x dominance and dominance x dominance types of epistatic variances.

Q. 15. What is classification of genetic variance according to Fisher (1918) ?

Ans. Fisher (1918) divided genetic variance into three components, viz. additive, dominance and epistatic variances.

Q. 16. What are components of genetic variance as per Wright (1935) ?

Ans. Wright (1935) classified genetic variance into two components, viz. additive variance and non-additive variance. Non-additive variance includes dominance variance and AxD and DxD types of epistasis.

Q. 17. What are components of genetic variance as per Mather (1949) ?

Ans. Mather (1949) divided genetic variance into two components, viz heritable fixable and heritable non-fixable. The heritable fixable variance is additive variance and heritable non-fixable variance refers to non-additive variance.

Q. 18. What is the genetic constitution of natural breeding populations ?

Ans. In natural plant breeding populations, additive variance is predominant followed by dominance variance and epistatic variance.

Q. 19. What is genetic constitution of F_2 population ?

Ans. F_2 population has ½ additive genetic variance and ¼ dominance variance.

Q. 20. What type of gene action is found in random mating populations ?

Ans. Random mating populations such as composite and synthetic varieties have both additive and non-additive gene actions.

Q. 21. Which breeding methods would be rewarding when a character is governed by additive gene action ?

Ans. In self pollinated species, if a character in governed by additive gene action three breeding procedures, viz. pure line selection, mass selection and progeny selection would be rewarding.

Q. 22. Which breeding methods would be rewarding in cross pollinated species, if a character is governed by additive gene actor ?

Ans. Under such situation recurrent selection for gca, synthetic and composite breeding methods would be rewarding.

Q. 23. Why selection is not effective in a pure line ?

Ans. In a pure line, further improvement through selection is not possible because additive genetic variance is completely depleted. The pure line population has exactly the same genetic constitution. The variation is due to environmental factors.

Q. 24. How variation is generated in a pure line ?

Ans. In a pure line over a period of time, additive variance is regenerated by natural out-crossing and spontaneous mutation.

Q. 25. When additive and non-additive gene actions are of equal magnitude, which breeding method would be effective ?

Ans. When a character is governed by both additive and non- additive gene actions, reciprocal recurrent selection would be rewarding for population improvement.

Q. 26. Which breeding method would be rewarding when a character is governed by non-additive gene action ?

Ans. When a character is governed by non-additive gene action, heterosis breeding would be rewarding both in self- and cross-pollinated species.

Q. 27. In F_1 population, what type of gene action is expected ?

Ans. In F_1 population, usually non-additive (dominance and epistasis) gene action is expected.

Q. 28. What type of gene action is expected in cross pollinated species ?

Ans. In a cross pollinated species, both additive and non-additive (additive, dominance and epistatic variances) gene actions are expected.

Q. 29. What type of gene action is expected in self pollinated species ?

Ans. In self pollinated species, additive gene action (additive variance and additive x additive type of epistasis) are expected.

Q. 30. What biometrical techniques are used for the study of gene action ?

Ans. The gene action can be studied with the help of following biometrical techniques.

(i) Diallel cross analysis
(ii) Partial diallel analysis
(iii) Line x tester analysis
(iv) Generation mean analysis
(v) Triallel cross analysis
(vi) Quadriallel cross analysis
(vii) Triple test cross analysis
(viii) Biparental cross analysis.

Q. 31. What are the factors affecting gene action ?

Ans. The gene action is influenced by the following factors :

(i) Type of genetic material used for study
(ii) Mode of pollination of a species
(iii) Mode of inheritance of the character
(iv) Presence or absence of linkage
(v) Sample size and sampling method
(vi) Method of calculation.

Q. 32. What is the relationship between gene action and heritability ?

Ans. Narrow sense heritability has positive association with additive gene action. Higher the value of additive genetic variance, higher the estimates of heritability in narrow sense and vice versa.

Q. 33. What is the relationship of gene action with heterosis ?

Ans. There is positive association between non-additive variance and magnitude of heterosis. The presence of significant amount of non-additive variance is a pre-requisite for commercial exploitation of heterosis.

Q. 34. How gene action is affected by mode of pollination ?

Ans. The additive gene action is associated with homozygosity, and therefore, it is expected to be the maximum in self pollinated species. The non-additive gene action is associated with heterozygosity and, therefore, it is expected to be maximum in cross pollinated species and minimum in self pollinated species.

Q. 35. What is the role of gene action in plant breeding ?

Ans. The knowledge of gene action to a plant breeder is useful in following three main ways.

(i) In the selection of suitable parents for hybridization.

(ii) In the choice of breeding procedures for the genetic improvement of various quantitative characters.

(iii) In the estimation of some other genetic parameters such as heritability, degree of dominance etc.

13

INTRODUCTION AND SELECTION

[PURELINE, MASS AND CLONAL SELECTION]

Q. 1. What are breeding methods ?

Ans. Various procedures (selections, hybridization, mutation, etc.) which are used for genetic improvement of crop plants are known as breeding methods.

Q. 2. How will you classify plant breeding methods ?

Ans. Plant breeding methods are classified on the basis of mode of pollination and reproduction, application and hybridization as follows :

(i) Based on mode of pollination and reproduction, crop plants are divided into three groups, *viz.* self pollinated species, cross pollinated species and asexually propagated species. Their breeding methods differ to some extent, though some methods are common.

(ii) Based on applications, breeding methods are of three types, *viz.* (*a*) general breeding methods, (*b*) special breeding methods, and (*c*) population improvement approaches.

(iii) Based on hybridization, breeding methods are classified into two broad groups as follows :

(a) Methods which involve hybridization such as pedigree, SSD, bulk and backcross breeding, heterosis breeding, population improvement approaches and molecular breeding.

(b) Methods which do not involve hybridization such as Introduction, Mass selection, Progeny selection, Pureline selection, Clonal selection, Mutation breeding and Transgenic breeding.

Q. 3. What are methods of breeding self-pollinated crops ?

Ans. In self pollinated species, the following breeding methods are used :

(i) Plant Introduction
(ii) Pureline selection
(iii) Mass selection
(iv) Pedigree method

(v) Bulk Breeding method
(vi) Single Seed Descent
(vii) Heterosis breeding
(viii) Mutation breeding
(ix) Polyploidy breeding
(x) Transgenic breeding
(xi) Molecular breeding etc.

Q. 4. What are methods of breeding cross pollinated crops ?

Ans. In cross pollinated crops, following breeding methods are used :

(i) Plant introduction
(ii) Mass and Progeny selection
(iii) Heterosis breeding
(iv) Synthetic breeding
(v) Composite breeding
(vi) Backcross method
(vii) Polyploidy breeding
(viii) Transgenic breeding
(ix) Molecular breeding etc.

Q. 5. What are methods of breeding asexually propagated species ?

Ans. In asexually propagated species, following breeding methods are used :

(i) Plant introduction
(ii) Clonal selection
(iii) Heterosis breeding
(iv) Mutation breeding
(v) Polyploidy breeding
(vi) Backcross method
(vii) Distant hybridization
(viii) Transgenic breeding
(ix) Molecular breeding.

Q. 6. What are breeding methods which are common for self pollinated, cross-pollinated and asexually propagated crops ?

Ans. The following breeding methods are used for genetic improvement of all three groups of crop plants :

(i) Plant introduction
(ii) Backcross method
(iii) Heteroses breeding
(iv) Palyploidy breeding

(v) Transgenic breeding
(vi) Molecular breeding.

Q. 7. What are general breeding methods ?

Ans. Various breeding approaches that are more commonly used for genetic improvement of crop plants are known as general breeding methods. They are listed below :

(i) Plant introduction
(ii) Pureline selection
(iii) Mass selection
(iv) Progeny selection
(v) Synthetic and composite breeding etc.
(vi) Pedigree Method
(vii) Bulk method
(viii) Backcross method
(ix) Heterosis breeding

Q. 8. What are special breeding methods ?

Ans. Those breeding methods that are rarely or less commonly used for crop improvement are known as special breeding methods. Such methods are listed below :

(i) Mutation breeding
(ii) Polyploidy breeding
(iii) Transgenic breeding
(iv) Molecular breeding.

Q. 9. What are breeding methods which do not involve hybridization ?

Ans. The following breeding methods do not involve hybridization :

(i) Plant introduction
(ii) Pureline selection
(iii) Mass selection
(iv) Progeny selection
(v) Mutation breeding
(vi) Clonal selection
(vii) Transgenic breeding.

Q. 10. What is plant introduction ?

Ans. Plant introduction refers to transposition of crop plants from the place of their cultivation to such areas where they were never grown earlier. It is of two types, *viz.* direct introduction and indirect introduction.

Q. 11. What is direct introduction ?

Ans. Introductors which are immediately adapted to the changed environment are called direct introductions.

Q. 12. What is indirect introduction ?

Ans. Introductions which require few years for adaptation are known as indirect introductions.

Q. 13. What is primary introduction ?

Ans. Introduction that can be used for commercial cultivation without any change in the original genotype is called primary introduction.

Q. 14. What is secondary introduction ?

Ans. Introduction that can be used as a variety after selection from the original genotype or used for transfer of some desirable genes to the cultivated variety is known as secondary introduction.

Q. 15. What is exotic variety ?

Ans. A foreign variety which is directly recommended for cultivation is called exotic variety.

Q. 16. What are examples of primary introduction ?

Ans. Examples of primary introduction in wheat and rice are given below :

(i) Wheat : Sonora 64 and Lerma Rojo.

(ii) Rice : Taichung Native 1, IR 8, IR 20 and IR 36.

Q. 17. Cite examples of secondary introduction.

Ans. Examples of secondary introduction are given below :

(*i*) **Wheat :** Kalyan sona and Sonalika. These varieties were released after selection from the material received from Mexico.

(*ii*) **Cotton :** PRS 72, SB 289 E, American Nectariless etc. Many such examples can be cited from other field crops.

Q. 18. How plants were introduced earlier ?

Ans. Earlier plants used to be introduced through travellers, traders, invaders and merchants.

Q. 19. How plants are collected now ?

Ans. Now plants are introduced/collected by specific organizations at international and national level :

(i) At international level, crop plants or gemplasms are collected by International Plant Genetic Resources Institute (IPGRI), Rome, Italy.

(ii) In India germplasm collections are made by following organizations.

  (a) National Bureau of Plant Genetic Resources (NBPGR), New Delhi.

  (b) Forest Research Institute (FRI), Dehradun.

  (c) Botanical Survey of India (BSI), Kolkata.

Q. 20. What are means of Plant Introduction ?

Ans. Plant material is collected by following methods (ways) :

(i) Through expeditions
(ii) Through personal visit
(iii) Through correspondence
(iv) By exchange of material
(v) By purchase, and
(vi) As a gift.

Q. 21. What are purposes of plant introduction ?

Ans. Main purposes of plant introduction are given below :

(i) To use as new source of food, vegetable, fibre fruit, timber plant etc.
(ii) For conservation of biodiversity.
(iii) For genetic improvement of crop plants.
(iv) For aesthic interest—flowering plants.
(v) For study of origin and evolution of crop plants.

Q. 22. What are sources of plant collections ?

Ans. Important sources of plant collections are given below :

(i) Centres of diversity
(ii) Gene banks
(iii) Gene sanctuaries
(iv) Seed companies
(v) Farmers fields
(vi) Markets.

Q. 23. What are uses of Plant introduction ?

Ans. The introduced plant material can be used in following main ways :

(i) It can be used directly as a variety.
(ii) As a new variety after selection.
(iii) As a parent in developing new hybrids.
(iv) As a parent in hybridization for developing new crop cultivars.

Q. 24. What are advantages of Plant introduction ?

Ans. Main advantages of plant introduction are given below :

(i) It is useful in introducing new crop plants.
(ii) It helps in conserving plant biodiversity.
(iii) It is useful in protecting those species that are threatened by danger of extinction.
(iv) It is a simple and rapid method of crop improvement.
(v) It is applicable to all crops *i.e.* self pollinated, cross pollinated and asexually propagated species.

Q. 25. What are demerits of Plant introduction ?

Ans. The main demerit of plant introduction is that sometimes new (*a*) diseases, (*b*) insects, and (*c*) weeds get entry in the country alongwith introduced material.

Q. 26. What are the crops introduced in India from other countries ?

Ans. Various important crops such as maize, potato, tomato, brinjal, groundnut, papaya, pineapple, triticale etc. have been introduced in India from other countries.

Q. 27. What is domestication ?

Ans. The process of bringing wild and weedy species under human management is called domestication.

Q. 28. What is acclimatization ?

Ans. Adaptation of an introduced variety to the new environment is known as acclimatization.

Q. 29. What is selection ?

Ans. The process which favours survival and further propagation of some plants having more desirable characters than others is called selection. It is of two types, *viz. natural* and *artificial.*

Q. 30. What is natural selection ?

Ans. Selection which operates in nature without human interference is called natural selection. It favours those plant characters which are essential for survival (adaptation) of a species.

Q. 31. What is artificial selection ?

Ans. Selection made by human is called artificial selection. It favours those characteristics of plants which are related to yield and quality.

Q. 32. What are different types of artificial-selection ?

Ans. Artificial selections are of different types as given below :

(i) Mass selection
(ii) Pureline selection
(iii) Progeny selection
(iv) Clonal selection
(v) Recurrent selection
(vi) Directional selection
(vii) Disruptive selection
(viii) Stabilizing selection.

Q. 33. Who coined the term Selection ?

Ans. The term *selection* was coined by Darwin in 1858.

Q. 34. What is pureline ?

Ans. Progeny of a self pollinated homozygous plant obtained by selfing is called pureline.

Q. 35. Who coined the term pureline ?

Ans. The term pureline was first used by Johannsen in 1903.

Q. 36. Who coined terms genotype and phenotype ?

Ans. Terms genotype and phenotype were coined by Johannsen in 1909.

Q. 37. Who developed the concept of pureline theory ?

Ans. The concept of pureline theory was developed by Johannsen in 1926 working with Princess variety of common bean (*Phoseolus vulgaris*).

Q. 38. What are main features of purelines ?

Ans. Main features of purelines are given below :

(i) Purelines are homozygous and homogeneous.
(ii) Purelines are relevant to self pollinated crops.
(iii) Purelines are highly uniform.
(iv) There is no heritable variation in a pureline variety.
(v) Purelines have narrow genetic base.
(vi) Purelines have narrow adaptation.

Q. 39. What are sources of variation in a pureline variety ?

Ans. Main sources of variation in a pureline variety are given below :

(i) Natural outcrossing
(ii) Spontaneous mutations and
(iii) Mechanical admixtures.

Q. 40. Why purelinels have poor adaptability ?

Ans. Purelines have pure adaptability due to narrow genetic base.

Q. 41. Why a pureline variety is more prone to the attack of new disease ?

Ans. A pureline variety is more prone to the attack of new race of a pathogen due to genetic uniformity and narrow genetic base.

Q. 42. What is pureline selection ?

Ans. Development of a new variety through identification and isolation of a single best plant progeny from mixed population of self pollinated species is called pureline selection.

Q. 43. What are advantages of pureline selection ?

Ans. Advantages of pureline selection are as follows :

(i) It helps in isolating best genotypes for disease resistance, insect resistance, yield, quality, earliness etc. from a mixed population of an old variety.
(ii) Pureline varieties are uniform and more attractive than mass selected varieties.
(iii) This is an easy and cheap method of crop improvement.

Q. 44. What are demerits of pureline selection ?

Ans. Main demerits of pureline selection are given below :

(i) Pureline varieties have danger of uniformity.

(ii) Pureline varieties have narrow genetic base.

(iii) Pureline varieties have narrow adaptation.

Q. 45. What steps are involved in development of a pureline variety ?

Ans. Pureline selection consists of following steps :

(i) Selection of a heterogeneous material as base population.

(ii) Isolation of purelines by individual plant selection.

(iii) Evaluation of purelines in field trials.

(iv) Release of the best pureline as a variety.

Q. 46. What is mass selection ?

Ans. A method of crop improvement in which individual desirable plants are selected on the basis of phenotype from a mixed population, their seeds are bulbed and used to grow next generation is called mass selection. It is of two types, *viz. positive* and *negative.*

Q. 47. What is positive mass selection ?

Ans. Selection of desirable plants from a mixed population and growing of next generation from the bulk seed of selected plants is called positive mass selection.

Q. 48. What is negative mass selection ?

Ans. Removal of off type plants from a mixed population and allowing rest of the plants further is called negative mass selection.

Q. 49. Who proposed grid method of mass selection ?

Ans. The grid method of mass selection was suggested by Gardner in 1962.

Q. 50. What are main features of mass selection ?

Ans. Main features of mass selection are given below :

(i) It is applicable to both self and cross-pollinated species.

(ii) In self pollinated species, mass selected varieties are homozygous but heterogeneous (mixture of several purelines).

(iii) Mass selected varieties have broad genetic base and wide adaptaion than purelines.

(iv) Mass selected varieties have heritable variation.

(v) The produce of mass selected varieties is lesser uniform than that of pureline varieties.

(vi) Mass selected varieties are less prone to the attack of new race of a pathogen due to genetic diversity.

Q. 51. What are main drawbacks of mass selection ?

Ans. Main drawbacks of mass selection are given below :

(i) The selection is based on phenotype which is not always indication of a superior genotype.

(ii) In cross pollinated species, there is no control on the pollination. Selected plants are pollinated by both superior and inferior pollen parents.

(iii) The produce of mass selected variety is lesser uniform than that of pureline variety.

Q. 52. What are modifications of mass selection ?

Ans. The following modifications of mass selection have been suggested by various workers. :

(i) Rejection of inferior pollen parents.

(ii) Use of composite pollen for pollinating selected plants.

(iii) Gardner (1962) suggested stratified method of mass selection in which field is divided into small plots and 40-50 superior plants are selected in each plot. It reduces the effect of soil heterogeneity.

Q. 53. What is progeny selection ?

Ans. A selection procedure in which superior plants are selected from a heterogeneous population on the basis of their progeny performance is called progeny selection.

Q. 54. What is progeny test ?

Ans. The test of genotypic value of an individual based on performance of its progeny is called progeny test.

Q. 55. Who suggested progeny test ?

Ans. Progeny test was first suggested by Louis de Vilmorin. It is also known as Vilmorin principle.

Q. 56. What is a clone ?

Ans. Progeny of a single vegetatively propagated plant obtained by asexual reproduction is called clone. Main features of a clone are given below.

(i) The progeny of a clone is homogenous but heterozygous.

(ii) Selfing of a clone leads to high inbreeding depression.

(iii) Heterosis can be easily conserved in clonally propagated crops.

(iv) The progeny of a clone has wider adaptability due to heterozygosity.

Q. 57. What are sources of variation in a clone ?

Ans. Main sources of variation in a clone are given below :

(i) Bud mutations

(ii) Mechanical mixture, and

(iii) Occasional sexual reproduction.

Q. 58. What is clonal selection ?

Ans. A procedure of selecting superior clones from the mixed populations of asexually propagated crops such as sugarcane, potato etc. is called clonal selection.

Q. 59. Who coined the term clone ?

Ans. The term clone was first used by Webber in 1903.

Q. 60. Why clones are more prone to new race of a pathogen ?

Ans. Clones are more prone to the incidence of new race of a pathogen due to uniformity (homogenous nature).

Q. 61. Why a cross between two clones exhibit segregation in F_1 ?

Ans. Clones are heterozygous, hence F_1 generation of a cross between two clones exhibits segregation.

Q. 62. Compare a pureline and a clone.

Ans. Comparison of a pureline and a clone is presented below in Table13.1.

Table 13.1. Comparison of a Pureline and a Clone

Particulars	A pureline	A clone
1. Relevant to	Autogamous species.	Asexually propagated species.
2. Definition	Progeny of a single homozygous self fertilized plant.	Progeny of single asexually propagated plant obtained by asexual means.
3. Genetic constitution	Homozygous and homogeneous.	Heterozygous and homogeneous.
4. Adaptation	Narrow	Wide
5. Component genotypes	Identical.	Identical
6. Produce/Product	Highly uniform.	Highly uniform.
7. Variation	Entirely environmental	Entirely environmental
8. Sources of genetic variation	Natural outcrossing spontaneous mutations, and mechanical mixtures.	Bud mutations, mechanical mixtures and occasional sexual reproduction.
9. Equilibrium.	Trivial, stable.	Heterozygous, stable.
10. Vulnerability to disease	Highly vulnerable to new races of a disease.	Highly vulnerable to new races of a disease.
11. Exact reconstitution	Not possible.	Not possible.
12. Maintenance	By natural self pollination.	By asexual propagation.

Q. 63. Compare mass selection and pureline selection.

Ans. Comparison of mass selection and pureline selection is presented below in Table 13.2.

Table 13.2. Differences between Mass Selection and Pureline Selection

Particulars	Mass Selection	Pureline Selection
1. Application	Used in both cross and self pollinated species.	Used in self pollinated species.
2. Genetic constitution of variety	Homozygous but heterogenous in self pollinated species. In cross pollinated species, heterozygous and heterogenous.	Homozygous and homogenous.
3. Component lines	Mixture of several purelines in self pollinated species. In cross pollinated species, mixture of several open pollinated genotypes.	It is a progeney of a single homozygote.
4. Genetic variation	Present	Absent
5. Adaptation	Mass selected variety has wide adaptation.	Pureline variety has narrow adaptation.
6. Genetic base	Mass selected variety has broad genetic base.	Pureline variety has narrow genetic base.
7. Produce of variety	Less uniform.	Highly uniform.
8. Progeny testing	Not required.	Progeny testing is essential.
9. Basis of Selection	Selection is based on the phenotype of plants.	Selection is based on the progeny performance of plants.
10. Time required	Release of new variety takes 6-7 year.	Release of new variety takes 9-10 years.
11. Vulnerability to new race of pathogen	Low	High

HYBRIDIZATION METHODS

[PEDIGREE, BULK, SSD AND BACKCROSS METHODS]

Q. 1. What are breeding methods involving hybridization ?

Ans. The following breeding methods involve hybridization :

(i) Pedigree breeding method.

(ii) Bulk breeding method.

(iii) Single seed descent method.

(iv) Heterosis breeding, and

(v) Population improvement approaches.

Q. 2. What are uses of hybridization ?

Ans. Hybridization is useful in following main ways :

(i) In combining desirable characters from different sources into one genotype.

(ii) In creating vast genetic variability for various economic characters in a population.

(iii) For developing hybrid varieties for commercial cultivation.

(iv) In developing simple varieties.

(v) In transferring desirable gene from wild species to the cultivated one.

Q. 3. Define hybridization.

Ans. Crossing between genetically dissimilar genotypes is called hybridization. After hybridization, the segregating material is handled in three main ways, *viz.* pedigree method, bulk method and backcross method.

Q. 4. What is pedigree ?

Ans. Record of the ancestry of an individual selected plant for its various generations is called pedigree.

Q. 5. What is pedigree breeding ?

Ans. A selection procedure which is used in segregating population of self pollinated species, and keeps proper records of plants and progeny selected in each generation is known as pedigree breeding.

Q. 6. What is mass pedigree method ?

Ans. It is a modified form of pedigree method in which segregating material is handled by bulk (mass) method when conditions are unfavourable for selection and by pedigree method when conditions are favourable for selection.

Q. 7. Who proposed mass pedigree method ?

Ans. Mass pedigree method was proposed by Harrington in 1937 and it was first used by him in cotton.

Q. 8. What are main features of pedigree method ?

Ans. Main features of pedigree method are given below :

(i) In self pollinated species, it is used for developing new varieties.

(ii) In cross pollinated species, it is used for developing inbreds.

(iii) Pedigree records are maintained for each selected plant in various generations.

(iv) A variety developed by this method is homozygous and homogenous because it is a progeny of single homozygote.

(v) Selection is based on progeny test (genotypic value).

Q. 9. What are advantages of pedigree method ?

Ans. Main advantages of pedigree method are given below :

(i) It offers chances of obtaining transgressive segregants.

(ii) It provides information about mode of inheritance of various qualitative characters.

(iii) The selection is based on genotypic value rather than phenotypic value.

Q. 10. What are demerits of pedigree method ?

Ans. Main demerits of pedigree method are given below :

(i) Breeder has to handle huge breeding material which sometimes becomes difficult.

(ii) Breeder has to maintain pedigree records for all selected plants for different generations which takes lot of valuable time of the breeder.

Q. 11. What varieties of rice have been developed by pedigree method?

Ans. In rice, varieties Krishna, Ratna, Sabarmati, Padma, Jaya, Bala, Kaveri etc. have been developed by pedigree method.

Q. 12. What varieties of wheat have been developed by pedigree breeding?

Ans. In wheat several varieties such as HD 2281, HD 2285, HD 2380, HD 2402, Janak, Pratap, Raj 2535, DWR 39, WH 331 and WL 616 have been developed by pedigree method.

Q. 13. What are pedigree bred varieties of cotton ?

Ans. In cotton several varieties such as LH 900, LH 1556, F 1054, F 1378, HS 6, H 1098, RST 9, Vikas, Khandwa 2, Sharda, Abhadita, Sahana, MCU 9, MCU 11,

LRA 5166, Anjali, Surabhi and Suvin have been developed by pedigree method.

Q. 14. What are pedigree bred varieties of chick pea ?

Ans. In chick pea, varieties T_1, T_2, T_3, T_5 and Radhey are outcome of pedigree breeding.

Q. 15. What are pedigree bred varieties of mungbean ?

Ans. In greengram (mungbean), varieties T2, T44, T51 and Sheela have been developed by pedigree method.

Q. 16. What is bulk breeding method ?

Ans. A selection method (procedure) used in segregating population of self pollinated species in which material is grown in bulk plots from F_2 to F_5 with or without selection, next generation is grown from bulk seeds and individual plant selection is practised in F_6 or later generations, is known as bulk breeding method.

Q. 17. Who developed the concept of bulk breeding method ?

Ans. The concept of bulk breeding method was developed by Nilsson Ehle in 1908.

Q. 18. What is evolutionary method of crop improvement ?

Ans. Bulk breeding method leads to significant evolutionary changes in the gene frequencies in a population. Hence, it is also known as evolutionary method of crop improvement.

Q. 19. What are main features of bulk breeding method ?

Ans. Main features of bulk breeding method are given below :

(i) It is used in self-pollinated species.

(ii) Natural selection operates during bulking period.

(iii) Selection in final stages is based on progeny test.

(iv) The variety developed by bulk method is homogenous and homozygous.

Q. 20. What are advantages of bulk breeding method ?

Ans. Main advantages of bulk breeding method are given below :

(i) It improves adaptation of the population, because natural selection operates during bulking period.

(ii) Chances of getting transgressive segregants are more in this method than pedigree breeding.

(iii) There is no need of maintaining pedigree records.

Q. 21. What are demerits of bulk method ?

Ans. Main demerits of bulk breeding method are given below :

(i) Information about mode of inheritance of various oligogenic characters can not be obtained.

(ii) It takes 2-3 years more for release of a variety than pedigree method of breeding.

(iii) Sometimes, natural selection favours undesirable alleles.

Q. 22. What is the variety of brown mustard developed by bulk breeding method ?

Ans. In brown mustard, variety "Narendra Rai" has been developed by bulk breeding method. In USA, more than 50 varieties of self pollinated crops have been developed by this method.

Q. 23. What is single seed descent method ?

Ans. A breeding procedure which is used with segregating population of self pollinated species in which plants are advanced by selecting single seed per plant from F_2 generation onward is called single seed descent method.

Q. 24. Who developed concept of single seed descent ?

Ans. Single seed descent (SSD) method was first suggested by Goulden in 1939 for advancing segregating population of self pollinated species.

Q. 25. What are main features of single seed descent method ?

Ans. Main features of SSD method of breeding are given below :

(i) It is used in self pollinated species.

(ii) Artificial selection is practised.

(iii) It is modified form of bulk breeding method.

(iv) The variety developed by this method is homozygous and homogeneous.

Q. 26. Who first used SSD method in oats ?

Ans. In oats, single seed descent method was first used by Grafius in 1965.

Q. 27. Who first used SSD method in soybean ?

Ans. In soybean, single seed descent method was first used by Brim in 1966.

Q. 28. What are advantages of SSD method ?

Ans. Main advantages of SSD method are given below :

(i) This method requires less space and labour. Hence large number of crosses can be evaluated by this method.

(ii) This is a simple, time saving and less expensive method.

(iii) It retains considerable variability in the population.

Q. 29. What are demerits of SSD method ?

Ans. Main demerits of SSD method are given below :

(i) Artificial selection can not be practised till F_5 generation.

(ii) Inheritance of oligogenic characters can not be studied.

(iii) Transgressive segregants can not be obtained due to selection of single seed per plant.

Q. 30. What is backcross ?

Ans. Crossing of F_1 with either of its parents is termed as backcross. When F_1 is crossed with its homozygous recessive parent, it is called test cross.

Q. 31. Who coined the term test cross ?

Ans. The term test cross was coined by Bridges in 1934.

Q. 32. What is donor parent ?

Ans. The parent which donates desirable character is called donor parent. It is also known as non-recurrent parent because donor parent is used only once in the crossing programme.

Q. 33. What is recipient parent ?

Ans. The parent which receives a desirable character is known as recipient parent, also called recurrent parent, because it is repeatedly used in backcrossing.

Q. 34. What are main features of backcross method ?

Ans. Main features of backcross method are given below :

(i) It is used in both self and cross pollinated species.

(ii) It is used to improve specific character of a well adapted variety.

(iii) It includes donor and recipient parents in crossing.

(iv) New variety is similar to parent variety except for the character under transfer.

Q. 35. What are basic requirements to start a backcross programme?

Ans. Basic requirements for starting a backcross programme are given below :

(i) A well adapted variety—deficient in one character.

(ii) A donor parent for desired character.

(iii) High heritability of the character under transfer.

Q. 36. How many backcrosses are required for transfer of oligogenic character ?

Ans. Generally, 5-6 backcrosses are sufficient to retain the genotype of original variety with new character.

Q. 37. What are uses of backcross method ?

Ans. Backcross method is used for various purposes as given below :

(i) It is used for transfer of oligogenic characters such as disease resistance in cultivated varieties.

(ii) It is used for developing isogenic lines and multiline variety.

(iii) Also used for transfer of male sterility and restorer genes to various agronomic bases.

(iv) Very much useful in interspecific gene transfer.

Q. 38. What are drawbacks of backcross method ?

Ans. Main drawbacks of backcross method are given below :

(i) It is more successful for transfer of monogenic and oligogenic characters than polygenic traits.

(ii) It involves crossing work for 5-6 generations.

(iii) Some times there is tight linkage between some desirable and undesirable characters which are also transferred alongwith desirable character.

Q. 39. What are cotton varieties developed by backcross technique ?

Ans. In cotton, varieties V 797, Digvijay, Vijalpa and Kalyan belonging to *Gossypium herbaceum* have been developed by backcross method.

Q. 40. What are multilines ?

Ans. The deliberate seed mixtures of isolines, closely related lines or unrelated lines is called multiline. Multilines are homozygous but heterogeneous.

Q. 41. Who first suggested use of multilines ?

Ans. The use of multilines was first suggested by Jensen in 1952 in oats.

Q. 42. Who outlined method of developing multilines ?

Ans. The method of developing multilines was outlined by Borlang and Gibler in 1953.

Q. 43. What are main features of multilines ?

Ans. Main features of multilines are given below :

(i) Multilines are relevant to self pollinated species.

(ii) Multilines are homozygous but heterogeneous.

(iii) Multilines are mixtures of several isogenic lines, related lines or unrelated lines.

(iv) Multilines are more adaptable to environmental variations than purelines.

(v) Multilines have broad genetic base than purelines.

Q. 44. What are advantages of multilines ?

Ans. There are following two main advantages of multilines.

(i) Multilines have wider adaptability due to genetic diversity.

(ii) They provide better protection from the infestation of new race of a pathogen.

Q. 45. What are drawbacks of multilines ?

Ans. Main drawbacks of multilines are given below :

(i) The yield of multilines is lesser than the most productive cultivar of a pureline under disease free conditions. However, reverse is true under adverse conditions.

(ii) The produce of a multiline variety is less uniform and less attractive than that of a pureline.

Q. 46. Give names of multiline varieties of wheat.

Ans. In India, three multiline varieties of wheat *viz.* KSML 3, MLKS 11 and KML 7404 have been released from PAU, Ludhiana.

Q. 47. How many isogenic lines are mixed to constitute a multiline cultivar ?

Ans. Generally 6-10 isogenic lines are mixed together to constitute a multiline cultivar.

Q. 48. Compare pedigree, bulk and backcross methods of breeding.

Ans. Comparison of pedigree, bulk and backcross methods of breeding is presented below in Table 14.1.

Table 14.1. Comparison of Pedigree, Bulk and Backcross Methods

Pedigree Method	Bulk Method	Backcross Method
1. Used in self and cross pollinated crops	Used in self pollinated crops	Used in self and cross pollinated crops
2. Crossing is done only once	Done only once	Done repeatedly with recurrent parent
3. Human selection plays important role	Both natural and human selections are involved	Human selection is involved
4. Smaller F_2 than bulk method is grown	Larger than pedigree method	Smaller than both pedigree and bulk method
5. Pedigree records are maintained	Pedigree records are not maintained	Pedigree records are not maintained
6. Equally effective with oligogenic and polygenic traits	Equally effective with oligogenic and polygenic traits	More effective with oligogenic traits such as disease resistance
7. Extensive testing is required before release of new variety	Extensive testing is required before release of new variety	Extensive testing is not required, because new variety differs from old variety in one trait only

8. It takes 14-15 years to release new variety.	It requires longer period than pedigree method	It takes 7-8 years for release of new variety
9. Breeding procedure is same for dominant and recessive characters	Same for dominnt and recessive characters	Differs for dominant and recessive characters
10. Variety released by this method has narrow adaptation	Variety released by this method has wide adaptation·	Adaptation is like that of parent variety
11. This method is widely used	Less popular	Widely used
12. New variety is different from both the parents	New variety is different from both the parents	New variety is identical to recurrent parent except for the character transf erred.

POPULATION IMPROVEMENT APPROACHES

Q. 1. What is population improvement ?

Ans. Accumulation of desirable genes in a population through various breeding techniques is called population improvement. And various breeding procedures which are used for accumulation of desirable alleles in a population are known as population improvement approaches.

Q. 2. What are population improvement approaches ?

Ans. In plant breeding following breeding approaches are used for population improvement :

(i) Recurrent selection
(ii) Disruptive selection
(iii) Diallel selection mating
(iv) Biparental mating.

Q. 3. What is recurrent selection ?

Ans. Reselection generation after generation with intermating of selected plants to provide for genetic recombination is known as recurrent selection. It is a cyclic selection which is used to improve the frequency of desirable genes for a particular character in a breeding population.

Q. 4. What are main features of recurrent selection ?

Ans. Main features of recurrent selection are given below :

(i) It is a modified form of progeny selection.
(ii) A heterozygous base population is required to start recurrent selection.
(iii) It is used to improve the frequency of desirable alleles in a population.
(iv) Intermating is done among selected plants.
(v) It can be used both in cross and self-pollinated species.

Q. 5. What are basic assumptions for recurrent selection ?

Ans. Recurrent selection is based on following three basic assumptions :

(i) Absence of epistasis

(ii) Absence of multiple alleles, and

(iii) Absence of linkage disequilibrium.

Q. 6. What type of basic population is used for recurrent selection ?

Ans. In cross pollinated species, any of the following populations can be used to start recurrent selection :

(i) An open pollinated variety.

(ii) A synthetic or composite variety.

(iii) Progeny of intercrosses among selected inbreds.

(iv) A double cross or single cross or multiple cross.

Q. 7. What steps are involved in recurrent selection ?

Ans. A simple recurrent selection scheme consists of following important steps :

(i) Selection of superior plants from base population.

(ii) Selfing of selected plants.

(iii) Growing progeny of selected plants in next generation from selfed seed.

(iv) Intermating among progeny.

(v) Bulking of crossed seed in equal quantity.

This completes original cycle of recurrent selection. The bulk seed is used for next cycle of selection which also involves above five steps.

Q. 8. What are uses of the end product of recurrent selection ?

Ans. The end product of recurrent selection can be used in following main ways :

(i) In producing homozygous inbred lines by selfing.

(ii) In developing hybrid varieties.

(iii) In developing synthetic varieties.

Q. 9. What are types of recurrent selection ?

Ans. There are four types of recurrent selection as follows :

(i) Simple recurrent selection.

(ii) Recurrent selection for gca.

(iii) Recurrent selection for sca.

(iv) Reciprocal recurrent selection.

Q. 10. What is simple recurrent selection ?

Ans. A type of recurrent selection which does not include a tester is called simple recurrent selection, also known as phenotypic recurrent selection. Main features of simple recurrent selection are given below :

(i) The tester is not used.
(ii) In does not measure combining ability.
(iii) Selection is based on phenotype.
(iv) It is used for those characters having high heritability.
(v) One selection cycle is completed in two years.

Q. 11. What is recurrent selection for gca ?

Ans. A form of recurrent selection which is used to improve general combining ability (*gca*) of a population for a character using heterozygous tester is called recurrent selection for *gca*. It is also known as half-sits recurrent selection with heterozygous tester.

Q. 12. What are main features of recurrent selection for gca ?

Ans. Main features of recurrent selection for *gca* are given below :
(i) A heterozygous tester is used.
(ii) It is used for improving gca of a character.
(iii) It is more effective with additive genes.
(iv) Selection is based on test cross performance.
(v) It is used for improving polygenic characters.

Q. 13. What is recurrent selection for *sca* ?

Ans. A form of recurrent selection which is used for improving specific combining ability (*sca*) of a population for a character using homozygous tester is called recurrent selection with homozygous tester.

Q. 14. What are main features of recurrent selection for sca ?

Ans. Main features of recurrent selection for *sca* are as follows :
(i) A homozygous tester is used.
(ii) It is used for improving sca of a character.
(iii) It is effective with non-additive genes.
(iv) Selection is based on test cross performance.
(v) It is used for improving polygenic characters.

Q. 15. What is reciprocal recurrent selection ?

Ans. A form of recurrent selection which is used for improving both *gca* and *sca* of a population for a character using two heterozygous testers is called reciprocal recurrent selection. It is also known as recurrent reciprocal half-sits selection.

Q. 16. What are main features of reciprocal recurrent selection ?

Ans. Main features of reciprocal recurrent selection are as follows :
(i) Two heterozygous testers are used.

(ii) It is used for improving both gca and sca of a character.
(iii) It is equally effective with additive and non-additive genes.
(iv) Selection is based on test cross performance.
(v) It is used for improving polygenic traits.

Q. 17. What are advantages of recurrent selection ?

Ans. Main advantages of recurrent selection are given below :

(i) It is useful in improving frequency of desirable alleles in a population for various polygenic characters.
(ii) It is useful in breaking repulsion phase linkages.
(iii) It creates vast variability in a population for various traits.

Q. 18. What is main drawback of recurrent selection ?

Ans. This method can not be used directly for developing new varieties. The end product can be used in hybridization for developing new varieties/hybrids.

Q. 19. Who gave initial idea of recurrent selection ?

Ans. The initial idea of recurrent selection was independently given by Hayes and Garber in 1919 and East and Jones in 1920.

Q. 20. Who outlined the procedure of recurrent selection ?

Ans. The procedure of recurrent selection was outlined by Jenkins in 1940.

Q. 21. Who coined the term Recurrent Selection ?

Ans. The term Recurrent selection was coined by Hull in 1945 working with maize.

Q 22. Who proposed recurrent selection for *sca* ?

Ans. Recurrent selection for specific combining ability was proposed by Hull in 1945.

Q. 23. Who proposed reciprocal recurrent selection ?

Ans. The reciprocal recurrent selection was proposed by Comstock *et al.* in 1949 for improving both *gca* and *sca* of a character in a population.

Q. 24. Under what conditions reciprocal recurrent selection is effective ?

Ans. Reciprocal recurrent selection is effective under following conditions :

(i) When additive and non-additive gene actions are of equal magnitude.
(ii) When both *gca* and *sca* of a character have to be improved.
(iii) It is equally effective with incomplete, complete and overdominance.

Q. 25. In which conditions recurrent selection for *gca* is more effective ?

Ans. Recurrent selection for *gca* is more effective with incomplete dominance and less effective with overdominance.

Q. 26. In what conditions recurrent selection for *sca* is more effective ?

Ans. Recurrent selection for *sca* is more effective with overdominance and less effective with incomplete dominance. It is used when a character is governed by non-additive gene action.

Q. 27. What is broad based tester ?

Ans. A heterozygous tester is known as broad based tester. In cross pollinated species, an open pollinated variety is used as a broad based tester.

Q. 28. What is narrow based tester ?

Ans. A homozygous tester is known as narrow based tester. An inbred is example of narrow based tester.

Q. 29. In which scheme of recurrent selection a broad based tester is used ?

Ans. In recurrent selection for *gca*, a broad based tester is used. In reciprocal recurrent selection two broad based testers (heterozygous testers) are used.

Q. 30. What is disruptive selection ?

Ans. A breeding approach in which both the extremes for a character are selected in a segregating population and then intermated is referred to as disruptive selection, also known as centrifugal selection.

Q. 31. Who developed the concept of disruptive selection ?

Ans. The concept of disruptive selection was developed by Mather in 1953. It was elaborated by Thoday in 1958 and 1960.

Q. 32. What are main features of disruptive selection ?

Ans. Main features of disruptive selection as follows :

(i) It can be used both in self and cross-pollinated species.

(ii) It is considered as a form of recurrent selection.

(iii) It is effective in breaking undesirable linkages.

Q. 33. What is difference between disruptive and directional selections ?

Ans. Main differences between these two approaches are given below :

(i) In case of directional selection, plants are selected from one extreme and intermated for further selection. In case of disruptive selection, plants are selected from two extremes and intermated for further selection.

(ii) Disruptive selection does not change mean of a population, whereas mean performance is improved by directional selection.

Q. 34. What are advantages of disruptive selection ?

Ans. Main advantages of disruptive selection are as follows :

(i) It is useful in breaking undesirable linkages.

(ii) It is very effective in creating genetic variability.
(iii) It is also effective in improving adaptation of a population.

Q. 35. What are demerits of disruptive selection ?

Ans. There are two main demerits of disruptive selection as given below :

(i) It involves repeated intermating.
(ii) Population of limited crosses can be handled by this technique at a time.

Q. 36. What are the crops where disruptive selection has been used ?

Ans. Disruptive selection has been used in sorghum, brown mustard, cotton and some other crops.

Q. 37. What is directional selection ?

Ans. Selection in favour of an extreme phenotype such as (i) earliness or lateness, and (ii) dwarfness or tallness is known as directional selection. Directional selection leads to improvement in population mean. It is also known as progressive selection and linear selection or dynamic selection.

Q. 38. What is stabilizing selection ?

Ans. Selection in favour of intermediate type is called stabilizing selection. It is also known as centripetal selection. In case of stabilizing selection population mean remains constant.

Q. 39. What are types of selection given by Mather (1953) ?

Ans. Mather (1953) gave following three types of selections :

(i) Disruptive selection.
(ii) Directional selection.
(iii) Stabilizing selection.

Q. 40. Who coined the terms centrifugal and centripetal selections ?

Ans. Terms centrifugal selection and centripetal selection were first used by Simpson in 1953.

Q. 41. Who coined the term progressive selection ?

Ans. The term progressive selection was first used by Schmalhausen in 1949.

Q. 42. Who used disruptive selection in sorghum ?

Ans. In sorghum, disruptive selection was used by Doggett and Majisu in 1972.

Q. 43. Who used disruptive selection in mustard ?

Ans. In brown mustard, disruptive selection was first used for improvement of yield by Murty *et al.* in 1972.

Q. 44. Who used disruptive selection in cotton ?

Ans. In upland cotton, disruptive selection was used for improvement of yield and fibre properties by Narayanan *et al.* in 1987.

Q. 45. What is difference between progeny and recurrent selections ?

Ans. There are following differences between progeny and recurrent selections :

(i) In recurrent selection, selected plants are self pollinated, whereas in progeny selection selected plants are open pollinated.

(ii) In recurrent selection, progeny selected plants are intermated, whereas in progeny selection they are allowed to open pollinate.

Q. 46. What is diallel selective mating ?

Ans. A method of pupulation improvement of autogamous species which involves parental diallel series, F_1 diallel series and selective mating series is called diallel selective mating. It is considered as a form of recurrent selection.

Q. 47. Who developed the concept of diallel selective mating ?

Ans. The concept of diallel selective mating was originally developed by Jensen in 1970 working with wheat.

Q. 48. What are uses of the end product of a diallel selective mating ?

Ans. The end product of diallel selective mating (DSM) can be used in following ways :

(i) As a pureline variety.

(ii) As mass selected variety

(iii) As a parental line in further breeding programme.

Q. 49. What are main features of DMS ?

Ans. Main features of DSM are given below :

(i) It is considered as a form of recurrent selection.

(ii) It is useful in improving small grained self-pollinated crops.

(iii) It permits inclusion of multiple parents in the breeding programme.

(iv) New germplasm can be included in the breeding programme at any stage.

(v) It broadens genetic base of the population, etc.

Q. 50. What are advantages of DSM ?

Ans. Advantages of DSM are given below :

(i) It is useful in breaking undesirable linkages.

(ii) It creates vast genetic variability in a population.

(iii) It is useful in broadening the genetic base of the population.

Q. 51. What is main drawback of DSM ?

Ans. DSM is less effective in improving those characters which have low heritability.

Q. 52. What is biparental mating ?

Ans. A method of pupulation improvement which involves crossing among randomly selected plants in F_2 or later generation to provide for genetic recombination is called biparental mating.

Q. 53. Who developed the concept of biparental mating ?

Ans. The concept of biparental mating was originally developed by Comstock and Robinson (1948, 1952).

Q. 54. What are advantages of biparental mating ?

Ans. Main advantages of biparental mating are given below :

(i) It is very effective in breaking undesirable linkages.
(ii) It creates vast genetic variability in a population.
(iii) It is simple and effective method of population improvement.

Q. 55. What are main drawbacks of biparental mating ?

Ans. Main drawbacks of biparental mating are given below :

(i) Limited populations can be handled by this technique.
(ii) Repeated selection and intermating have to be made for 3-4 generations.

Q. 56. Who coined the term inter-population selection ?

Ans. The term inter-population selection was first used by Grant in 1963.

Q. 57. Compare recurrent selection for *gca* and recurrent selection for *sca*.

Ans. Comparison of recurrent selection for gca and recurrent selection for *sca* is presented below in Table 15.1.

TABLE 15.1. Comparison of recurrent selection for *gca* and recurrent selection for *sca*

Particulars	Recurrent selection for gca	Recurrent selection for sca
1. Application	Used to improve polygenic traits.	Also used to improve polygenic traits.
2. Basis of selection	Test cross performance.	Test cross performance.
3. Tester used	Heterozygous.	Homozygous.
4. Effectiveness	More effective with incomplete dominance.	More effective with complete and overdominance.
5. Condition of use	Used when additive gene action is important.	Used when non-additive gene action is important.
6. Impact	It improves *gca* of a character.	It improves *sca* of a character.

❑❑❑❑❑

16

HETEROSIS BREEDING

Q. 1. What is heterosis ?

Ans. The superiority of F_1 hybrid in one or more characters over its parents is called heterosis ; also known as hybrid vigour.

Q. 2. Who coined the term heterosis ?

Ans. The term heterosis was coined by Shull in 1914 working with maize crop.

Q. 3. What is luxuriance ?

Ans. The increase of F_1 over parents in vegetative growth, but not in yield and adaptation is called luxuriance.

Q. 4. Who coined the term luxuriance ?

Ans. The term luxuriance was coined by Renner in 1929.

Q. 5. What are main features of heterosis ?

Ans. The main features of heterosis are listed below :

(i) The hybrid F_1 exhibits superiority over both parents in yield, adaptation, quality, maturity etc.

(ii) The superiority is confined to F_1 generation.

(iii) The superiority (heterosis) is under genetic control.

(iv) The superiority is reproducible in specific environment.

(v) The heterosis has association with specific combining ability and heterozygosity.

(vi) In hybrids, recessive deleterious genes are concealed in heterozygous condition.

(vii) The frequency of desirable heterosis is very low.

Q. 6. What are genetic bases of heterosis ?

Ans. There are three possible genetic causes of heterosis, *viz.* dominance, overdominance and epistasis. In other words, there are three genetic explanations of heterosis.

Q. 7. Explain dominance theory of heterosis.

Ans. According to dominance hypothesis, heterosis is the result of accumulation of dominant genes in F_1 from both the parents. Suppose the genetic constitution of female parent is AABBccdd and that of male parent as aabbCCDD. A hybrid between these two parents will have four dominant genes (Aa Bb Cc Dd) and exhibit superiority over both the parents having two dominant genes each.

AABBccdd Female parent with 2 dominant genes	×	aabbCCDD Male parent with 2 dominant genes
	↓ Aa Bb Cc Dd Hybrid with 4 dominant genes.	

Q. 8. Who proposed the dominance hypothesis of heterosis ?

Ans. The dominant hypothesis of heterosis was proposed by Davenport (1910), Bruce (1910) and Keeble and Pellow (1910).

Q. 9. What are objections to dominant theory of heterosis ?

Ans. There are two objections to dominance hypothesis of heterosis as follows :

(i) If the dominance hypothesis is true, it should be possible to obtain pure heterotic individuals in F_2 which are homozygous for all the dominant genes. However, such individuals are never obtained.

(ii) If the heterosis is due to dominance, the F_2 curve should be skewed towards dominant genes, but the curve of F_2 is found always smooth and symmetrical.

Q. 10. Why homozygous heterotic individuals with all dominant genes are not recovered in F_2?

Ans. Jones (1917) suggested that due to linkage between some dominant and recessive genes, it is not possible to obtain true breeding individuals with all dominant genes in F_2 generation.

Q. 11. Why the curve of F_2 generation of a hybrid is always smooth and symmetrical ?

Ans. Collins (1921) provided explanation for this. He suggested that trait like yield is governed by large number of genes or polygenes which exhibit continuous variation resulting in symmetrical distribution of genes.

Q. 12. What is overdominance hypothesis of heterosis ?

Ans. According to this hypothesis, heterosis is due to superiority of heterozygote over its both homozygous parents. The superiority of heterozygote over both homozygotes may arise due to following reasons :

(i) Production of superior hybrid substance in heterozygote which is completely different from either of the homozygous products or due to.

(ii) Greater buffering capacity in the heterozygote resulting from cumulative action/stimulation of divergent alleles.

Q. 13. Who proposed overdominance hypothesis of heterosis ?

Ans. The overdominance hypothesis of heterosis was independently proposed in 1908 by Shull and East and it was supported by East (1936) and Shull (1945).

Q. 14. Who coined the term overdominance ?

Ans. The term overdominance was coined by Hull in 1945 working with maize crop.

Q. 15. What is the main difference between dominance and overdominance theories of heterosis ?

Ans. According to dominance hypothesis, heterosis results due to masking effect of dominant desirable alleles over harmful recessive alleles. According to overdominance, heterosis results due to complementation between divergent alleles.

Q. 16. What is the role of epistasis in heterosis ?

Ans. Epistasis is of three types, *viz.* additive × additive, additive × dominance and dominance × dominance. The last type of epistasis (dominance × dominance) may contribute to heterosis.

Q. 17. Which is the most widely accepted theory of heterosis ?

Ans. Dominance hypothesis is the most widely accepted genetic explanation of heterosis.

Q. 18. What are different methods of fixation of heterosis in crop plants ?

Ans. In crop plants, heterosis can be conserved in following three ways :

(i) By asexual reproduction. This method can be used for conserving heterosis in vegetatively propagated crops such as sugarcane, potato, sweet potato, banana etc.

(ii) By apomixis. The fixation of heterosis by apomixis is common in citrus fruits, black berries, roses, blue grasses and many other flowering plants.

(iii) By polyplody. This method can be used for fixation of heterosis in interspecific and intergeneric hybrids.

Q. 19. How heterosis is conserved in *Oenothera* ?

Ans. In evening primerose (*Oenothera* spp.), the heterosis is conserved by balanced lethal system. Homozygotes are lethal, hence they die. Only heterozygotes survive.

Q. 20. List genetic factors which affect level of heterosis in crop plants.

Ans. In crop plants, the level of heterosis has been reported to be associated with genetic diversity, genetic base and adaptability of parental lines involved in hybridization programme. Moreover, the level of heterosis is generally higher in cross pollinated species than in self-pollinated crops.

Q. 21. How will you estimate heterosis ?

Ans. Heterosis is estimated in three ways, *viz.* over the average value (mean value) of both the parents, over better parent and over the value of commercial cultivar/hybrid.

Q. 22. How would you classify heterosis ?

Ans. The heterosis can be classified as follows :

(i) On the basis of estimation, heterosis is of three types, *viz.* average heterosis, heterobeltiosis and useful heterosis or economic heterosis.

(ii) On the basis of direction, heterosis is of two types *viz.* positive heterosis and negative heterosis.

(iii) Heterosis is also classified as real heterosis and pseudo-heterosis (Luxuriance).

Q. 23. How would you estimate average heterosis ?

Ans. The average heterosis is estimated over the mean value of both the parents as follows :

Average heterosis = $\frac{F_1 - MP}{MP} \times 100$ where F_1 is the mean value of F_1 hybrid and MP is the mean value of two parents involved in the hybrid or cross.

Q. 24. Who coined the term useful heterosis ?

Ans. The term useful heterosis was coined by Meredith and Bridge in 1972 working with cotton crop.

Q. 25. How would you estimate useful heterosis ?

Ans. Useful heterosis also known as economic heterosis is estimated over the standard commercial check variety of a region as follows :

Useful heterosis = $\frac{F_1 - CC}{CC} \times 100$ where CC is the mean value of commercial cultivar.

Q. 26. How heterobeltiosis is estimated ?

Ans. Heterobeltiosis is estimated over the better parent of the hybrid as follows :

Heterobeltiosis = $\frac{F_1 - BP}{BP} \times 100$

There BP is the value of better parent.

Q. 27. What are main effects of heterosis ?

Ans. Heterosis leads to increase in yield, reproductive ability, adaptability, general vigour, quality etc.

Q. 28. What are the traits for which negative heterosis is desirable ?

Ans. Negative heterosis is desirable for some characters such as earliness, height in cereals, micronaire value in cotton and toxic substances like neurotoxin in *Lathyrus sativus.*

Q. 29. What are the factors which help in commercial exploitation of heterosis ?

Ans. In crop plants, the following factors help in the commercial exploitation of heterosis :

(i) Enough magnitude of heterosis.

(ii) High percentage of outcrossing.

(iii) Floral biology emanable for large scale production of hybrid seed with less expenditure.

(iv) Availability of male sterility and self incompatibility systems for easy hybrid seed production.

Q. 30. What is hybrid ?

Ans. The progeny of a cross between genetically different (dissimilar) plants is called hybrid.

Q. 31. What is hybrid variety ?

Ans. The F_1 population which is used for commercial cultivation is called hybrid variety or hybrid cultivar.

Q. 32. What are main features of hybrids ?

Ans. The main features of hybrid cultivars are as follows :

(i) Hybrid varieties are vigorous and highly productive due to full exploitation of hybrid vigour.

(ii) Hybrid varieties are heterozygous but homogeneous populations.

(iii) Hybrids have wider adaptability than inbreds and purelines due to high inherent buffering capacity.

(iv) Hybrids are more tolerant to biotic and abiotic stresses than inbreds and purelines.

Q. 33. What are various types of hybrids ?

Ans. Commercially cultivated hybrids are of two types, *viz.* intraspecific hybrids and interspecific hybrids.

Q. 34. What is intraspecific hybrid ?

Ans. A hybrid between genetically different genotypes of the same species is called intraspecific or intervarietal hybrid. Such hybrids are commonly used for commercial cultivation in several cross pollinated crops and some self pollinated crops. Intraspecific hybrids are always fertile. Based on type of cross, intraspecific hybrids are of three types, *viz.* (a) single cross hybrid, (b) three way cross hybrid, and (c) double cross *hybrid.*

Q. 35. What is single cross hybrid ?

Ans. A hybrid between two inbreds or purelines is called single cross hybrid. In cross pollinated crops, such hybrids are developed from a cross between two inbred lines, whereas in self pollinated species such hybrids are developed from a cross between two purelines.

Q. 36. What are uses of single crosses ?

Ans. Single crosses are useful in following ways :

(i) In developing single cross hybrid.
(ii) In developing three-way and double cross hybrids.
(iii) In predicting the performance of double cross hybrids.
(iv) In developing varieties in self pollinated specises from segregating generations.

Q. 37. What is three way cross hybrid ?

Ans. The hybrid progeny between a single cross and an inbred line is called three-way cross hybrid. The single cross is used as female and inbred as male. Such hybrids are developed when only three good inbreds are available.

Q. 38. What is double cross hybrid ?

Ans. The hybrid progeny from a cross between two single crosses is called double cross hybrid. Such hybrids are commonly used in maize and sugarbeet.

Q. 39. How many single crosses can be made using *n* inbred lines ?

Ans. The following formula is used to decide number of all possible single crosses among *n* inbred lines.

All possible single crosses = $n(n-1)/2$

Q. 40. What is the formula used to decide all possible three-way crosses among *n* inbreds ?

Ans. The number of all possible three-way crosses among *n* parents is determined by the following formula :

All possible three-way crosses = $(n -1) (n -2) / 2$

Q. 41. How many double crosses can be made using n inbreds ?

Ans. The number of all possible double crosses among *n* inbreds is determined by the following formula :

All possible double crosses = $n (n - 1)(n - 2) (n -3)/ 8$

Q. 42. What is top cross ?

Ans. A cross between an inbred line and an open pollinated variety is known as top cross or inbred variety cross. Top crosses are generally made for testing the combining ability of inbreds.

Q. 43. What is double top cross ?

Ans. The hybrid progeny between a single cross and an open pollinated variety is called double top cross. It is represented as follows.

Double top cross = (A × B) × open pollinated variety :

Q. 44. What is multiple cross ?

Ans. A cross involving more than four inbred lines is called multiple cross or composite cross. Such crosses are made to combine monogenic characters from different sources into a single genotype. Multiple crosses generally have wider adaptation.

Q. 45. What is polycross ?

Ans. Polycross refers to open pollination of a group of selected genotypes in isolation from other compatible genotypes to promote random mating among selected genotypes.

Q. 46. How the performance of a double cross hybrid is predicted ?

Ans. Single crosses are used to predict the yield performance of a double cross hybrid. The mean yield of the four non- parental single crosses is used to predict the yield performance of the double cross hybrid.

Predicted yield of the double cross (A × B) × (C × D)

= (A × C) + (A × D) + (B × C) + (B × D) / 4

Q. 47. List important hybrids of maize.

Ans. Important hybrids of maize include Ganga 2, Ganga 5, Ganga 11, Deccan 103, Makka 1, VL 42 etc.

Q. 48. Give names of some important hybrids of pearlmillet.

Ans. Important hybrids of pearlmillet include HB 3, BJ 104, MBH 110, BK 560, MH 179 etc.

Q. 49. List important hybrids of sunflower.

Ans. Important hybrids of sunflower are BSH 1, HSFH 1, MSFH 10, KBSH 1 and KBSH 11.

Q. 50. Name important hybrids of castor.

Ans. Important castor hybrids are GCH 2, GCH 3, Gauch 1 and GCH 4.

Q. 51. List important hybrids of sorghum.

Ans. Important sorghum hybrids are CSH 1, CSH 5, CSH 6, CSH, CSH 10 and CSH 11 etc.

Q. 52. Give names of important intra-hirsutum cotton hybrids.

Ans. Important intra-hirsutum cotton hybrids include H 4, H 6, H 8, H 10, JKHy 1, JKHy 2, PKVHy 2, PKVHy 3, PKVHy 4, PKVHy 5, NHH 44, CICRHH 1, Savita, Surya, DHH 11, Fateh, Maruvikas, Dhanlaxmi, LHH 144, Omshankar etc.

Q. 53. What are interspecific tetraploid cotton hybrids ?

Ans. Important interspecific tetraploid cotton hybrids include Varalaxmi, DCH 32, HB 224, TCHB 213, DHB 105, Sruthi etc.

Q. 54. Give names of diploid cotton hybrids.

Ans. Diploid cotton hybrids are of two types *viz.* intra arboreum and interspecific. Intra-arboreum hybrids include LDH 11, AAH 1, AKDH 7 and G. Cot MDH 11. Interspecific diploids hybrids include DH 7, DH 9, DDH 2 and Pha 46.

Q. 55. What are male sterility based diploid cotton hybrids ?

Ans. The male sterility based diploid cotton hybrids include AAH 1, AKDH 7, and G. Cot MDH 11. All these hybrids have been developed using genetic male sterility.

Q. 56. What are male sterility based hybrids of upland cotton ?

Ans. The male sterility based hybrids of upland cotton include Suguna, PKVHy 3, DKVHy 4, PKVHy 5 and MECH 11. The first hybrid is genetic male sterility based and rest are cytoplasmic genic male sterility based.

Q. 57. What are advantages of hybrid varieties ?

Ans. Hybrid varieties are more productive, more uniform and attractive than synthetics, composites and open pollinated varieties.

Q. 58. What are demerits of hybrid varieties ?

Ans. The main demerits of hybrid cultivars in comparison to synthetics, composites and open pollinated cultivars are as follows :

(i) The seed is expensive.

(ii) Fresh seed is to be purchased by farmers every year.

(iii) The cultivation requires more inputs.

Q. 59. What is inbreeding ?

Ans. Mating between closely related individuals such as sefling and sib mating is called inbreeding. Inbreeding is required to maintain the genetic purity of the inbreds.

Q. 60. What is inbreeding depression ?

Ans. The decline in fitness and vigour due to inbreeding is called inbreeding depression.

Q. 61. What is inbred ?

Ans. In cross pollinated species, a true breeding line obtained by continuous inbreeding is called inbred.

Q. 62. Compare dominance and overdominance theories of heterosis.

Ans. Comparison of dominance and overdominance theories of heterosis is presented below in Table 16.1.

TABLE 16.1. Comparison of dominance and overdominance theories of heterosis

Particulars	Dominance hypothesis	Overdominance hypothesis
Similarities		
1. Effect of inbreeding	Inbreeding tends to decline in vigour.	Also leads to decline in vigour.
2. Outbreeding	Outbreeding restores vigour.	Outbreeding restores vigour.
3. Parental diversity	Heterosis is associated with parental diversity.	Also associated with parental diversity.
4. Control	Heterosis is controlled by genes.	Also controlled by genes.
Difference		
5. Cause of heterosis	Heterosis results due to masking effect of dominant esirable alleles over harmful recessive alleles.	Heterosis results due to complementation between divergent alleles.

SYNTHETIC AND COMPOSITE VARIETIES

Q. 1. What is inbred line ?

Ans. In allogamous species, a true breeding line obtained by continuous inbreeding is called inbred line. Main features of inbred line are given below.

(i) It is relevant to cross pollinated species

(ii) It is true breeding

(iii) It is homozygous and homogeneous

(iv) It is obtained by continuous inbreeding

Q. 2. What is synthetic variety ?

Ans. A variety which is developed by intermating in all possible combinations a number of inbred lines with good general combining ability and mixing their seeds (seeds of F1s) in equal quantity is refered to as synthetic variety. It is relevant to cross pollinated species.

Q. 3. What are base materials used for developing synthetic variety ?

Ans. A synthetic variety can be developed from inbreds, clones and open pollinated varieties. The end products of recurrent selection which are already tested for gca are used to constitute synthetic variety.

Q. 4. Why synthetic varieties are developed ?

Ans. The basic concept in the development of synthetic variety is exploitation of heterosis. In the development of synthetics, heterosis is partially exploited, because some degree of inbreeding takes place due to open pollination in later generations.

Q. 5. What is genetic constitution of a synthetic variety ?

Ans. Initially, a synthetic variety consists of several heterozygotes. A synthetic variety is maintained by open pollination, some degree of selfing occurs resulting in fixation of some genes. As a result in later generations, a synthetic variety consists of several heterozygotes and homozygotes.

Q. 6. What type of variance is exploited in synthetic varieties ?

Ans. Synthetic varieties exploit more of additive gene action than dominance gene action.

Q. 7. How many inbreds are used to constitute a synthetic variety ?

Ans. Generally, 5-8 good general combining inbred lines are used to constitute a synthetic variety.

Q. 8. After how many years a synthetic variety is reconstituted ?

Ans. A synthetic variety can safely be grown for a period of 4-5 years without reduction in yield potential. After a period of five years, it would be desirable to reconstitute the synthetic variety. Farmers can use their own seeds upto five years.

Q. 9. What are main features of synthetic variety ?

Ans. Main features of synthetic variety are given below :

(i) They are relevant to cross pollinated species .

(ii) They are maintained by open pollinaton.

(iii) They exploit more of additive variance.

(iv) In later generations, they consist of several heterozygotes and homozygotes.

(v) Synthetics have wider adaptability.

Q. 10. What is yield level of synthetic variety ?

Ans. The yield of synthetic variety is always higher than open pollinated variety but lower than the yield of single and double cross hybrids.

Q. 11. What is difference between mass selected variety and synthetic variety ?

Ans. Main difference between mass selected and synthetic varieties are given below :

(i) In mass selected variety, component genotypes are selected on the basis of phenotypic performance, where as component lines of synthetic variety are selected on the basis of gca.

(ii) Development of synthetics involves intermating among selected lines which is not in ease of mass selection.

(iii) Synthetics have better buffering capacity to environmental changes.

(iv) Synthetics are more tolerant to insects and diseases.

Q. 20. What are demerits of synthetic varieties ?

Ans. Main demerits of synthetic varieties are given below :

(i) The produce of synthetic varieties is generally less uniform and less attractive than hybrids.

(ii) The yield of synthetic varieties is poorer than single or double cross hybrids.

Q. 21. Why yield of synthetics is poorer than hybrids ?

Ans. The yield of synthetics is poorer than hybrids, because in synthetics heterosis is partially exploited, whereas in hybrids heterosis is fully exploited.

Q. 22. Why heterosis is partially exploited in synthetics ?

Ans. In synthetics, heterosis is partially exploited because some level of inbreeding takes place due to open pollination in later generations.

Q. 23. Why synthetics have better resistance to diseases ?

Ans. Synthetic varieties have better resistance to plant diseases due to their heterogeneous nature and broad genetic base.

Q. 24. What is a composite variety ?

Ans. In cross pollinated species, a variety which is developed by mixing the seed of various genotypes (in equal quantity) which are similar in maturity, height, seed size, colour etc. is called composite variety.

Q. 25. What are main features of composites ?

Ans. Main features of composites are given below :

(i) Component genotypes are open pollinated.

(ii) They are relevant to cross pollinated species.

(iii) They are maintained by open pollination.

(iv) The exploit heterosis partially.

(v) Exact reconstitution is not possible.

(vi) Highly adaptable to environmental changes.

Q. 26. Is it possible to reconstitute a composite variety ?

Ans. It is not possible to reconstitute a composite variety, because such variety is developed from heterozygous genotypes.

Q. 27. Cite examples of composite variety.

Ans. Composite varieties have been developed in maize. Important composite varieties of maize are given below :

Jawahar	Sona	Vikas
Vijay	Vikram	Tarun
Kiran	Protina	Renuka
Kisan	Ratan	Kanchana
Amber	Shakti	Co1.

Q. 28. What is composite variety of Brassica campestris ?

Ans. The composite variety of *Brassica campestris* is composite 1.

Q. 29. What are similarities between synthetics and composites ?

Ans. Main similarities between synthetics and composites are given below. Both are :

(i) Polymorphic and stable
(ii) Relevant to allogamous species
(iii) Mixture of several homozygotes and heterozygotes
(iv) Both have broad genetic base.

Q. 30. What steps are involved in the development of composite variety ?

Ans. Development of a composite variety consists of the following important steps.

(i) Selection of heterozygous base material.
(ii) Intermating among selected genotypes to get F_1s.
(iii) Evaluation of F_1 crosses.
(iv) Mixing in equal quantity parental seeds of superior crosses.

Q. 31. What are similarities and dissimilarities between synthetic and composite varieties ?

Ans. Comparison of synthetic and composite varietes is presented below in Table 17.1.

TABLE 17.1. Similarities and dissimilarities between Synthetic and Composite Cultivars

Particulars	Synthetic cultivars	Composite cultivars
Similarities		
1. Equilibrium	Polymorphic stable	Polymorphic stable
2. Relevant to	Allogamous Species	Allogamous Species
3. genetic constitution	Heterogeneous, mixture of several homozygotes and heterozygotes	Heterogeneous, mixture of several homozygotes and heterozygotes
4. Adaptation	Wide due to broad genetic base	Wide due to broad genetic base
Dissimilarities :		
5. Base material	Inbreds or clones	Open pollinated varieties or any other heterozygous source
6. Component genotypes	Two to eight	Two to many
7. Evaluation for gca	Always done	Usually not done
8. Seed replacement	After 4-5 years	After 3-4 years
9. Exact Reconstitution	Possible	Not possible

18

MUTATION BREEDING

Q. 1. What is mutation ?

Ans. Mutation refers to sudden heritable change in the phenotype of an individual. In molecular terms, mutation is defined as the permanent and relatively rare change in the number or sequence of nucleotide.

Q. 2. Who first discovered mutation ?

Ans. Mutation was first discovered by Wright in 1791 in male lamb who had short legs.

Q. 3. Who first discovered mutation in plants ?

Ans. In plants, mutation was first discovered in 1900 by de Vries in *Oenothera*.

Q. 4. Who coined the term mutation ?

Ans. The term mutation was coined/first used by de Vries in 1901.

Q. 5. Who first discovered mutation in Drosophila ?

Ans. In Drosophila, the white eyed mutatant was first discovered in 1910 by Morgan.

Q. 6. What are different types of mutation ?

Ans. Mutations have been classified in various ways as follows :

Basis of Classification	Types of Mutation
Origin	Spontaneous and induced
Direction of change	Forward mutation and reverse mutation
Tissue involved	Somatic and germinal mutation
Site	Point mutation and cytoplasmic mutation
Character involved	Morphological and biochemical mutation
Visibility/Phenotypic	Macro-mutation and micro-mutation effect
Survival	Lethal, sub-lethal, sub-vital and vital mutations

Q. 7. What are macro-mutations ?

Ans. Induced mutations are of two types, viz. macro-mutations and micro-mutations. Mutations with distinct morphological changes in the phenotype are called macro-mutations. Such mutations are found in qualitative character and, therefore, are also called oligogenic mutations. Identification of such mutations is easy.

Q. 8. Who coined the term macro-mutation ?

Ans. The term macro-mutation was first used by Gold Schmidt in 1940.

Q. 9. What are micro-mutations ?

Ans. Mutations with invisible phenotypic changes are called micro-mutations. Such mutations are observed in quantitative characters and, hence are also known as polygenic mutations. Identification of such mutations is very difficult. Micro-mutations are of economic value in plant breeding.

Q. 10. Define lethal mutation ?

Ans. A mutation which kills all the individuals that carry it is called lethal mutation.

Q. 11. What is sub-lethal mutation ?

Ans. A mutation which kills more than 50% of individuals that carry it is termed sub-lethal mutation.

Q. 12. Define sub-vital mutation.

Ans. A mutation which kills less than 50% of individuals that carry it is known as sub-vital mutation.

Q. 13. What is vital mutation ?

Ans. A mutation in which all the mutants survive is called vital mutation.

Q. 14. What is point mutation ?

Ans. Gene mutation or mutation at molecular level is called point mutation. The term point mutation was first used by Bridges in 1923.

Q. 15. What is mutagen ?

Ans. Physical or chemical agents which greatly enhance the frequency of mutation are called mutagens.

Q. 16. What are physical mutations ?

Ans. Physical mutagens include various types of radiations as given below :

(1) X-rays (2) Gamma rays

(3) Alpha rays (4) Beta particles

(5) Fast neutrons (6) Thermal (slow) neutrons

(7) Ultraviolet rays.

Q. 17. What are sparsely ionizing radiations ?

Ans. Sparsely ionizing radiations include X-rays, Gamma rays and Beta particles.

Q. 18. What are densely ionizing radiations ?

Ans. Fast and thermal neutrons and alpha particles are densely ionizing radiations.

Q. 19. Who discovered X-rays ?

Ans. X-rays were first discovered by Roentgen in 1895.

Q. 20. Who first used X-rays for induction of mutation ?

Ans. Muller first used X-rays for induction of mutation in 1927 in *Drosophila.*

Q. 21. Who first used X-rays for induction of mutation in crop plants ?

Ans. Stadler in 1928 first used X-rays for induction of mutation in barley.

Q. 22. Who won Nobel prize for induction of mutation ?

Ans. Muller, a US genelicist, was awarded Nobel prize in 1946 for experimental induction of mutation in *Drosophila.*

Q. 23. How X-rays induce mutations ?

Ans. X-rays induce mutations by forming free radicals and ions. X-rays cause both chromosomal and gene mutations.

Q. 24. How gamma rays induce mutations ?

Ans. Gamma rays induce chromosomal and gene mutations by ejecting atoms from the tissue.

Q. 25. How gamma rays are generated ?

Ans. Gamma rays are generated from radioactive decay of some elements such as 14C, 60C, radium etc. The Cobalt-60 is commonly used for the production of gamma rays.

Q. 26. How alpha particles induce mutation ?

Ans. Alpha particles induce chromosomal mutations through ionization and excitation.

Q. 27. How beta particles are generated ?

Ans. Beta particles are generated from radioactive decay of heavier elements such as 3H, 32P, 35 S etc.

Q. 28. How ultraviolet rays are generated ?

Ans. Ultraviolet rays are generated from mercury vapour lamps or tubes. They are also present in solar radiation. They are non-ionizing and low penetrating.

Q. 29. What are chemical mutagens ?

Ans. There is a long list of chemicals which are used as mutagens. Chemical mutagens are divided into following four groups.

(1) Alkylating agents (3) Acridine dyes

(2) Base analogues (4) Others.

Q. 30. What are alkylating agents ?

Ans. Chemical mutagens which cause mutation by adding alkyl group at various positions in DNA are called Alkylating agents. Such chemicals cause mutation through transitions. Examples of alkylating agents are given below :

(i) Ethyl methane sulphonate
(ii) Methyl methane sulphonate
(iii) Ethyl ethane sulphonate
(iv) Ethylene imines

Q. 31. What are base analogues ?

Ans. Chemical compounds which are similar to DNA bases are called base analogues. Base analogues include 5 Bromo Uracil and 2 Amino Purine.

Q. 32. What are acridine dyes ?

Ans. Acridine dyes are very effective mutagens. They lead to addition or deletion and cause frameshift mutations. Acridine dyes which used for induction of mutation are given below :

Proflavin — Acriflavin
Acridine orange — Ethadium bromide
Acridine yellow

Acridine dyses are also known as frameshift mutagens.

Q. 33. What are frameshift mutations ?

Ans. Mutations which arise due to addition or deletion of nucleotides in mRNA are called frameshift mutations. Such mutations are induced by acridine dyes.

Q. 34. Who first used the term frameshift mutation ?

Ans. The term frameshift mutation was first used by Brenner *et al.* in 1961.

Q. 35. Define transition.

Ans. Substitution of one purine by another purine or one pyrimidine by another pyrimidine is called transition. The term transition was coined by Freese in 1959.

Q. 36. What is transversion ?

Ans. Substitution of a purine by a pyrimidine or vice-versa is known as transversion. The term transversion was coined by Freese in 1959.

Q. 37. What are nonsense mutations ?

Ans. Mutations which have nonsense codons are called nonsense mutations. The term nonsense mutation was coined by Brenner *et al* in 1961.

Q. 38. What are missense mutations ?

Ans. Mutations which have missense codons are termed missense mutations. The term missense mutation was coined by Brenner *et al.* in 1961.

Q. 39. What is tautomerization ?

Ans. The process of shifting hydrogen atoms from one position to another in a purine or pyramidine base in called tautomerization. The product of this process is known as tautomer.

Q. 40. What is mutator gene ?

Ans. Gene which enhances the natural mutation rate of other genes in the same genome is called mutator gene. Example is dotted gene is maize. The term mutator gene was coined by Demerec in 1937.

Q. 41. What is mutable gene ?

Ans. A gene which exhibit higher mutation rate than other genes is called mutable gene.

Q. 42. What is antimutator gene ?

Ans. A gene which reduces the frequency of spontaneous mutation of other genes in the same genome is called antimutator gene. Such genes have been reported in bacteria and bacteriophages.

Q. 43. Who coined the term antimutator gene ?

Ans. The term anti-mutator gene was first used by Drake and Allen in 1968.

Q. 44. What is mutation load ?

Ans. Decrease in the relative fitness of a genotype due to presence of deleterious mutation is called mutation load. The term mutation load was coined by Muller in 1950.

Q. 45. What is mutation pressure ?

Ans. Increase of a mutant allele in a population due to its recurrent production is called mutation pressure. The term mutation pressure was first used by Wright in 1921.

Q. 46. What is muton ?

Ans. A subdivision of gene which is the site of mutation is called muton. The term muton was coined by Benzer in 1957.

Q. 47. What are hot spots ?

Ans. Highly mutable sites within a gene are called hot spots. The term mutational hot spot was coined by Benzer in 1955.

Q. 48. What is LD 50 ?

Ans. A dose of mutagen that kills 50% of treated individuals is called LD 50.

Q. 49. Define mutability.

Ans. The property of any gene or genotype to undergo mutation is called mutabiltiy.

Q. 50. What is mutagenicity ?

Ans. Any event that leads to alteration of DNA is called mutagenicity.

Q. 51. Who coined the term mutation repair ?

Ans. The term mutation repair was first used by Helsings *et al* in 1976.

Q. 52. What is mutation breeding ?

Ans. The genetic improvement of crop plants for various economic characters through the use of induced mutations is called mutation breeding. Mutation breeding is used when desired variability is not found in the germplasm of cultivated species.

Q. 53. What steps are involved in mutation breeding ?

Ans. The process of mutation breeding consists of following important steps :

(1) Choice of plant material
(2) Choice of mutagen
(3) Mutagen treatment
(4) Handling of treated material
(5) Identification of desired mutant
(6) Evaluation of mutant

Q. 54. What are main achievements of mutation breeding ?

Ans. The main achievements of mutation breeding are given below :

(1) Development of improved varieties
(2) Induction of male sterility
(3) Production of haploids
(4) Creation of genetic variability
(5) Overcoming self incompatibility
(6) Improvement in adaptation.

Q. 55. What are limitations of induced mutations ?

Ans. Main limitations of induced mutations are as follows :

(1) Most of the mutation are deleterious.
(2) The frequency of desirable mutants is very low (0.1%).
(3) Identification of polygenic mutations is difficult.
(4) It requires screening of large populations.

Q. 56. Write the characters which have been improved through mutation breeding ?

Ans. Mutation breeding has been used for improvement of various characters such as yield, quality, disease resistance, insect resistance, drought resistance, dwarfness, earliness, plant type and adaptation.

Q. 57. What are mutatant varieties of wheat ?

Ans. Mutant varieties of wheat released in India include NP 836, JRC 7447, Pusa Lerma and Sharbati Sonora. However, now these varieties are not under cultivation.

Q. 58. List mutant varieties of rice released in India ?

Ans. In rice several mutant varieties have been developed through X-rays, gamma rays and Ethyl methane sulphonate (EMS) as given below :

(1) X-rays : Biraj, CNM 20, LNM 25, CNM 31, CNM 6 and Jagannath.
(2) Gamma rays : AU1, AU4, K84, Mohan, Padmini, Rasmi, Sattari, Hybrid mutant
(3) EMS : PL 56, IIT 60, Indira and Prabhavati.

Q. 59. Name the mutant varieties of cotton released in India.

Ans. In cotton, some mutant varieties have been released through X-rays and Gamma rays as given below :

(1) X-rays : Indore 2, MA 9 and MCU 7
(2) Gamma rays : MCU 10, Rasmi, DS 1 and Pusa Ageti.

Q. 60. What are mutant varieties of barley.

Ans. Mutant varieties of barley include DL 253 (gamma rays), PL 56 (EMS) and RDB (N).

Q. 61. What are mutant varieties of sugarcane ?

Ans. The mutant varieties of sugarcane include Co 6608, Co 8153 and Co 997. All these varieties have been developed through gamma rays.

Q. 62. What are mutant varieties of chickpea ?

Ans. Mutant varieties of chickpea released in India include Pusa 408, Pusa 413, Pusa 417 (gamma rays) and Kiran (TN).

Q. 63. What are mutant varieties of groundnut ?

Ans. In ground nut, mutant varieties have been released through X-rays, Gamma rays and EMS as given below :

(1) X-rays : TG3, TG4, TG17 and Vikram
(2) Gamma rays : BP1 and BP2
(3) EMS : CO_2

Q. 64. What are mutant varieties of cowpea ?

Ans. The mutant varieties of cowpea include V16, V37, V38 and V240, all developed through chemical mutagen (dMS).

Q. 65. List mutant varieties of mung bean.

Ans. In mung bean some mutant varieties have been developed through gamma irradiation. These varieties are Co4, ML 26-10-3, Pant Mung 2 and TAP 7.

Q. 66. What are mutant varieties of pigeonpea ?

Ans. Important mutant varieties of pigeonpea released in India are CO_3 (EMS), TAT5, Trombay (FN) and Co_5 (gamma rays).

Q. 67. What are mutant varieties of tomato ?

Ans. Mutant varieties of tomato include PKM1, S 12 and Pusalal Meeruthi.

Q. 68. What is difference between mutant and segregant ?

Ans. There are two main differences between mutant and segregant as follows :

(1) The frequency of mutant is very low, whereas the frequency of segregant is quite high.

(2) The mutant differs from parent variety in one or few characters, whereas the segregant will differ from the parent variety in several characters.

Q. 69. Give the names of organizations where mutation breeding work is carried out in India.

Ans. In India, mutation breeding work is carried out in the following organizations :

(1) Bhapba Atomic Research Centre, Trombay, Mumbai.

(2) ICAR Crop Research Institutes

(3) State Agricultural Universities, and

(4) In some general universities.

Q. 70. Where is the IAEA located ?

Ans. The International Atomic Energy Agency is located at Vienna, Austria.

Q. 71. Give names of some renound mutation breeders.

Ans. Names of some renound mutation breeders are given below :

(1) H.J. Muller (USA)

(2) L.J. Stadler (USA)

(3) C. Aureback

(4) Brock, R.D.

(5) AU Harten

(6) M.S. Swaminathan

Q. 72. Compare hybridization method and mutation breeding.

Ans. Comparison of combination (hybridization) breeding and mutation breeding is presented below in Table 18.1

TABLE 18.1. Comparison of Combination (hybridization) breeding and Mutation breeding

Particulars	Hybridization method	Mutation breeding
1. Material required	Minimum two genetically different genotypes are required.	One genotype or variety is sufficient.
2. Purpose	To combine desirable genes from two or more different sources.	To induce new variability for various characters.
3. Crossing	Crossing is involved.	Crossing is not involved.
4. Breeding method	It comes under general method of plant breeding.	It comes under special method of plant breeding.
5. Use	Widely used in the genetic improvement of both self and cross pollinated species.	Less popular and used in self pollinated crops and asexually propagated crops.
6. Time taken	It takes 14-15 years for release of a new variety by pedigree method and more by bulk method.	It takes lesser time for release of a new variety.
7. Mutagenic treatment	Not required	Required.
8. Skill	It requires less skill in handling of material.	It requires more skill in handling of mutagens and treated material.
9. Creation of variability	New variability is created by recombination and segregation.	New variability is created by gene mutation.
10. Effectiveness	Equally effective for the improvement of both oligogenic and polygenic traits.	More effective with oligogenic traits.
11. Selection	Selection is practised in F_2 or later generations.	Selection is practised in M_2 or later generations.
12. Features of end product	New variety is different from both the parents except in backcross breeding.	New variety is identical to parent variety except for the character improved.

POLYPLOIDY BREEDING

Q. 1. What is genome ?

Ans. A basic set of chromosomes in which each type of chromosome is represented only one is called genome. Haploid individuals have single genone, diploid species two, triploid species three and so on.

Q. 2. Who coined the term genome ?

Ans. The term genome was first used by Winkler in 1920 and genome analysis was first proposed by Kihara in 1930.

Q. 3. What is heteroploidy ?

Ans. Any change in chromosome number from the diploid state is termed heteroploidy and such individuals are called heteroploids. Heteroploidy is of two types, viz. euploidy and aneuploidy.

Q. 4. What coined the term heteroploid ?

Ans. The term heteroploid was first used by Winkler in 1916.

Q. 5. What is euploidy ?

Ans. Euploidy refers to numerical change in the entire genome. It includes haploids and polyploids.

Q. 6. What is aneuploidy ?

Ans. Numerical change in one or few chromosomes of the genome is called aneuploidy. It includes monosomics, mullisomics and polysomics.

Q. 7. Define haploidy.

Ans. An individual with gametic chromosome number is called haploid and such condition is known as haploidy. Haploid chromosome number is denoted by *n*.

Q. 8. Who coined the term haploid ?

Ans. The term haploid was first used by Strasburger in 1905.

Q. 9. What is monoploid ?

Ans. An individual with basic chromosome number of a species is called monoploid and such condition is known as monoploidy. The term monoploid was first used by Kotayam in 1935.

Q. 10. What is basic chromosome number ?

Ans. The gametic chromosome number of a true diploid species is called basic number. It is represented by x.

Q. 11. What is difference between haploid and monoploid ?

Ans. The main differences between haploid and monoploid are given below.

(1) Monoploids have single copy of genome, whereas haploids may contain one or more copies of genome.

(2) Monoploids are denoted by x and haploids are represented by n.

(3) Monoploids represent gametic chromosome number of a true diploid species, whereas haploids represent gametic chromosome number of any species.

(4) Monoploids are always haploids but haploids cannot always be monoploids.

Q. 12. What are different types of haploids ?

Ans. Depending upon the origin, haploids are of two types, viz. euphapolids and aneuhaploids. Euhoploids develop from a euploid species and have complete chromosome set. Aneuhaploids develop from aneuploid species and have either one additional or missing chromosome.

Q. 13. What are euhaploids ?

Ans. Euhaploids develop from euploid species and are of following types :

(1) Monhaploids–develop from a true diploid species

(2) Polyhoploids–develop from a polyploid species

(i) Allophaloids–develop from allopolyploid species

(ii) Dihaploids–develop from a tetraploid species.

Q. 14. What are aneuhaploids ?

Ans. Haploids that develop from aneuploid species are called aneuhaploids. Aneuhaploids are of following types :

(1) Disomic haploids–have one extra chromosome ($n + 1$)

(2) Nullisomic haploids–have one missing chromosome ($n–1$)

(3) Substitution haploids have one missing and one extra chromosome ($n -1 + 1$).

(4) Misdivision haploids–have an isochromosome.

Q. 15. How haploids originate ?

Ans. In crop plants, haploids may originate in following ways :

(1) By haploid parthenogenesis

(2) From interspecific crosses

(3) As a member of twin embryos
(4) By semigamy
(5) From alien cytoplasm
(6) By anther culture.

Q. 16. What are uses of haploids in crop Improvement ?

Ans. Haploids are used in plant breeding in several ways as given below :

(1) For development of pure lines through chromosomal doubling.

(2) For development of inbred lines in dioecious species.

(3) For obtaining disease free plants from anther culture as has been used in tobacco.

(4) For developing nullisomics, monosomics and trisomics which in turn are used for locating genes on different chromosomes.

Q. 17. What are monosomics ?

Ans. An individual lacking one chromosome from a diploid set ($2n-1$) is called monosonic and such condition is known as monosomy. Monosomics are of two types as follows :

(1) Single monosomic–Loss of one chromosome from one pair ($2n-1$).

(2) Double monosomic–Loss of one chromosome each from two different pairs ($2n-1-1$).

Q. 18. How monosomics originate ?

Ans. Monosomics may originate in the following three main ways :

(1) They may originate spontaneously from diploids due to non-disjunction.

(2) From union of nullisomic gamete ($n-1$) with normal gamete.

(3) From trisomics due to non-disjunction of three homologous chromosomes.

Q. 19. What are nullisomics ?

Ans. An individual lacking one pair of chromosome from a diploid set ($2n-2$) is called nullisomic and such condition is known as nullisomy. Nullisomics are inviable in true diploid species. Among polyploids, hexaploids can tolerate loss of one pair of chromosomes more than tetraploids, because they have two other pairs of similar chromosomes in other genomes.

Q. 20. What are polysomics ?

Ans. An individual having either single chromosome or one pair of extra chromosome in the diploid complement is called polysomic and such condition is known as polysomy. Polysomics are of two types, viz. trisomics and tetrasomics.

Q. 21. What are trisomics ?

Ans. An individual having addition of one chromosome to one pair in a diploid set is called trisomic and such condition is known as trisomy. Trisomics are of following two types

(1) Simple trisomic–Addition of one chromosome to one pair only ($2n + 1$).

(2) Double trisomic–Addition of one chromosome to two different pairs ($2n + 1 + 1$).

Depending on the nature of extra chromosome, simple trisomics are of following three types :

(a) Primary trisomic–additional chromosome is normal.

(b) Secondary trisomic–additional chromosome is isochromosome.

(c) Tertiary trisomic–additional chromosome is translocated one.

Q. 22. What are tetrasomics ?

Ans. An individual having addition of two chromosomes to one pair in a diploid set is called tetrasomic and such condition is known as tetrasomy. Tetrasomics are of following two types :

(1) Simple tetrasomic–Addition of two chromosomes to one pair ($2n + 2$)

(2) Double tetrasomic–Addition of two chromosomes to two different pairs ($2n + 2 + 2$).

Q. 23. Who coined the terms monosomic, nullisomic and tetrasomic ?

Ans. Terms monosomic, nullisomic and tetrasomic were coined by Blakeslee in 1921.

Q. 24. Who coined the terms polysomic and trisomic ?

Ans. Terms polysomic and trisomic were first used by Blakeslee in 1921.

Q. 25. What are Hypoploids ?

Ans. Monosomics and nullisomics are combinedly known as hypoploids, which refer to loss of one or two chromosomes from the diploid set.

Q. 26. Who coined the term Hypoploid ?

Ans. The term hypoploid was first used by Belar in 1928.

Q. 27. What are Hyperploids ?

Ans. Polysomics are also known as hyperploids, which refer to addition of one or two chromosomes to single or two different pairs. The term hyperploid was coined by Belar in 1928.

Q. 28. What are uses of aneuploids in Crop improvement ?

Ans. Aneuploids are useful in crop improvement in following ways.

(1) In locating genes on different chromosomes. Monosomics and nullisomics are used for this purpose.

(2) In transferring desirable genes from one species to another through transfer of chromosome.

(3) In developing alien addition and alien substitution lines in different crops.

(4) In identification of chromosomes involved in translocations.

Q. 29. What are limitations of aneuploids ?

Ans. Main limitations of aneuploids are given below :

(1) In diploid species, monosomies are inviable. They can survive only in polyploid species *i.e.* tetroploid and hexaploid.

(2) Similar is the ease with nullisomics.

(3) Trisomics can survive in diploid species.

(4) Maintenance of nullisomics and monosomics is difficult.

Q. 30. What is polypolidy ?

Ans. An individual having more than two basic sets of chromosome is called polyploid and such condition is known as polyploidy. It is estimated that about one third species of flowering plants are polyploids. Polyploidy is of two types, viz. autopolyploidy and allopolyplody.

Q. 31. Who coined the term polyploidy ?

Ans. The term polyploidy was first used by Strasburger in 1910.

Q. 32. Who coined the terms autopolyploidy and allopolyploidy ?

Ans. Terms autopolyploidy and allopolyploidy were first used by Kihara and Ono in 1926.

Q. 33. What is autopolyploidy ?

Ans. Polyploids which originate by multiplication of the chromosome of a single species are known as autopolyploids or autoploids and such condition is called autopolyploidy. Autopolyploids are also known as simple polyploids or single species polyploids. Autopolyploids are of following types

(1) Autotriploids–have three copies of the same genome (3X)

(2) Autotetraploids–have four copies of the same genome (4X)

(3) Autopentaploids–have five copies of the same genome (5X)

(4) Autohexaploids–have six copies of the same genome (6X).

Q. 34. What are practical uses of autotriploids ?

Ans. Triploid cultivars have been developed for commercial cultivation in banana, apple, sugarbeet and watermelon.

(1) Banana–triploid banana is seedless and has larger fruits than diploids.

(2) Watermelon–triploid watermelon is seedless.

(3) Sugarbeet–triploid sugarbeets have higher sugar contants their diploids.

(4) Apples–triploid apples are seedless and are propagated by budding and grafting.

Q. 35. What are practical uses of autotetraploids ?

Ans. Autotetraploid cultivars have been developed for commercial cultivation in Rye, grapes and alfalfa.

Q. 36. What is allopolyploidy ?

Ans. A polyploid individual which originates by combining complete chromosome sets from two or more species is known as allopolyploid or alloploid and such condition is called allopolyploidy. Allopolyploids are also known as hybrid polyploids. Allopolyploidy can be developed by interspecific crosses and fertility can be restored by chromosome doubling with colchicine treatment.

Q. 37. What is amphidiploid ?

Ans. An allotetraploid which arises by combining genomes of two diploid species is called allotetraploid or amphidiploid.

Q. 38. Who coined the term Amphidiploid ?

Ans. The term amphidiploid was coined by Navoshin in 1927.

Q. 39. Give examples of some natural allopolyploids.

Ans. Natural allopolyploid crops include wheat, cotton, tobacco, oats etc.

Q. 40. What are the sources of A, D and B genomes of bread wheat.

Ans. The sources of different genones of wheat are as follows :

A–Genome from Triticum monococcum

D–Genome from Triticum tauchi

B–Genome from unknown source–now extinct.

Q. 41. What are the species involved in the development of tobacco ?

Ans. There are two cultivated species of tobacco, viz. *Nicotiana tabacum* and *N.rustica.* The former has originated from a cross between *N. sylvestris* and *N.tomentosa* and the latter from a cross between *N.paniculata* and *N.undulata.*

Q. 42. Explain genetic origin of American cotton ?

Ans. American cotton (*Gossypium hirsutum*) has originated from a cross between *G.africanum* and *G. rainmondii.*

Q. 43. What are three basic species of Brassica ?

Ans. Following are three basic species of *Brassica*

(1) Brassica nigra (BB, $n = 8$)

(2) Brassica oleracea (CC, $n = 9$)

(3) Brassica campestris (AA, $n = 10$).

Q. 44. What are amphidiploid species of Brassica ?

Ans. The following are amphidiploid species of *Brassica.*

(1) Brassica corinata (BBCC, $2n = 34$)

(2) Brassiaa napus (AACC, $2n = 38$)

(3) Brassic juncea (AABB, $2n = 36$)

Q. 45. How Brassica carinata has originated ?

Ans. The *Brassica carinata* has originated from a cross between *B. nigra* and *B.oleracea.*

Q. 46. How Brassica napus has evolved ?

Ans. The *Brassica napus* has originated from a cross between *B. compestris* and *B.oleracea.*

Q. 47. What is the genetic origin of Brassica juncea ?

Ans. The *Brassica juncea* has evolved from a cross between *B.compestris* and *B. nigra.*

Q. 48. Who traced the genetic origin of tetraploid species of Brassica ?

Ans. The genetic origin of tetraploid species of Brassica was traced by Nagaharu U in 1935. He proposed a triangle for the origin of tetraploid species of *Brassica.*

Q. 49. What is *Raphanobrassica* ?

Ans. Raphanobrassica is an amphidiploid between radish (*Raphanus* sativus n=9) and cabbage (Brassica oleracea, n= 9).

Q. 50. Who developed *Raphanobrassica* ?

Ans. The Raphanobrassiea was developed by Karpechenko in 1928 in Russia from a intergeneic cross between radish and cabbage ($2n = 36$).

Q. 51. Who synthesized nicotiana digluta species of tobacco ?

Ans. The new species of tobacco *i.e.* Nicotiana digluta was synthesized by Clasuen and Goodspeed in 1925 from a cross between *N. tabacum* and *N. glutinosa.*

Q. 52. Who first developed hexaploid wheat ?

Ans. The hexaploid wheat was first synthesized by Mc Fadden and Sears in 1947 from a cross between *Triticum tungidum* and *T. tauschi.*

Q. 53. Who proposed genetic origin of American cotton ?

Ans. The genetic origin of American cotton was proposed by Harland (1940) and Phillips (1963). According to Harland, American cotton has evolved from a cross between *G. arboreum* and *G.thurberi* . However, Philips is of the opinion that American cotton has originated from a cross between *G.africanum* and *G.raimondii.* The second view is widely accepted.

Q. 54. What is colichicine ? How it induces polyploidy ?

Ans. Colchicine is an alkaloid which is obtained from the seed of a plant called *Colchicum autumnale*. Colchicine induces polyploidy by inhibiting the formation of spindle fibres. Thus chromosomes divide without moving to opposite poles resulting in doubling of chromosomes. Colchicine does not inhibit spindle formation in *Calchicum* because this plant has an anticolchicine substance.

Q. 55. What is Colchiploidy ?

Ans. The colchicine induced polyploidy is known as Colchiploidy. For induction of polyploidy colchicine is applied to growing tips, meristematic cells, seeds and auxillary buds.

Q. 56. What is the chemical formula of colchcine ?

Ans. The chemical formula of colchicine is $C_{22}H_{25}O6N$.

Q. 57. Who discovered Colchicine ?

Ans. The colchicine was discovered by Blackslee and Avery in 1937 and used for induction of polyploidy.

Q. 58. Who coined the term colchiploidy ?

Ans. The term colchiploidy was first used by Dermen in 1957.

Q. 59. Who first made cross between wheat and rye ?

Ans. The intergeneric cross between wheat and rye was made by Rimpu in Sweden around 1890.

Q. 60. What is triticale ?

Ans. Triticale is a new plant species which has originated from a cross between wheat (*Triticum aestivum*) and rye (*Secale cereale*).

Q. 61. What is primary triticale ?

Ans. Triticales which have originated from a cross between wheat (tetraploid or hexaploid) and rye are called primary triticales.

Q. 62. What is secondary triticale ?

Ans. Triticales which develop from a cross either between two primary triticales or between primary triticale and wheat are called secondary triticales.

Q. 63. What is quasidiploid ?

Ans. An individual having one additional chromosome to a pair and lacking one chromosome from a pair is called quasidiploid. It is represented as $(2n + 1 - 1)$.

Q. 64. Who coined the term quasidiploid ?

Ans. The term quasidiploid was coined by Hsu in 1957.

Q. 65. What is diploidzation ?

Ans. The process by which a polyploid species behaves like a diploid species is called diploidization. The cotton, wheat and tobacco are polyploids but behave like diploid species.

Q. 66. What is autosyndesis ?

Ans. In allopolyploids, the pairing between the chromosomes of same species *i.e.* A with A and B with B genome, also called intragenomal pairing or homologous pairing.

Q. 67. What is allosyndesis ?

Ans. In allopolyploids, the pairing between the genomes of two different species is called allosyndesis; also called intergenomal pairing and homeologous pairing.

Q. 68. Who coined the term tetraploid ?

Ans. The term tetraploid was first used by Nemec in 1910.

Q. 69. Why hexaploid wheat behaves like a diploid species.

Ans. The hexaploid wheat behaves like a diploid species because of 5 B system. A mutant gene is present on the long arm of chromosome number 5 of B genome which promotes homologous pairing and inhibits homeologous pairing. As a result hexaploid wheat behaves like a diploid species due to intragenomal pairing.

Q. 70. Who discovered 5 B system in wheat ?

Ans. In wheat, the 5B system was discovered by Riley and Chapman in 1958.

Q. 71. Who coined the term double reduction ?

Ans. The term double reduction was first used by Darlington in 1929. In refers to production of gametes with AA alleles in a simplex polyploid (Aaaa)

Q. 72. What are applicators of polyploidy in crop improvement ?

Ans. The following are main uses of polyploidy in crop improvement.

(1) It helps in tracing the origin of crop species.

(2) It is useful in creating new crop species.

(3) It is useful in interspecific gene transfer.

(4) It is useful in making bridge crosses.

Q. 73. What are limitations of polyploidy ?

Ans. Following are main limitations of polyploidy :

(1) Maintenance of monoploids and triploids is not possible in seed propagated species.

(2) Autopolyploidy has limited applications.

(3) It sometimes leads to undesirable combination such as *Raphanobrassica*.

(4) Polyploids have low fertility, late maturity, slow growth rate etc.

Q. 74. What are differences between monoploids and haploids ?

Ans. Main difference between monoploids and haploids are presented below in Table 19.1.

TABLE 19.1. Differences between Monoploids and Haploids

Monoploids	Haploids
1. Represent gametic chromosome number of a diploid species.	Represent gametic chromosome number of any species.
2. Denoted by *x*.	Denoted by *n*.
3. Monoploids are always haploids.	Haploids cannot always be monoploids.
4. Contain single set of genome.	May contain one or more copies of genome.

DOUBLED HAPLOIDS IN CROP IMPROVMENT

Q. 1. What do you mean by doubled haploids ?

Ans. Doubled haploids refer to the diploid lines which are obtained by doubling the chromosome number of a haploid line by colchicine treatment. Such condition is referred to as doubled haploidy.

Q. 2. What are main features doubled haploids ?

Ans. Main features of doubled haploids are briefly presented below :

1. Doubled haploids are represented as DH.
2. Double haploids are developed from haploids by doubling the chromosome number of haploids by colchicine treatment.
3. Double haploids are completely homozygous lines.
4. Doubled haploids can be produced in both self pollinated and cross pollinated species.
5. Double haploid method allows to fix recombinant gametes directly as fertile homozygous lines.
6. This is a rapid method of producing pure lines. The pure lines can be produced from a heterozygous parents in one generation. The development of pure lines or inbred lines by conventional methods take 3-5 years.
7. Doubled haploids can be released as cultivars after testing or can be used as parents of hybrids.
8. Doubled haploid method reduces the time needed to develop and release new cultivars by 3-5 years in comparison with conventional breeding techniques.
9. Doubled haploids are homozygous and homogeneous populations and hence their produce is highly uniform.
10. The doubled haploids obtained from recombinant lines generally have broad genetic base due to gene combination from two different sources.

11. The doubled haploid method opens a unique way for the fixation hybrid vigour or heterosis in homozygous lines, which would avoid all problems associated with the production of hybrid seed. For example, from a cross between AAbb x aaBB genotypes we can get a haploid with AB genotype which can be fixed as doubled haploid (AABB).
12. Doubled haploids segregate much simpler in the progeny of crosses being true breeding individuals.

Q. 3 What are differences between doubled haploid and conventional methods of producing pure lines ?

Ans. There are several differences in the doubled haploid and conventional breeding methods used for developing pure lines/inbred lines.

Table 20.1 Differences between doubled Haploid and conventional methods of producing pure lines

Particulars	Doubled Haploid Method	Conventional Method
1. Time required for developing pure lines.	One year or one crop season	3-5 years.
2. Time required for developing cultivars	Two to three years	Seven-eight years
3. Fixation of heterosis	Possible	Not possible
4. Expenditure/cost involved	More than conventional method	Lesser than DH method.
5. Identification of recessive mutants	Very easy	Difficult

Q. 4. What are advantages of doubled haploid method ?

Ans. The double haploid method of breeding has several advantages which are briefly discussed below :

1. **A Rapid Method.** It is a rapid method of developing pure lines as compared to conventional method. It takes one year for developing pure lines and 2-3 years for developing Cultivars.
2. **Fixation of Heterosis.** The DH method permits fixation of hybrid vigour. The recombinant gametes with heterotic effects can be fixed by this method. Thus DH method is useful in conserving heterosis.
3. **Gene Mapping.** Doubled haploid populations are commonly used for gene mapping and construction of linkage maps.
4. **Identification of Mutants.** Doubled haploid method is very much useful in the instant identification of recessive mutants which is very difficult by conventional methods.

5. **Instant homozygosity.** DH method is useful in obtaining instant homozygosity. The homozygosity can be obtained in a single step.
6. **Uniform Produce.** Doubled haploid cultivars are homozygous and homogeneous, hence their produce are highly uniform and attractive.

Q. 5. What are disadvantages of doubled haploid method ?

Ans. The doubled haploid method as some limitations or disadvantages which are briefly presented below:

1. Difficult to Obtain

The doubled haploids are developed from haploids. It is difficult to obtain haploid in all crops. The production of haploids varies from crop to crop. In some crops such as cereals and rapseed they can be easily obtained while in other crops they are obtained with difficulty .

2. Expensive Method

It is an expensive method of developing pure lines. The cost is 2-3 times higher than conventional method of developing purelines. However, high cost of this method is compensated from the results achieved. This method requires well equipped tissue culture laboratory and well trained manpower, because haploids are obtained from anther culture or interspecific crosses.

Q. 6. What do you mean by haploid ?

Ans. An individual having gametic chromosome number in the somatic cells is known as haploid and such condition is referred to as haploidy. Main features of haploids are presented below:

1. Haploids occur spontaneously as well as can be induced artificially. The spontaneous haploids have been reported in 10 families covering 26 genera and 36 species. Main crops in which spontaneous haploids have been reported include tomato, cotton, coffee, beets, barley, flax, pearl millet, rapeseed, asparagus and wheat.
2. Haploids are sterile due to lack of pairing partners of chromosomes.
3. Haploids are represented by n. Haploids differ from monoploids. Monoploids have single copy of the genome and are represented by x. In a true diploid species both monoploid and haploid chromosome number is the same ($n = x$). Thus a monoploid is haploid but all haploids cannot be monoploids.
4. Haploids are of different types, their detailed description is beyond the scope of this discussion.

Q. 7. How would you induce haploids ?

Ans. Haploids can be induced in four principal ways, viz. by *(i)* wide crossing, (ii) irradiration and - chemical treatment, *(iii)* selection of twins, and *(iv)* by anther and pollen culture. These are briefly discussed below:

1. Wide Crossing

It includes interspecific and intergeneric hybridization and is also known as distant hybridization. Interspecific crosses have resulted in development of haploids in wheat and barley. Use of maize as pollen parent for wheat has produced high frequency of haploids in wheat.

2. Irradiation and chemical treatment

The pollination with irradiated pollen stimulates the unfertilized eggs to parthenogenic development. Such haploids have been developed in tabacco, wheat, snapdragon and **oenothera**. Application of a dye **toluene blue** (TB) to the pistils has induced haploids in **Vinca rosea**, tomato, maize and poplar.

3. Selection of Twins

In some species haploids can be obtained from the screening of polyembryonic seeds. Polyembyonic seeds can produce three types of seedlings viz. haploid-haploid; diploid-diploid and haploid-diploid. This has been reported in **capsicum** and in some horticultural species.

4. Anther and Pollen culture

This method is widely used for production of haploids. Haploids develop directly from pollen grains in culture, either through direct formation of embryos from pollen grains or formation of callus and subsequent plant regeneration.

Q. 8. What are advantages of haploids ?

Ans. Haploids have several advantages or applications in crop improvement which are presented below:

1. Haploids are used for rapid development of homozygous diploids or polyploids. In other words, haploids are used for developing doubled haploids which in turn are used in various ways as discussed earlier .
2. Haploids reduce the time needed for developing new cultivars.
3. Haploids are useful in identification of recessive mutants.
4. Haploids can be used to produce a male or female plant from a dioecious species such as asparagus.
5. Haploids can also be used for fixation of heterosis in recombinant lines.
6. Haploids can be used for gene mapping, transformation, somatic hybridization, germplasm storage etc.

Q. 9. What are disadvantages of haploids ?

Ans. There are some disadvantages of haploids which are briefly presented below:

1. The crop cultivars respond differently to nutrient or culture medium.

The culture medium suitable for regeneration of one variety may not be suitable for another variety.

2. The response to colchicine treatment also varies from genotype to genotype.
3. Sometimes unwanted traits such as albinism may be fixed with doubled haploid method.
4. The cost of raising doubled haploids for plant breeding is expensive.
5. Doubled haploid lines often appear inferior to inbred lines obtained by conventional method. This may be due to very high level of homozygosity.
6. Moreover, colchicine treatment, gametoclonal and somaclonal variation may have negative effect on the agronomic performance of doubled haploid plants.

Q. 10. What is dihaploid ?

Ans. A haploid individual which is produced from a tetraploid species is referred to as dihaploid. It has haploid or gametic chromosome number of a tetraploid species. In other words, it has two copies of the genome.

Q. 11. Define double haploid.

Ans. A haploid individual with haploid chromosome sets of two diploid species is referred to as double haploid. Such individuals have two sets of genome, one from each species. The term double haploid was coined by Warmke and Blakeslee in 1939.

Q. 12. What do you mean by deploidization ?

Ans. The process by which a polyploid species behaves like a diploid species is referred to as diploidization. There are several polyploid species such as wheat, oats, Brassica, tobacco and cotton which behave like a diploid species. The diploidization is the result of intragenomal pairing or homologous pairing. This pairing behaviour of chromosomes in these polyploid species is under genetic control. In wheat the homologous pairing is controlled by a mutant gene present in the long arm of chromosome 5 of B genome. This system in wheat is known as 5B system which prevents homeologus (intergenomal) pairing and permits only homologous pairing.

Q. 13. What is double reduction ?

Ans. In polyploids, a condition in which a heterozygous individual (simplex-aa aA) produces homozygous gametes with dominant alleles (AA) is referred to as double reduction. It occurs when sister alleles migrate to the same gamete during meiosis. It generates variability and causes decrease in the equilibrium frequencies of deleterious alleles. Such gametes are formed under special situations.

Q. 14. What do you mean by quasi-diploid ?

Ans. An individual having trisomic-monosomic constitution *i.e.* 2n+I-I is referred to as quasi-diploid. Such individuals originate from a cross between trisomic and nullisomic parents.

Q. 15. Describe briefly practical achievements of doubled haploid method.

Ans. Doubled haploids (DH) are presently being used in breeding of a number of crop species. Doubled haploid cultivars have been developed in different countries such as Canada, USA, UK, Spain, China, Hungry , Italy etc. The maximum DH cultivars have been developed in Barley followed by Rapeseed, wheat, melon, pepper, Rice, Asparagus, Tobacco and Eggplant (Table 20.2). Moreover, maximum doubled haploid cultivars have been released by anther culture.

Table 20.2 Number of doubled haploid cultivars released for commercial cultivation in different crops.

Name of Crop	Number of Cultivars Released	Developed
Barley (Hordeum vulgare)	115	Anther culture
Rapeseed (Brassica campestris)	47	Anther culture
Wheat (Triticum aestivum)	21	Anther culture
Melon (Cucurbita marchata)	9	Irradiated pollen
Pepper (Capsicum annum)	8	Anther culture
Rice (Oriza sativa)	8	Anther culture
Asparagus (Asparagus officinalis)	7	Irradiation
Tobacco (Nicotiana tabacum)	6	Irradiation
Eggplant (Solanum melongena)	5	Anther/Irradiation

Source: Modified from Euphytica 2004.

Q. 16. What are uses of doubled haploids ?

Ans. The use of doubled haploid method is gaining increasing importance these days looking to its various advantages. Doubled haploids are used in :

1. Development of purelines
2. Development of cultivars
3. Development of hybrids as parents
4. Construction of genetic maps
5. Gene tagging/locating genes
6. Identification of molecular markers for trait selection.

DISTANT HYBRIDIZATION

Q. 1. What is distant hybridization ?

Ans. Crossing between different species of the same genes or different genera of the same family is called distant hybridization; also known as wide crossing. It is of two types, viz. interspecific hybridization and intergeneric hybridization.

Q. 2. What is interspecific hybridization ?

Ans. Crossing between two different species of the same genus is termed interspecific hybridization or intra-generic hybridization. Such crosses are called interspecific crosses and progeny of such cross is called interspecific hybrid.

Q. 3. Who first used interspecific hybridization ?

Ans. Interspecific hybrid was first developed by Thomas Fairchild in 1717 between sweet william and carnation species of *Dianthus* (*Dianthus barbatus* × D.Canyophyllus).

Q. 4. When will you use interspecific hybridization ?

Ans. Interspecific hybridization is used when the desirable character is not found within the species of a crop.

Q. 5. What are types of interspecific crosses ?

Ans. Interspecific crosses are of following three types :

(*i*) *Fully fertile crosses.* Such crosses are obtained between those species which have complete chromosome homology.

(*ii*) *Partially fertile crosses.* Such crosses are obtained between those species which differ in ploidy level but have some chromosomes in common.

(*iii*) *Fully sterile crosses.* Such crosses are obtained between those species which donot have chromosome homology.

Q. 6. What is intergeneric hybridization ?

Ans. Crossing between two genera of the same family is called intergeneric hybridization. Triticale and Raphanobrassica are outcome of intergeneric crosses.

Q. 7. What is introgression ?

Ans. Transfer of some genes from one species into the genome of another species is called introgression.

Q. 8. Who coined the term Introgression ?

Ans. The term introgression was first used by Anderson and Hubricht in 1938.

Q. 9. Who made first cross between wheat and rye ?

Ans. The first intergeneric cross between bread wheat and rye was made by Rimpu in 1890 in Sweden.

Q. 10. Who made first intergeneric cross between radish and cabbage ?

Ans. The first intergeneric cross between radish and cabbage was made by Karpechenko in 1928 in Russia.

Q. 11. What are parents of *Raphanobrassica* ?

Ans. The parents of Raphanobrassica are radish (*Raphanus stativus*) and cabbage (*Brassica oleracea*).

Q. 12. What are basic diploid species of *Brassica* ?

Ans. Following are three basic diploid species of *Brassica* :

(i) Rapeseed (Brassica campestris $2n = 20$, AA)

(ii) Black mustard (Brassica nigra $2n = 16$, BB)

(iii) Cabbage (Brassica oleracea $2n = 18$, CC)

Q. 13. What are three tetraploid species of Brassica ?

Ans. Three tetraploid species of *Brassica* are given below :

(i) Brassica juncea ($2n = 32$, AABB)

(ii) Brassica carinata ($2n = 34$, BBCC)

(iii) Brassica napus ($2n = 38$, AACC)

Q. 14. Who proposed origin of tetraploid species of *Brassica*?

Ans. The origin of tetraploid species of *Brassica* was proposed by Nagaharu, U in 1935 using a triangle which is popularly know as U's triangle.

Q. 15. Who first synthesized amphidiploid in cotton ?

Ans. Harland (1940) made a cross between *Gossypium arboreum* and *G.thurberi* and synthesized an amphidiploid which was similar to upland cotton (*G. hirsutum*).

Q. 16. Who traced genetic origin of upland cotton ?

Ans. Phillips (1963) based on morphological and cytological studies, reported that *Gossypium hirsutum* (upland cotton) has originated from a cross between *G.africanum* (AA) and *G.raimondii* (DD).

Q. 17. Who traced genetic origin of bread wheat ?

Ans. McFadden and Sears (1947) reported that *Triticum turgidum* and *T.tauchi* are invalved in the evolution of bread wheat (*T.aestivum*).

Q. 18. Who synthesized *Nicotiana digluta species* of *tobacco* ?

Ans. The new species *Nicotina digluta* of tobacco was synthesized by Clausen and Goodspeed in 1925 from a cross between *N.tabacum* and *N.glutinosa.*

Q. 19. Who traced origin of Nicotiana tabacum ?

Ans. The origin of cultivated tobacco (Nicotiana tobacum) was traced by Goodspeed and Clausen in 1928. They established that cultivated tobacco (*N.tabacum*) is amphidiploid between *N.sylvestris* and *N.tomentosa.*

Q. 20. What are progenitors of Nicotiana rustica ?

Ans. Goodspeed and Clausen (1928) established that *Nicotiana rustica* is an amphidiploid between *N. paniculata* and *N.undulata.*

Q. 21. What are main barriers of distant hybridization ?

Ans. The main barriers of distant hybridization are :

(*i*) Cross incompatibility (*ii*) Hybrid inviability

(*iii*) Hybrid sterility, and (*iv*) Hybrid breakdown.

Q. 22. What is cross incompatibility ?

Ans. Inability of the functional pollen of one species or genus to effect fertilization of female gametes of another species or genus of the same family is known as cross incompatibility.

Q. 23. What is hybrid inviability ?

Ans. Inability of a hybrid zygote to grow into a normal embryo under normal conditions of development is known as hybrid inviability. The main reasons of hybrid inviability are given below :

(i) Unfavourable interaction between chromosomes of two species.

(ii) Dishormony between cytoplasm and nuclear genes, and

(iii) Unfavourable interaction among embryo, endosperm and maternal tissues.

Q. 24. What are reasons of cross incompatibility ?

Ans. Main reasons of cross incompatibility are given below :

(i) Lack of pollen germination.

(ii) Insufficient growth of pollen tube to reach ovule, and

(iii) Inability of male gamete to unite with egg cell.

Q. 25. How will you overcome problem of hybrid inviability ?

Ans. Following techniques can be used to overcome the problem of hybrid inviability.

(i) Proper choice of parents
(ii) Making reciprocal crosses
(iii) Application of growth hormones
(iv) Transfer of embryo to culture medium.

Q. 26. What is hybrid sterility ?

Ans. Inability of a hybrid to produce viable offspring is called hybrid sterility. The problem of hybrid sterility is more actute in intergeneric crosses than in interspecific crosses. The main cause of hybrid sterility is lack of structural homology between the chromosomes of two species. This can be overcome by chromosome doubling.

Q. 27. What is cryptic structural hybridity ?

Ans. In interspecific hybrids, sometimes sterility is due to small structural changes in chromosomes which are not detectable during meiosis. Stebbins termed it as cryptic structural hybridity.

Q. 28. Who coined the term cryptic structural hybridity ?

Ans. The term cryptic structural hybridity was coined by stebbins in 1958.

Q. 29. What is hybrid breakdown ?

Ans. A condition in which F_1 plants of an interspecific cross are vigorous and fertile but their F_2 progeny is weak and sterile is called hybrid breakdown. It may result due to following reasons.

(i) Linkage of dominant genes in one species and recessive in other species.
(ii) Small structural differences in the chromosomes of two species.

Q. 30. What is alien addition ?

Ans. Addition of one chromosome of wild species to the normal complements of a cultivated species is known as alien addition.

Q. 31. What is alien addition line ?

Ans. A line having one chromosome of wild species to its somatic chromosome complements is called alien addition line.

Q. 32. Who coined the term alien addition ?

Ans. The term alien addition was coined by Leighty and Taylor in 1924.

Q. 33. What is alien substitution ?

Ans. Replacement of one pair of chromosomes of cultivated species with those of wild donor species is known as alien substitution.

Q. 34. What is alien substitution line ?

Ans. A line in which one pair of chromosome is from wild donor species is called alien substitution line.

Q. 35. Who coined the term alien substitution ?

Ans. The term alien substitution was first used by Kattermann in 1938.

Q. 36. How will you make wide crosses successful ?

Ans. The following techniques are useful in making distant crosses successful.

(i) Selection of compatible parents
(ii) Making reciprocal crosses
(iii) Manipulation of ploidy level
(iv) Making bridge crosses
(v) Use of pollen mixtures
(vi) Use of growth regulators (hormones)
(vii) Use of embryo rescue technique
(viii) Use of protoplast fusion technique
(ix) Cutting of long style to proper size etc.

Q. 37. What is the role of wide crossing in crop improvement ?

Ans. Distant hybridization has played significant role in the improvement of crop plants for various characters as follows :

(i) Imparting disease and insect resistance
(ii) Improvement in quality
(iii) Improvement in adaptation
(iv) Improvement in yield
(v) Development of interspecific hybrids
(vi) Development of new crop species etc.

Q. 38. What are main limitations of wide crossing ?

Ans. Main limitations of wide crossing are given below :

(i) Cross incompatibility, hybrid inviability, hybrid sterility and hybrid breakdown.
(ii) Linkage of some undesirable genes with desirable one.
(iii) It is a long term programme.

Q. 39. What is somatic hybridization ?

Ans. Crossing of plants through fusion of somatic cells, *i.e.* protoplasts is called somatic hybridization.

Q. 40. Who coined the term somatic hybridization ?

Ans. The term somatic cell hybridization was coined by Barski *et al.* in 1960.

Q. 41. What is protoplast ?

Ans. Naked cells or cells without cell wall are referred to as protoplasts.

Q. 42. Who coined the term protoplast ?

Ans. The term protoplast was first used by Hanstein in 1880.

Q. 43. What is bridge cross ?

Ans. A cross which involves third species (which is compatible with both species) to make gene transfer possible between two cross incompatible species in called bridge cross.

Q. 44. What is bridging species ?

Ans. A species which can cross with both cross incompatible species and is used to make bridge between such species is called bridging species.

Q. 45. What are varieties of cotton developed through interspecific hybridization ?

Ans. Varieties of upland cotton developed through interspecific hybridization include MCU 2, MCU 5, Deviraj, Devitej, Khandwaj 1, Khandwa 2, Bodnawar 1, PKV 081, PKV Rajat, and Arogya.

Q. 46. What are interspecific tetraploid hybrids of cotton ?

Ans. In cotton several interspecific tetraploid hybrids have been released for commercial cultivation in India. Such hybrids include Varalaxmi, JKHY 11, CBS-156, DCH 32, NHB 12, TCHB 213, DHB 105, NHB 302, Sruthi etc.

Q. 47. What are interspecific diploid cotton hybrids ?

Ans. In diploid cotton, some interspecific hybrids have been released for commercial cultivation in India. Such hybrids include DH 7, DH 9, DDH 2 and Pha 46.

Q. 48. What are differences between interspecific and intergeneric hybridization ?

Ans. Main differences between interspecific and intergeneric hybridization are presented below in Table 21.1.

TABLE 21.1 Differences between Interspecific and Intergeneric hybridization

Particulars	Interspecific Hybridization	Intergeneric Hybridization
1. Parents involved	Involves two different species of the same genes.	Involves two different genera of the same family.
2. Fertility	Such hybrids vary from completely fertile to completely sterile.	Hybrids are always sterile.

3. Seed setting	More than intergeneric crosses.	Low
4. Use in crop improvement	More than intergeneric crosses.	Less than interspecific crosses.
5. Release of Hybrid varieties	Possible in some crops.	Not possible.
6. Evolution of new crops	Not possible, but evolution of new species is sometimes possible.	Sometimes possible, example is *Triticale*.

PLANT TISSUE CULTURE AND SOMATIC HYBRIDIZATION

Q. 1. What is biotechnology ?

Ans. Biotechnology is the application/use of biological organisms/ processes for mass production of useful substances/products for industry, medicine and agriculture. For example, Penicillium and Streptomyces fungi are used for mass production of antibiotics penicillin and streptomycin. Similarly fermenting yeasts are used for manufacturing beer and wine and cheese is produced from milk by bacteria. Thus biotechnology is the controlled use of biological agents (micro-organisms or cellular components) for beneficial use.

Q. 2. What are different branches of biotechnology ?

Ans. Based on applications, there are five main branches of biotechnology as given below :

(1) Animal biotechnology (2) Medical biotechnology (3) Industrial biotechnology (4) Environmental biotechnology and (5) Plant biotechnology

Q. 3. Define animal biotechnology.

Ans. It deals with development of transgenic animals for increased milk, meat, egg and wool production with resistance to various diseases. It also deals with invitro fertilization and transfer of embryo in animals including human.

Q. 4. Explain medical biotechnology.

Ans. It deals with diagnosis of various diseases and large scale production of various drugs, hormones and vaccines as given below :

(1) Production of various antibiotics (2) Production of hormones such as insulin and interferon (3) Vaccines for chicken pox, rabies, polio etc. (4) Blood serum proteins, etc.

Q. 5. What is industrial biotechnology ?

Ans. It is a branch of biotechnology which deals with commercial production of various useful organic substances such as acetic acid, citric acid, acetone, glycerine etc. It also deals with production of antibiotics like penicillin, streptomycin, mitomycin etc. The micro-organisms such as fungi and bacteria are used for production of these substances.

Q. 6. Explain environmental biotechnology.

Ans. It is a branch of biotechnology which deals with detoxification of wastes and industrial effluents, treatment of sewage water and control of plant diseases and insects through the use of biological agents such as viruses, bacteria and fungi.

Q. 7. What is plant biotechnology ?

Ans. It is a branch of biotechnology which deals with development of transgenic (genetically engineered) plants for various purposes such as having resistance to a particular insect or disease. Transgenic plants are also developed with increase herbicide resistance and with improved quality. Plant biotechnology consists of the application of two basic techniques, viz. tissue culture and genetic engineering. It bypasses sexual process in the development of new crop cultivars. It makes distant crosses (interspecific and intergeneric) practically feasible.

Q. 8. What are various applications of plant biotechnology ?

Ans. Plant biotechnology has various practical applications as listed below :

(1) Development of transgenic plants with disease and insect resistance
(2) Development of herbicide resistant plants
(3) Development of plants with improved quality
(4) Development of male sterility
(5) Production of tetraploids in single step
(6) Distant hybridization through protoplast fusion etc.

Q. 9. Define plant tissue culture.

Ans. Plant tissue culture refers to development of whole plant from regeneration of living plant tissues in the nutrient medium (in vitro) or culture medium.

Q. 10. What is culture medium ?

Ans. Culture medium is a nutrient medium which contains all essential micro and macro nutrients, carbohydrate (glucose), vitamins and hormones. The pH of culture medium should be 5.5. However, the culture medium differs from species to species.

Q. 11. What is explant ?

Ans. Any plant part which is used for regeneration in tissue culture technique is called explant. It may be a cell, a protoplast (naked cell), a tissue or an organ. The effective explant may differ from species to species.

Q. 12. What is totipotency ?

Ans. The regeneration capacity or ability of a plant cell to develop into a whole plant is called totipotency. It reveals that each cell is capable of giving rise to a whole plant.

Q. 13. What is callus ?

Ans. A mass of unorganized regenerated cells in culture medium is called callus (plural *calli*). In some species like carrot and sandal wood, somatic embryos are developed from the callus, while in wheat, rice, barley and tobacco, roots and shoots develop directly from calli.

Q. 14. What are basic requirements of tissue culture ?

Ans. There are some basic requirements of tissue culture techniques which are listed below :

(1) A well equipped pathogen free laboratory.
(2) The laboratory should be air conditioned (18–25°C).
(3) Proper culture medium.
(4) Sub-culturing.

Q. 15. Who first developed the culture medium ?

Ans. The culture medium was first developed by Murashige and Skoog in 1962. This culture medium with modifications is widely used for tissue culture technique.

Q. 16. What are important steps in tissue culture techniques ?

Ans. The tissue culture technique generally involves following four important steps :

(1) Isolation of tissues/organ to be used as explant.
(2) Regeneration and callus formation in culture medium.
(3) Formation of somatic embryos from the callus (embrogenesis).
(4) Differentiation of root and shoot from the somatic embryos (organogenesis).

Q. 17. What is embryogenesis ?

Ans. In the tissue culture technique, the process of formation of somatic embryos from the callus is called embryogenesis. Sometimes, somatic embryos are not formed from the callus and somatic buds are formed which after germination give rise to whole plant.

Q. 18. Define organogenesis.

Ans. In tissue culture technique, the process of differentiation of roots and shoot from the somatic embryos is called organogenesis. Sometimes, the whole plant develops directly from the somatic bud. In such cases somatic embryos are not formed.

Q. 19. What are different techniques of plant tissue culture ?

Ans. Depending upon the plant part used as explant, plant tissue culture techniques are of following five main types :

(1) Meristem Culture
(2) Embryo Culture
(3) Anther Culture
(4) Cell Culture
(5) Ovule Culture

Q. 20. What is meristem culture ?

Ans. Meristem culture refers to regeneration of whole plant in culture medium from tissues of an actively dividing plant part such as stem tip, root tip or auxillary bud. Generally shoot apical meristem is used for regeneration. This technique has been widely used in vegetatively propagated crops such as sugarcane, potato, banana and several fruit trees and timber species.

Q. 21. What are uses of meristem culture in crop improvement ?

Ans. Important applications of meristem culture in crop improvement are given below :

1. It is used for micro propagation in banana, strawberries, citrus and some timber trees such as *Delbergia sissoo.*
2. It is used for obaining virus free plants.
3. It can be used for exchange of germplasm.
4. The germplasm can be conserved in the form of meristems at −196°C in liquid nitrogen for long term storage.

Q. 22. What is embryo culture ?

Ans. Regeneration of whole plant from an embryo in culture medium is called embryo culture. Embryos of appropriate stage are excised from the seed and are transferred to culture medium. This technique is used when there is disharmony between the embryo and endosperm. This technique is used to make distant crosses successful. The embryo in distant crosses is removed before abortion and cultured in culture medium.

Q. 23. How embryo culture is useful in crop improvement ?

Ans. The main uses of embryo culture in crop improvement are given below :

(1) It is useful in making distant crosses (interspecific and intergeneric) successful.
(2) It is useful in obtaining haploids from distant crosses.
(3) Embryo culture is useful in orchid breeding where seeds lack stored food and are unable to propagate.
(4) It is effective in breaking long seed dormancy in *Prunus, Iris* and *Taxus.*

Q. 24. What is anther culture ?

Ans. Regeneration of whole plant from anther or pollen in the culture medium is called anther culture. Anthers of appropriate stage are taken from the plant. The optimum stage may differ from species to species. This technique is used to obtain haploid plants. It has been used in crops like *Brassica*, Tobacco, Rice, Barley, Wheat, Tomato, Potato, Triticale etc.

Q. 25. Define cell culture.

Ans. Regeneration of whole plant from cells in culture medium is called cell culture. Protoplasts are also used for regeneration of whole plants. Cell culture technique has been used in different crop plants such as wheat, rice, barley, maize, tobacco etc. Main uses of this technique are given below :

(1) It is useful in mass clonal production of different species.

(2) Protoplast culture is useful in somatic hybridization and overcoming barriers of cross incompatibility in distant crosses.

(3) Cell culture and protoplast culture are useful in the development of transgenic (genetically engineered) plants.

(4) Embryogenic cell cultures are used for cryopreservation in liquid nitrogen at —196°C for long term preservation.

Q. 26. What is ovule culture ?

Ans. Regeneration of whole plant from ovule in culture medium is called ovule culture. This is also called ovary culture. This technique is used for embryo rescue in distant crosses. Ovules are of two types, viz. fertilized and unfertilized. Haploid plants are obtained from culture of unfertilized ovules and diploid from fertilized ovules. This technique has been used to limited extent.

Q. 27. What is tissue culture induced variation ?

Ans. Depending upon the explant, the tissue culture induced variation is of three types, viz. (1) gametoclonal variation, (2) somaclonal variation, and (3) protoclonal variation.

Q. 28. What is somaclonal variation ?

Ans. The variation that is observed among the plants which are regenerated from callus cultures of somatic explants such as meristems is called somaclonal variation. Such variation has been observed in sugarcane, potato, rice, wheat, maize etc.

Q. 29. What do you mean by gametoclonal variation ?

Ans. The variation which is observed among the plants which are regenerated from gametic culture (anther or ovule culture) is called gametoclonal variation. Such variation has been observed in tobacco, rice and *Brassica*.

Q. 30. Define protoclonal variation.

Ans. The variation which is observed among the plants which are regenerated from callus culture of protoplasts is calld protoclonal variation. Such variation has been observed in potato, tobacco, lucern and many other crops.

Q. 31. What are main features of tissue culture induced variation ?

Ans. Main features of tissue culture induced variation are briefly presented below :

(1) The variation is genetic in origin and hence heritable.

(2) It occurs in both sexually and asexually propagated species. However, the frequency of somaclonal variation is very high in vegetatively propagated species (upto 75% in potato and sugarcane) and low in seed propagated species (0.8–1.2% in maize).

(3) It occurs in both oligogenic and polygenic characters.

(4) Generally, somaclonal variation is heterozygous in origin and rarely homozygous.

(5) Tissue culture induced variation results due to structural chromosomal changes, viz. deletion, duplication, translocations and inversions.

Q. 32. How somaclonal variation is useful in crop improvement ?

Ans. The somoclonal variation is useful in various ways. It is useful in isolation of plants with :

(1) Salinity resistance
(2) Drought resistance
(3) Herbicide resistance
(4) Tolerance to metal toxicity
(5) Early maturity and
(6) Resistance to various insects and diseases.

Disease resistant somaclones have been identified in sugarcane, potato, tobacco and ornamental plants.

Q. 33. What are applications of tissue culture in crop improvement ?

Ans. Main applications of tissue culture in crop improvement are :

(1) Creation of variability
(2) Development of haploids
(3) Embryo rescue
(4) Somatic hybridization
(5) Development of transgenic plants
(6) Drought resistance
(7) Disease resistance
(8) Salinity resistance
(9) Micropropagation
(10) Cryopreservation.

Q. 34. What is somatic hybridization ?

Ans. Crossing of plants through fusion of somatic cells is known as somatic hybridization. In somatic hybridization, the sexual process is bypassed. The fusion of cells takes place through protoplasts. Somatic hybridization permits gene transfer between any two species irrespective of their taxonomic relationship. It overcomes the barriers of cross incompatibility and makes incompatible crosses compatible.

Q. 35. What is protoplast ?

Ans. Protoplasts are nothing but naked cells or cells without cell wall. Protoplasts are isolated from mesophyll cells and callus culture. Protoplasts can be isolated mainly by two methods *i.e.* mechanical method and enzymatic method.

Q. 36. What are the steps involved in somatic hybridization ?

Ans. Somatic hybridization consists of several steps which are listed below :

(1) Isolation of protoplasts
(2) Fusion of protoplasts
(3) Selection of hybrid cells
(4) Culture of hybrid cells
(5) Regeneration of plants from hybrid tissues
(6) Characterization of hybrid plants.

Q. 37. What are advantages of somatic hybridization ?

Ans. There are several potential uses or advantages of somatic hybridization which are given below :

(1) It permits hybridization between distantly related species and unrelated species.
(2) It permits modification of cytoplasmic genes.
(3) It is useful in production of somatic hybrids.
(4) It helps in production of allotetraploid in single step.
(5) It is useful in conservation of heterosis.
(6) It is useful in development of transgenic hybrids.
(7) It is useful in transfer of male sterility from one species to another.

Q. 38. Who coined the term somatic hybridization ?

Ans. The term somatic cell hybridization was first used by Barski *et al* in 1960.

Q. 39. Who first used the term protoplast ?

Ans. The term protoplast was first used by Hanstein in 1880.

Q. 40. What are different types of somatic hybrids ?

Ans. Based on taxonomic relationship of species involved in the hybridization, somatic hybrids are of following three types :

(1) Interspecific somatic hybrids
(2) Intergeneric somatic hybrids
(3) Intertribal somatic hybrids.

Based on combination of chromosomes and cytoplasm from parental species, somatic hybrids are again of three types as given below :

(1) Symmetrical hybrids (2) Asymmetrical hybrids, and (3) Cybrids.

Q. 41. What do you mean by symmetrical hybrids ?

Ans. Those somatic hybrid cells or plants that contain all chromosomes of both the species involved in the fusion are called symmetrical hybrids. Such hybrids have been obtained in tobacco, carrot, petunia, etc.

Q. 42. What are asymmetrical hybrids ?

Ans. Those somatic hybrid cells or plants that contain complete chromosome complement of one species and only a part of somatic complement of another species are called asymmetrical hybrids. Such hybrids have been obtained in tobacco, potato, petunia and several other intergeneric and intertribal somatic hybrids.

Q. 43. What is cybrid ?

Ans. Those somatic hybrids that involve normal protoplast of one species and nucleusless protoplast of another species are known as cybrids. Such hybrids are useful in transfer of cytoplasmic male sterility from one species to other.

Q. 44. Who coined the term cybrid ?

Ans. The term cybrid was first used by Bunn *et al.* in 1974.

Q. 45. What is cytoplast ?

Ans. A protoplast either without nucleus or with inactive nucleus is called cytoplast.

Q. 46. What are homokaryons ?

Ans. Hybrid cells involving protoplasts of the same species are called homokaryones.

Q. 47. What are heterokaryons ?

Ans. Hybrid cells involving protoplasts of two different species are known as heterokaryons.

Q. 48. What are limitations of somatic hybridization ?

Ans. Main limitations of somatic hybridization are listed below :

(1) It can be used in those species where regeneration of plants is possible from protoplasts.
(2) Intergeneric and intertribal hybrids are usually sterile.

(3) Such hybrids have many undesirable traits of wild or unrelated species.

(4) Generally, intergeneric and intertribal hybrids exhibit high degree of sterility and cannot be backcrossed with cultivated species.

Q. 49. What is micro propagation ?

Ans. Mass multiplication of plant material by regeneration of somatic tissues is called micro propagation or micro-cloning. It can be used for mass multiplication of crop plants which are difficult to propagate sexually or those vegetatively propagated species in which rate of multiplication is slow.

Q. 50. What are advantages of micro propagation ?

Ans. Main advantages of micropropagation are given below :

(1) It helps in rapid multiplication of material.

(2) It helps in isolation of pathogen free propagules.

(3) The propagules developed by micro propagation are easy for transportation and require lesser space.

Q. 51. What are differences between somatic hybridization and sexual hybridization ?

Ans. Main differences between somatic hybridization and sexual hybridization are presented below in table 22.1.

TABLE 22.1 Differences between Somatic hybridization and Sexual hybridization

Somatic Hybridization	Sexual Hybridization
1.It involves fusion of somatic cell (protoplasts).	It involves fusion of male and female gametes.
2.It requires tissue culture technique.	It does not require tissue culture technique.
3.Cytoplasm is contributed by both parents.	Cytoplasm is strictly contributed by female parent.
4.Hybrid combines somatic chromosomes of both parents.	Hybrid combines haploid chromosomes of both parents.
5.Segregation does not occur.	Segregation occurs.
6.Crosses between unrelated species can be easily made.	Crosses between unrelated species are incompatible.

23

TRANSGENIC BREEDING

Q. 1. What is transgenic ?

Ans. Genetically engineered organisms are called transgenics. In other words, a genotype developed by the process of genetic engineering is called transgenic (plural '*transgenics*). It may be a plant, an animal or microbes such as fungi, viruses and bacteria.

Q. 2. Who coined the term transgenic ?

Ans. The term transgenic was first used in 1981 by Gordon and Ruddle.

Q. 3. In which crop first transgenic plant was developed ?

Ans. The first transgenic plant was developed in tobacco in USA by Fraley *et al* in 1983.

Q. 4. How transgenic plants are developed ?

Ans. Transgenic plants are developed by the techniques of genetic engineering. The sexual process is bypassed in developing transgenic plants.

Q. 5. What is transgene ?

Ans. Transgene refers to foreign gene or modified gene of the same species which is used for development of transgenic organisms. Transgene may be from the same species (in modified form), related wild species, unrelated species and microbes such as fungi, bacteria and viruses.

Q. 6. What are sources of transgenes ?

Ans. The main sources of transgenes are same species, related wild species, unrelated species, micro-organism and animals.

Q. 7. What is genetic transformation ?

Ans. The process of transfer, integration and expression of transgene in an organism is called genetic transformation.

Q. 8. What are main features of transgenic plants ?

Ans. The main features of transgenic plants are given below :

(1) Transgenic plants contain transgene.

(2) They are developed by the process of genetic engineering.

(3) The sexual process is bypassed in developing transgenic plants.

(4) The frequency of transgenic plants is extremely low (0.001%).

(5) Transgenic plants are developed to solve specific problem which can not be tackled by conventional methods.

Q. 9. What is transgenic breeding ?

Ans. Genetic improvement of crop plants, domestic animals and useful micro-organisms, through genetic engineering, in relation to their economic use for mankind is referred to as transgenic breeding.

Q. 10. What is transgenic plant breeding ?

Ans. Transgenic plant breeding refers to genetic improvement of crop plants in relation to various economic characters useful for human beings, through genetic engineering.

Q. 11. What are advantages of transgenic crop breeding ?

Ans. The main advantages of transgenic plant breeding are briefly presented below :

(1) It is a rapid method of crop improvement.

(2) It overcomes barriers of cross incompatibility.

(3) It is useful in solving those problems which can not be solved by conventional methods.

(4) It permits gene transfer from any source including animals.

Q. 12. What are methods of foreign gene transfer ?

Ans. There are four important methods of foreign gene transfer as given below :

(1) Plasmid method or biological method

(2) Particle bombardment method

(3) Direct DNA uptake method

(4) Micro-injection method.

Q. 13. What is plasmid ?

Ans. Plasmids are extra chromosomal elements which are found in bacterial cell. They can replicate independently of chromosomal DNA and are not essential for normal growth and function of a bacterium. Plasmids are used as cloning vectors in genetic engineering. The soil bone bacterium *Agrobacterium tumefaciens* is used for development of transgenic plants.

Q. 14. What is biological method of DNA delivery ?

Ans. The plasmid method is also known as biological method of DNA delivery. In this method, *Agrobaclerium tumefaciens* is used for gene transfer. This method of foreign gene transfer consists of following four main steps.

(1) Gene cloning through plasmids of *Agrobaclerium tumefacirens.*

(2) Mixing of cloned genes with cell or protoplast culture for transformation.

(3) Identification of transformed cells by screening in culture medium containing Kanamycin. In kanamycin medium only transformed cells will survive and rest will die.

(4) Regeneration of transformed cells into whole plant.

Q. 15. What are advantages of plasmid method of DNA delivery ?

Ans. There are two main advantages of Agrobacterium mediated DNA transfer method as given below :

(1) This method has same control over the copy number and site of integration of transgenes which is not possible in particle bombardment method.

(2) This is cheaper method of foreign gene transfer than particle bombardment method.

Q. 16. What are limitations of plasmid method of genetic transformation ?

Ans. There are three main limitations of plasmid method of foreign gene transfer as follows :

(1) It has host specificity. It can not infect monocotyledonous crops especially cereals.

(2) The somaclonal variation is very high which poses problems in identification of transformed cells.

(3) The regeneration of cells or protoplast into whole plant is very slow.

Q. 17. What is particle bombardment method of DNA delivery ?

Ans. In this method, foreign DNA is delivered into plant cells through high velocity metal particles. This method is also known as biolistic micro-projectile bombardment, microballestics and particle acceleration method.

Q. 18. Which plant material is used for particle gun method ?

Ans. In this method, regenerable tissues or organs are required for bombardment. Generally, meristems, immature embryos, embryogenic callus and suspension culture have been used for DNA delivery in different crop plants. Isolation of protoplast is not essential in this method.

Q. 19. What are the metals used for particle bombardment ?

Ans. Generally, gold and tungsten particles are used for bombardment. These metals do not react with plant tissues. Gold particles, being dense, can penetrate into deeper cell layers than tungsten particles. These two metals are commonly used for bombardment because they do not react with DNA and their fine particles can be made.

Q. 20. How DNA is delivered through particle gun ?

Ans. Minute particles (1.0–1.5 μ m) of tungsten or gold are coated with DNA of interest. The primary cloned form of DNA *i.e.* plasmid construction is used for coating with gold or tungsten particles. The DNA coated particles are accelerated with such force that they should penetrate the outer cell wall of target tissues. Some of the DNA of these particles enters in the nuclei of target cells and is integrated with the DNA of host cells resulting in transformation.

Q. 21. What are types of particle gun ?

Ans. The particle gun is of two types as follows :

(1) Gun powder driven device, and

(2) Helium particle inflow gun.

The gun powder device is more cumbersome as it requires more DNA loaded particles than helium particle inflow gun. Hence, the helium particle inflow gun is in more common use than gun powder device.

Q. 22. Name the crops in which particle acceleration method has been used for genetic transformation.

Ans. Particle acceleration method has been used for development of transgenic plants in several crops such as wheat, barley, oats, rice, corn, sugarcane, cotton, peanut, soybean, common bean, papaya, sunflower, tobacco, etc.

Q. 23. What are advantages of particle gun method of DNA delivery ?

Ans. There are two main advantages of particle gun method of genetic transformation as follows :

(1) This method does not exhibit host specificity. It can be effectively used for development of transgenic plants in monocots as well as dicots.

(2) This method is technically simpler than Agrobacterium method. Various plant parts such as meristems, embryos and leaves can be used for bombardment.

Q. 24. What are disadvantages of particle gun method ?

Ans. This method has following two main disadvantages :

(1) There is no control over the copy number and site of integration of foreign gene.

(2) The equipment is very costly which prohibits its use by many researchers.

Q. 25. Explain direct DNA transfer method of genetic transformation.

Ans. This method is used with protoplast. A suspension of protoplast with desired DNA is prepared. Then high voltage current is applied to the protoplast–DNA suspension. The electric current leads to formation of small temporary holes in the membrane of protoplasts through which the DNA can pass. After entry into the cells, the foreign DNA gets incorporated with the host genome resulting in genetic transformation.

Q. 26. What are limitations of direct DNA transfer method of genetic transformation ?

Ans. This method can be used only in those crop species in which regeneration from protoplasts is possible. This method has been used in rice and corn.

Q. 27. What is micro-injection method of foreign gene transfer ?

Ans. In this method, foreign DNA is delivered in the protoplasts through microscopic needles called micro-injections. This method does not have host range limitations. However, in this method also regeneration from protoplasts is a basic requirement which restricts application of this method only to those species that can regenerate from protoplasts.

Q. 28. What are the charactors that have been improved through transgenic breeding ?

Ans. In field crops, the following characters have been improved through transgenic breeding :

(1) Resistance to disease and insects
(2) Resistance to herbicides
(3) Resistance to drought conditions
(4) Resistance to metal toxicity
(5) Quality of useful products
(6) Development of male sterility.

Q. 29. Cite examples of transgenic insect resistance in crop plants.

Ans. Examples of transgenic insect resistance in crop plants are given below.

(1) In cotton, resistance to *Helicoverva* has been transferred from soil borne bacterium *Bacillus thuringiensis* by Monsanto Company in USA.
(2) In corn, resistance to European Corn Borer had been transferred from above bacterium.
(3) In tobacco, insect resistance has been achieved by transferring trypsin inhibitor gene from cow pea.

Q. 30. Cite examples of transgenic disease resistance in crop plants.

Ans. Transgenic virus resistant genotypes have been developed in various crops such as tobacco, tomato, potato and cucumbers.

Q. 31. Cite examples of cold resistance achieved through transgenic breeding.

Ans. Examples of cold resistance achieved through transgenic breeding are cited below :

(1) In tomato and tobacco, antifreezing gene has been transferred from fish (winter flounder).

(2) In tobacco, cold resistance has been achieved by transferring a gene from *Arabidopsis thaliana*.

Q. 32. Cite example of male sterility induced through transgenic breeding.

Ans. In rapeseed, male sterility has been induced by transferring a gene from *Bacillus amyloliquefaciens*.

Q. 33. Cite examples of quality improvement achieved through transgenic breeding in crop plants.

Ans. Examples of quality improvement achieved through transgenic breeding are cited below :

(1) In potato, protein quality has been improved by transferring serum albumin gene from human.

(2) In tobacco, protein quality has been improved by transferring glutenin gene of wheat.

(3) In alfalfa, protein quality has been enhanced by transferring ovalbumin gene from chicken.

Q. 34. Cite examples of herbicide resistance in crop plants achieved through transgenic breeding.

Ans. Herbicide resistant transgenic plants have been developed in several crops such as cotton, tobacco, wheat, maize, potato, tomato, soybean, flax, rapeseed, sugarbeet, alfalfa, cabbage and other crops. In majority of cases, herbicide resistant genes have been transferred from micro-organisms.

Q. 35. What are limitations of transgenic breeding ?

Ans. The main limitations of transgenic breeding are given below :

(1) Transgenic plants sometimes exhibit unstable performance for the character under consideration.

(2) Sometimes, the transgene has undesirable side effects.

(3) The position of integration of foreign gene in the host genome affects the expression of transgene in the host plant.

(4) The frequency of desirable transgenic is extremely row.

(5) Transgenic breeding is very expensive method of crop improvement. Moreover, development of transgenic plants requires high technical skill.

Q. 36. What is transgenic technology ?

Ans. The technique or process of developing transgenic individuals is reffered to as transgenic technology. It enables plant breeders to combine useful genes in one controlling specific characteristics in one kind of organism, and transferring these genes into another quite different organism, which will then also have those characteristics.

Q. 37. Define transformation technology ?

Ans. The transfer of a gene from one organism and incorporation into the genome of other organism/individual by gene technology is called transformation and a cell containing foreign gene is called transformant. Transformation technology leads to development of transgenic individuals and conveys the same meaning as transgenic technology.

Q. 38. What it genetic engineering ?

Ans. Genetic engineering is the process of removing, modifying or adding desirable genes into the recipient cell or into the DNA strand of recipient plant by molecular techniques. It also has the same meaning as gene technology.

Q. 39. What is recombinant DNA technology ?

Ans. Recombinant DNA is a hybrid or chimeric DNA which cam combine with DNA of any organism. Recombinant DNA technology refer to a set of techniques that are used to isolate the DNA from one organism, modify it and incorporate into the genome of another individual or organism. It is the same as genetic engiuneering.

Q. 40. What is genetic modification ?

Ans. Genetic modification refers to transferring single gene between different plants and animals, turning existing genes up (switch on) and down (switch off), or removing a gene from its original position and placing it in a new position in the same organism. It also means the same as genetic engineering.

Q. 41. Compare transgenic and conventional breeding methods.

Ans. Comparison of transgenic and conventional breeding methods in presented below in Table 23.1.

TABLE 23.1 Comparison of Transgenic and Conventional Breeding methods

Particulars	Transgenic breeding	Conventional breeding (hybridization)
1. Sexual process	Bypassed	Involved
2. Methods involved	Tissue culture and genetic engineering	Hybridization between two genotypes
3. Transfer of genes from microbes and animals	Possible	Not possible

4. Frequency of desirable plants	Very low	Adequate
5. Technical skill required	Very high	Moderate
6. Expenditure involved	Very high	Low
7. Time required for release of new variety	3-4 years	10-15 years
8. Equipments required	Sophisticated	Simple or not required
9. Improvement of polygenic traits	Not possible	Possible
10. Facilities required	Well equipped lab	Mainly field facilities

Q. 42. Compare different methods of foreign gene transfer.

Ans. Comparison of methods of foreign gene transfer (DNA delivery) is presented below in Table 23.2.

TABLE 23.2. Comparison of three methods of foreign gene (DNA) transfer

Particulars	Plasmid method	Particle bombardment method	Micro-injection method
1. Basic requirement	Regenerable cells or protoplasts.	Regenerable tissues or organs or cells.	Regenerable cells or protoplasts.
2. Host specificity	Exhibits severe host specificity.	Does not exhibit host specificity.	Does not exhibit host specificity.
3. Effectiveness	More effective with dicots than monocots.	Equally effective with dicots and monocots.	Equally effective with dicots and monocots.
4. Recovery of transgenic plants	Low to moderate.	Low to moderate.	Extremely low.
5. Expenditure involved	Lesser than particle bombardment method.	Higher than plasmid and micro-injection methods.	Lesser than particle bombardment method.
6. Procedure	Difficult.	Simple.	Difficult.

24

INTRODUCTION TO MOLECULAR PLANT BREEDING

Q. 1. What is plant breeding ?

Ans. Plant breeding is considered as the art, science and technology of improving crop plants in relation to their economic use for mankind. In plant breeding various techniques or technologies are used which are divided into two groups viz. *(i)* soft technologies, and *(ii)* hard technologies.

Q. 2. What are soft technologies ?

Ans. Those techniques which can be easily used or applied without much technical skill are referred to as soft techniques. Such techniques include conventional selection methods or phenotypic selection methods (pure line selection, mass selection, clonal selection, progeny selection etc.)

Q. 3. What are hard technologies ?

Ans. Those technologies which require sophisticated laboratory and high technical skill for use or practical application are referred to as hard technologies. Such technologies include non-conventional techniques such as molecular marker technologies (marker assisted selection, DNA fingerprinting) and transformation technology.

The hard technologies would not replace the soft technologies and a combination (fusion) of both would achieve the best results. The plant breeding techniques are also classified as (i) upstream technologies, and *(ii)* down stream technologies (Sokefun, 2003). These are defined below.

Q. 4. What are upstream technologies ?

Ans. Those techniques which generate additional variation in plant populations are referred to as upstream technologies. Such technologies include transformation technology, protoplast fusion technology and induced mutagenesis.

Q. 5. What are downstream technologies ?

Ans. Those techniques which are used to select superior variants in breeding populations are known as downstream technologies. Such technologies include conventional phenotypic selection, marker assisted selection (MAS) and DNA fingerprinting. These technologies donot create additional variation in plant populations.

Q. 6. What is molecular plant breeding ?

Ans. Molecular breeding is an important branch of plant breeding which is gaining increasing significance these days. Molecular breeding is of very recent origin. It developed In the 1980s with the evolution of DNA marker technologies. Molecular breeding is defined as a branch of plant breeding which utilizes molecular genetic tools and approaches for genetic improvement of crop plants. In other words, genetic improvement of crop plants for various economic traits using molecular marker and transformation technologies is referred to as molecular plant breeding.

Q. 7. Explain main points related to molecular plant breeding.

Ans. Main points related to molecular plant breeding are briefly presented below:

1. Molecular Plant Breeding (MPB) utilizes latest genetic technologies to develop better varieties of crop plants. Two molecular technologies, viz. molecular marker technology and transformation technology are used in molecular plant breeding.
2. Molecular breeding is applicable to both plant breeding and animal breeding. In other words MB can be used for genetic improvement of both plants and animals.
3. MPB is more precise, rapid and cost effective in comparison to conventional phenotypic selection. MPB reduces the time significantly required for the development of new cultivars. It takes 4-5 years for development of new cultivars against 10-12 years of conventional phenotypic selection method.
4. It offers possibility of achieving those goals which can not be achieved by conventional methods.
5. It permits gene transfer between unrelated organisms through transformation technology or genetic engineering.
6. It helps in identification of new genes of economic importance.
7. Now-a-days, molecular breeding is a vital part of plant breeding programmes.
8. MPB sometimes contributes novel traits of agronomic value in crop plants.

Q. 8. What are branches of molecular breeding ?

Ans. In agriculture or crop improvement, molecular breeding has two distinct branches, viz. *(i)* molecular marker technology and *(ii)* transformation technology.

Q. 9. What are advantages of molecular markers ?

Ans. A molecular marker is a unique sequence of nucleotide located close to the gene of interest. In other words, molecular markers are special segments of DNA which flag the presence of genes that control particular trait. They are located very close to the gene of interest and are easily identifiable. The important advantages of molecular markers in plant breeding are presented below.

(i) Molecular markers allow selection of desirable trait at any stage of plant growth. There is no need to wait till maturity of the plants.

(ii) They are not affected by environmental conditions. In other words molecular markers are stable.

(iii) They eliminate the need for phenotypic scoring.

(iv) They provide uniform method for scoring.

(v) A very small sample of plant, leaf or grain is required for scoring of molecular markers.

(vi) Much less number of breeding generations are required for evaluation of molecular markers.

(vii) They provide information about the percentage of genome contributed by each parent.

Q. 10. What are applications of molecular markers ?

Ans. There are several useful applications of molecular markers. Some Important applications of molecular markers are listed below:

(i) Molecular markers help in better understanding of quantitative trait loci (QTL).

(ii) They are highly reproducible and thus help in accurate selection of desired genotypes.

(iii) They are useful in gene pyramiding *i.e.* in incorporation of multiple target genes into a single cultivar.

(iv) They are useful in identification of cultivars and parents of hybrid.

Q. 11. What is molecular marker technology ?

Ans. A technology that utilizes molecular markers for genetic improvement of crop plants is referred to as molecular marker technology. It is also known as DNA marker technology. Important points related to this technology are briefly presented below:

1. It does not involve genetic transformation and hence does not require biosafety and bioethics measures.
2. It utilizes various DNA markers such as restriction fragment length polymorphisms (RFLP), amplified fragment length polymorphisms

(AFLP), random amplified polymorphic DNA (RAPD), simple sequence repeats (SSR), single nucleotide polymorphisms (SNP), expressed sequence tags (EST), sequence tagged site (STS), etc.

3. It makes use of two major technologies, viz. marker assisted selection (MAS) and DNA fingerprinting.
4. Molecular marker technology helps in early detection of cytoplasmic male sterility. By this technology, the cytoplasm type of unknown plants can be identified.
5. It is highly accurate and rapid method of crop improvement. Marker assisted selection (MAS) is a potential tool of crop improvement which should be deployed along with conventional breeding methods.
6. It is effective technology in achieving new goals which can not be easily achieved by traditional breeding such as pyramiding disease resistance genes with indistinguishable phenotypes.
7. The DNA fingerprinting is useful in identification of cultivars, pure lines and parents of a hybrid.
8. Molecular marker technology (MMT) is useful in mapping desirable quantitative trait loci (QTL).

Q. 12. What is transformation technology ?

Ans. The genetically engineered genes are known as transgenes. The technology that utilizes transgenes in developing new crop cultivars is referred to as transformation technology. It is also known as gene technology or genetic engineering. Genetic engineering is the process of removing, modifying or additing genes to a strand of DNA for specific purposes. For example rye grass produces a protein in its pollen that causes an allergic reaction in humans. By switching off the gene responsible for producing allergen, we can develop allergy free rye grass. Main points related to transformation technology are briefly presented below:

1. It involves transgenes and hence biosafety and bioethical measures are essential.
2. It leads to development of transgenic cultivars which are not permitted for organic farming. Moreover, development of transgenic cultivars is not permitted under organic plant breeding.
3. Transformation technology is useful in developing crop cultivars resistant to biotic (insects, diseases and parasitic weeds) and abiotic stresses and also in improvement of quality traits.
4. It is an accurate and rapid method of developing superior crop cultivars.

5. It utilizes mainly two methods of gene transfer, viz. biolistic method and agrobacterium mediated method. These methods are being applied to various field crops for developing transgenic cultivars.
6. This technology leads to over expression of novel genes conferring novel traits. It leads to inhibit expression of harmful genes.
7. It is useful in evaluation of gene promoters and also of the function of unknown genes.

Q. 13. Give a comparison of molecular marker technology and transformation technology.

Ans. There are some similarities and some dissimilarities between marker technology and transformation technology which are presented in Table 24.1

Table 24.1 Comparison of molecular marker technology , and transformation technology.

Sr. No. Particulars	Molecular Marker Technology	Transformation Technology
A. Similarities		
I. Well equiped Laboratory	Required	Required
2. Costly chemicals	Required	Required
3. Glasswares	Required	Required
4. High Technical Skill	Required	Required
5. Initial cost involved	High	High
6. Release of new variety	Possible	Possible
7. Improvement in resistance	Possible	Possible
8. Improvement in quality	Possible	Possible
9. Slow or rapid method	Rapid method	Rapid method
10. Accuracy	Very High	Very High
B. Dissimilarities		
I. Transgene	Not Involved	Involved
2. Biosafety measures	Not Required	Required
3. Bioethical measures	Not Required	Required
4. Sexual Process	Involved	Not Involved
5. End product	Non transgenic	Transgenic
6. QTL Mapping	Possible	Not Possible
7. Gene Pyramiding	Possible	Not Possible
8. Gene transfer between unrelated species	Not possible	Possible
9. Tissue culture	Not Involved	Involved

10. DNA markers	Used	Not used
11. Adoption of technology	Moderate	Very high
12. Cultivars Released	Very few	Many in different crops
13. Organic farming of cultivars released	Permitted	Not permitted

Q. 14. What are practical applications of molecular marker technology ?

Ans. Molecular marker technology has four principal applications in crop improvement programmes, viz., *(i)* Molecular selection, *(ii)* molecular backcrossing, *(iii)* molecular mapping, and *(iv)* molecular genotyping. These are briefly discussed below:

1. Molecular Selection

This method is used in the segregating populations (F_2 or F_3) for selection of plants carrylng the gene of Interest. It is used to select those genes which are difficult to select on the basis of phenotypes. It is very useful in selection of genes which are highly influenced by environmental conditions. It allows easy stacking of disease resistant genes. The selection can be practised at an early stage of plant growth.

2. Molecular Backcrossing

This is the most common application of molecular marker technology in crop improvement. It requires less material and less space for conducting DNA marker test. It allows recovery of the highest portion of the recurrent parent, with desired genes, in the shortest number of generations. It is possible to identify between homozygous and heterozygous plants through DNA markers. Thus it helps in rapid transfer of desirable gene.

3. Molecular Mapping

DNA marker technology plays important role in gene mapping. It is useful in mapping of both qualitative and quantitative traits. It identifies molecular markers linked to chromosomal regions affecting the desirable trait and thus helps in gene mapping. The gene mapping is possible even during early stages of plant growth. Various breeding populatlons, viz. recombinant inbred lines (RILs), near isogenic lines (NILs), bulk segregant populations, introgressed lines etc. are used for gene mapping.

4. Molecular Genotyping

It refers to molecular characterization of breeding material. It fullfils DNA fingerprinting needs. It is useful to identifying differences between varieties and hybrids. It is used for varietal verification, varietal purity and hybrid purity and thus is useful for patent protection.

The gene technology or genetic engineering has four principal applications in crop improvement which are listed below:

1. Development of transgenic crop cultivars in different crop plants resistant to biotic and abiotic stresses.
2. Development of herbicide resistant cultivars in different crop plants.
3. Improvement in nutritional quality , keeping quality and industrial quality in food, vegetable, fruits and other commercial crops.
4. Identification of new genes and gene systems. It enables the introduction of novel traits in crop plants such as golden rice and longer blooming flowers in flower plants such as petunia.

Q. 15. What are practical achievements of molecular plant breeding ?

Ans. The molecular plant breeding has been used for genetic improvement of various field crops (maize, barley, wheat, rice, sorghum, soybean, chick pea, pea, mungbean), vegetable crops, forage crops and fruit crops for various economic characters. The notable achievements and briefly presented below:

1. Bacterial blight and blast resistant cultivars have been developed in rice through marker assisted selection.
2. In soybean, cyst nematode resistant cultivars have been developed through MAS.
3. In maize, quality protein maize (QPM) lines have been developed through MAS.
4. In rye, bread making quality has been significantly improved by transferring glutenin gene of wheat through genetic engineering. .
5. In wheat and rice, yield has been increased by transferring a gene from maize through gene technology.
6. In cotton, bollworms and herbicide resistant cultivars have been developed through transgenic technology by using Bt. gene from soil bacterium. Several other examples can be cited from vegetable and forage crops.

Q. 16. What are advantages of molecular marker technology ?

Ans.

1. It helps in better understanding of quantitative trait loci and as a result in more effective breeding.
2. Rapid Introgression of simply inherited characters : The number of backcrosses required can be reduced drastically once the marker for the character to be introduced is identified.
3. Easy or early indirect character selection : It is useful for those genes

that can not be detected at an early stage of plant growth. For example lysline and tryptophan genes in malze.

4. It is useful in achieving new goals that cannot be achieved through conventional breeding such as pyramiding of disease resistance genes which cannot be detected by phenotypes. In other words, it offers possibility of achieving unattainable goals.
5. It is accurate, rapid and more reliable method of crop improvement.
6. The DNA marker test can be conducted at any stage of plant growth even for those traits which express at the maturity .
7. A very small amount of plant tissue i.e. leaf, stem, flower, seed etc. is required to conduct DNA marker test.
8. The DNA test is the only available practical technique for ensuring the presence of multiple beneficial genes in a single variety .
9. DNA based fingerprinting allows the reliable tracking of beneficial traits during varietal selection.
10. DNA based fingerprinting is the best currently available technology to differentiate among varieties for patent protection and plant variety protection Act Certification.
11. DNA based marker systems are useful in selection of disease or insect resistant material even in the absence of pest incidence.

Q. 17. What are advantages of gene technology ?

Ans.

1. It involves transgenes and leads to development of transgenic cultivars in different self and cross pollinated species.
2. It permits transfer of desirable gene from unrelated species *i.e.* from micro-organisms to crop plants and even from animals to plant.
3. It enhances the expression of novel genes conferring novel traits and switch off the expression of harmful traits such as allergen in rye grass.
4. It is precised and rapid method of crop improvement. Moreover, it requires less space than conventional breeding methods.
5. It is useful method of improving crop plants for resistance to biotic and abiotic stresses and quality traits.
6. It helps in evaluating the functionality of unknown genes.

Q. 18. What are limitations of molecular plant breeding ?

Ans. Molecular plant breeding has several advantages as discussed above. However, it has some limitations which are briefly presented below :

1. The initial cost of both the branches of molecular plant breeding *i.e.* molecular marker technology and gene technology is very high. Both these technologies require costly equipment, glasswares and chemicals.

2. Molecular breeding requireds team of well trained manpower for handling of costly equipments, chemicals and glassware.
3. Gene technology involves transgene and thus requires biosafety and bioethical measures.
4. The scoring of DNA or molecular markers is a laborious and time consuming task.
5. In gene technology, the frequency of desirable transformants is very low which restricts the use of this technology.

Q. 19. What are restriction enzymes ?

Ans. A group of endonuclease enzymes that cuts DNA only at specific sites or base sequences.

Q. 20. What are endonuclease enzymes ?

Ans. The enzyrnes which are able to make Internal cut in the DNA molecule are called endonucleases. *Most* of the endonucleases cut DNA molecule at random sites, but some cut only at specific sites. Endonucleases are found In mlcro-organisms and are also referred to as biochemical scissors. Endonucleases do not cut DNA of their own cell and protect it by methylation process.

Q. 21. What is recognition site ?

Ans. The specific base sequence or site recognized by restriction enzyme in DNA molecule to make a cut or cleavage Is called recognltion site; also called recognition sequence.

Q. 22. What is DNA Finger Printing ?

Ans. A method of genetic engineering in which a DNA probe is hybridized with DNA of plants or animals to determine similarities : and differences between two DNA molecules.

Q. 23. What are DNA probes ?

Ans. Small segments of DNA with known base sequences, origin and function are called DNA probes. DNA probes are useful in marker aided selection.

Q. 24. What is DNA library ?

Ans. A set of DNA probes is called DNA library.

Q. 25. Define gene amplification.

Ans. An increase in the copy number of a gene/DNA sequence is called gene amplification.

Q. 26. What is restriction fragment ?

Ans. A fragment of DNA created by cleavage at specific sites by a restriction endonuclease is called restriction fragment.

25

MOLECULAR MARKERS [DNA MARKERS]

Q. 1 What is marker character ?

Ans. Any property of an individual showing heritable variation is referred to as a character. It includes morphological, physiological and biochemical properties in plants. Plant characters are of two types viz. qualitative (governed by one or few genes) and quantitative (governed by several genes). Those characters which can be easily identified are referred to as marker characters.

Q. 2. What are types of markers ?

Ans. Markers are of three types, viz. morphological markers, biochemical markers and DNA markers. These are defined below:

1. **Morphological Markers.** These are related to shape, size, colour and surface of various plant parts. Such characters are used for varietal identification.
2. **Biochemical Markers.** Such markers are related to variations in protein and amino acid banding patterns. Gel electrophoretic studies are used for identification of biochemical markers.
3. **DNA Markers.** DNA markers are related to variations in DNA fragments generated by restriction endonuclease enzymes. DNA markers are also known as molecular markers.

Q. 3. What are features of an ideal DNA marker ?

Ans. A DNA marker refers to a unique sequence of nucleotides found on a strand of DNA. The DNA markers are also known as genetic markers. The main features or characteristics of an ideal marker are given below:

1. **Polymorphic.** It means that the markers tend to look slightly different in different individuals. This presents variability in the population which is essential for a geneticist or plant breeder to select superior genotypes.
2. **Codominant.** An ideal DNA marker should be codominant. Codominance refers to absence of intra-locus interaction. It helps in identification of heterozygotes from the homozygotes.

3. **Non-epistatic.** It refers to absence of inter locus interaction. It helps in predicting the genotype of an individual from its phenotype.
4. **Multi allelic.** It refers to presence of more than two alleles at a locus. It is useful in getting more variability or polymorphism for a character.
5. **Neutral.** It means that the allelic substitutions at the marker locus do not alter the phenotype of an individual. Almost all molecular markers have this property.
6. **Insensitive to Environment.** It means the molecular marker should not be affected by environmental conditions. This property is found in almost all DNA markers. It helps in predicting genotype from the phenotype of an individual.

Q. 4. What are differences between morphological and DNA markers ?

Ans. The DNA markers differ from morphological markers in several ways (Table 25.1)

Table 25.1 Difference between morphological and molecular markers

Particulars	Morphological Markers	Molecular Markers
1. Polymorphism	Insufficient	Very high
2. Expression	Generally dominant	Generally codominant
3. Epistasis	Generally present	Generally absent
4. Multiple alleles	Mostly present	Mostly absent
5. Allele substitution effect	Very clear	Generally neutral
6. Environmental effect	Very high	No effect

Q. 5. What are the limitations of biochemical markers ?

Ans. There are some limitations of biochemical markers (protein markers) as given below:

1. The biochemical markers are detected through isozyme analysis. Isozyme technique can detect only limited number of loci. For example it can detect 30-40 isozyme markers (protein markers) in segregating populations of rice or maize.
2. Secondly, all the enzymes are not present or active in all the organs of an individual.

The DNA markers, on the other hand, are numerous in number and uniformly present in all organs and stage of an individual. Thus main differences between DNA markers and protein markers are given In Table 25.2.

Table 25.2 Differences between biochemical (protein) and DNA markers

Particulars	Protein Markers	DNA Markers
I. Number of Markers	Limited	Numerous
2. Effect of organ of an individual	Variable	Uniform
3. Effect of stage of an individual	Yes	No

Q. 6. List various types of DNA markers.

Ans. There are several types of DNA markers. The list of DNA markers which are commonly used in plant breeding is given below :-

I. Restriction Fragment Length Polymorphisms (RFLPs)
2. Amplified Fragment Length Polymorphisms (AFLPs)
3. Random Amplified Polymorphic DNA (RAPD)
4. Cleaved Amplified Polymorphic Sequence (CAPS)
5. Simple Sequence Repeats (SSRs)
6. Single Strand Conformational Polymorphisms (SSCPs)
7. Single Nucleotide Polymorphisms (SNPs)
8. Heteroduplex Analysis (HA)
9. Expressed Sequence Tags (ESTs)
10. Sequence Tagged Sites (STSs)

Description, advantages, disadvantages and uses of all these DNA markers are briefly presented below:

Q. 7. What are restriction fragment length polymorphisms (RFLPs).

Ans. RFLPs refer to variations found within a species In the length of DNA fragment generated by specific endonuclease. RFLPs are useful as molecular markers. RFLPs are the first type of DNA markers developed to distinguish individuals at the DNA level. RFLP technique was developed before the discovery of PCR. The RFLP technique consists of following three main steps.

1. Cutting genomic DNA with specific restriction enzymes.
2. Separating different sized DNA fragments on electrophoresis gels (agarose), and
3. Hybridizing a labelled, radioactive DNA probe to complementary DNA fragment on the gel in order to visualize them (using X-ray film exposure).

Q. 8. What are advantages of RFLPs ?

Ans. RFLP technique has several advantages. Important advantages of this technique are presented below:

1. It is a simple and cheaper technique of DNA sequencing.

2. It does not require special instrumentation.
3. The majority of RFLP markers are co-dominant and highly locus specific.
4. RFLP markers are powerful tools for comparative and synteny mapping.
5. Other markers such as CAPS or INDEL can be developed from RFLP probe sequences.
6. Numerous samples can be simultaneously screened by this technique using different RFLP probes.
7. RFLP genotypes for single or low copy number genes can be easily interpreted and scored.

Q. 9 What are disadvantages of RFLPs ?

Ans. Disadvantages of RFLPs are given below :

1. Developing sets of RFLP probes and markers is labour intensive.
2. This technique requires large amount of high quality DNA.
3. The multiplex ratio is low (typically one per gel).
4. The genotyping through put is low (difficult to automate).
5. Mostly genotyping is done using radio active methods, despite availability of non- radio active methods.
6. RFLP finger prints for multigene families are often complex and difficult to score.
7. RFLP probes must be maintained physically and are thus a hassle to share between laboratories.

Q. 10. What are uses of RFLPs ?

Ans. RFLPs can be used in a variety of ways for various purposes. Important uses of RFLPs are given below:

1. RFLPs can be used in determining paternity cases.
2. RFLPs are also used in crimal cases to determine the source of DNA sample.
3. RFLPs can be used to determine the disease status of an individual.
4. In plant breeding and genetics RFLPs are used for gene mapping, germplasm characterization and marker assisted selection.
5. It is also used for diagnosis of pathogen in plants even if it is in latent stage.

Q. 11. What are amplified fragment length polymorphisms (AFLPs) ?

Ans. AFLPs are differences in restriction fragment length caused by SNPs (single nucleotide Polymorphism or INDELs (insertion-deletion) that create or abolish restriction endonuclease recognition sites. AFLP assay are performed by selectively amplifying a pool of restriction fragments using PCR. The RFLP technique was originally known as selective restrictive fragment amplification (SRFA). It was developed by Zabeau and Vas in 1993. The AFLP protocol consists of the following five main steps.

1. Digestion of DNA with two different restriction, enzymes (generally a rare and a frequent cutter).
2. Ligation of double stranded adapters to the end of restriction fragments.
3. Optional DNA-pre-amplification of ligated products directed by primers complementary to adapter and restriction site sequences.
4. DNA amplification of subsets of restriction fragments using selective AFLP primers and labelling of amplified products.
5. Finally the labelled reaction products are separated by electrophoresis (using denaturing poly acrylamide gels) and exposed to X-ray films to visualise the AFLP fingerprint (if radio active) or by running out the product on an automated DNA sequencer.

Q. 12. What are advantages of AFLPs ?

Ans. Advantages of AFLPs are as follows :

1. This technique provides very high multiplex ratio and genotyping through put.
2. This technique can be applied to virtually any organism with no formal marker development.
3. The results are highly reproducible across laboratories.
4. No special instrumentation is needed for performing AFLP assays except for co-dominant scoring.
5. The start up cost of AFLP is moderately low.
6. AFLP assay can be performed using very small DNA samples.

Q. 13. What are disadvantages of AFLPs ?

Ans. Disadvantages of AFLPs are given below :

1. High quality DNA is needed to ensure complete restriction enzyme digestion.
2. The homology of a restriction fragment can not be unequivocally ascertained across genotypes or mapping populations.
3. It is difficult to develop locus specific markers from individual fragments.
4. It generally involves radio active methods. Though non-radio active methods are available but they are rarely used.
5. In species with large genomes such as barley and sunflower, AFLP markers often densely cluster in telomeric regions.

6. The maximum polymorphic information content for any bi-allelic marker is 0.5.
7. Proprietary technology is needed to score heterzygotes and ++ homozygotes.

Q. 14. What are uses of AFLPs ?

Ans. This technique has been widely used In the construction of genetic maps containing high densities of DNA marker. In plant breeding and genetics AFLP markers are used in varietal identification, characterization of germplasm, gene tagging and marker assisted selection.

Q. 15. What is random amplified polymorphic DNA (RAPD) ?

Ans. RAPD refers to polymorphism (variations) found within a species in the randomly amplified fragments of DNA generated by restriction endonuclease enzyme. RAPDs are PCR based molecular (DNA) markers. RAPD marker assays are performed using single DNA primer of arbitrary sequence. This technique was proposed by Williams et. al. 1990. The RAPD markers are dominant. RAPD polymorphisms are produced by the presence or absence of PCR amp licons.

Q. 16. What are advantages of RAPDs ?

Ans. Advantages of RAPDs are as follows :

1. This technique is simple and quick as compared to RFLPs.
2. RAPD primers are readily available being universal.
3. It can be performed with any species using universal primers.
4. It provides more polymorphisms than RFLPs.
5. It does not need special equipment. Only PCR is required.
6. The startup cost is low.
7. RAPD marker assays can be performed using very small DNA samples (5-25 ng per individual).
8. RAPD markers are universal, can be commercially purchased and are easily shared between laboratories.
9. RAPD technique permits development of locus specific co-dominant PCR based markers from RAPD markers.
10. Large number of bands are produced per primer.

Q. 17. What are Disadvantages of RAPDs ?

Ans. Disadvantages of RAPDs are as follows :

1. The detection of polymorphism is limited as in case of RFLPs.
2. This technique only detects dominant markers.
3. The reproducibility of results across laboratories is low or inconsistent.
4. It is not feasible for use in marker assisted breeding program.

5. The homology of fragments across genotypes cannot be ascertained without mapping.

Q. 18. What are uses of RAPDs ?

Ans. A segregating population is used to detect polymorphism. RAPD markers are useful in following ways :

1. For assessment of genetic variation in breeding populations, germplasm and different species of a cross.
2. To study phylogenetic relationship among species and subspecies.
3. For gene tagging and construction of linkage maps.
4. For varietal identification.
5. For DNA fingerprinting of cultivars and germplasm.

RAPD markers have played a very important role in selection of genotypes with desired characteristics and also in the production of species specific, genome-specific and chromosome specific markers.

Q. 19. What is cleaved amplified polymorphic sequence (CAPS) ?

Ans. It refers to polymorphism (variations) that are found in restriction fragment lengths caused by SNPs or INDELs that create or abolish restriction endonuclease recognition sites in PCR amplicons produced by locus specific oligonucleotides primers (Tragonrung et al. 1992 ; Konieczny and Ausubel 1993). CAPs assays consist of following main steps.

1. Digesting locus specific PCR amplicons with one or more specific enzymes.
2. Separating the digested PCR amplicons with one or more specific enzymes.
3. Exposure to X-ray films to visualise CAPS fragments.

The modification CAPS has been proposed by some workers (Michaels and Amasino, 1998), called d CAPS (derived CAPS) which is based on SNPs. It is PCR based technique.

Q. 20. What are advantages of CAPS ?

Ans. Advantages of CAPS are as follows :

1. It is a simple PCR assay. The markers are developed from DNA sequences of previously mapped RFLP markers.
2. The most CAPS markers are co-dominant and locus specific.
3. It provides moderately high genotyping through put.
4. No special equipment is needed to perform manual CAPS marker assays.
5. The CAPS marker assays can be performed using semi-automated methods e.g. florescent assays on DNA sequencer.

6. The startup cost is low for manual assay methods.
7. The CAPS assays can be performed using very small DNA samples (50-100 ng per individual).
8. The most CAPS genotypes are easily scored and interpreted.
9. The CAPS markers are easily shared between laboratories.

Q. 21. What are disadvantages of CAPS ?

Ans. Disadvantages of CAPS are as follows :

1. Other methods better than CAPS such as SNPS assays are now available for screening locus specific DNA fragments for polymorphism.
2. Scoring and interpretation of multigene families is some times difficult by this technique.
3. A variety of restriction enzymes have to be tested to find polymorphism.

Q. 22. What are simple sequence repeats (SSRs) ?

Ans. SSRs also called microsatellites, are randomly repeated mono, di, tri, tetra, penta, and hexo nucleotide motifs. The SSR length polymorphisms are caused by differences in the number of repeats. The SSR loci are individually amplified by PCR using pairs of oligonucleotide primers specific to unique DNA sequences flanking the SSR sequence. Jeffreys (1985) showed that some restriction fragment length polymorphisms are caused by VNTRs (Variable Number of Tandem Repeats). These are known as minisatellites and have similarity to larger satellite DNA repeats.

Q. 23. What are advantages of SSRs ?

Ans. Advantages of SSRs are as follows :

1. The SSR markers tend to be highly polymorphic.
2. Most SSR markers are co-dominant and locus specific.
3. This is a simple PCR based technique.
4. SSRs technique does not require special equipments. However, sometimes DNA sequencer is required.
5. The startup cost is low for manual assay methods.
6. It provides high genotyping throughput.
7. The SSR assays can be performed using small DNA samples (100 ng per individual).
8. The SSR DNA markers can be easily shared between Laboratories.
9. The SSR markers can be multiplexed using PCR.

Q. 24. What are disadvantages of SSRs ?

Ans. Disadvantages of SSRs are given below :

1. The development of SSRs is Labour intensive.
2. The cost of developing SSR markers is very high.
3. The SSR markers are specific to plant species.
4. The start up cost is high for automated method of SSR assays.
5. The development of PCR multiplexer is difficult and expensive. Some markers may not multiplex.

Microsatellites or simple sequence repeats (SSRs), simple sequence length polymorphisms (SSL.Ps), short tandem repeats (STRs), simple sequence motifs (SSMs), sequence target microsatellites (STMs) is a class of repetitive sequence which are widely distributed in all eukaryotic genomes. They consist of arrays of tandemly repeated short nucleotide motifs of 1-4 bases, and are called mono, di, tri or tetranucleotide repeats respectively.

Q. 25. What are uses of SSRs ?

Ans. The use of microsatellites or genetic markers has been proposed for genetic mapping of eukaryotes (Backmann and Soller, 1990).

Q. 26. What are single strand conformational polymorphisms (SSCPs) ?

Ans. SSCPs refer to DNA polymorphism produced by differential folding of single stranded DNA harboring mutations (Orita et al. 1989). The conformation of follded DNA molecule is produced by intra-molecular interaction and thus is a function of DNA sequence. The SSCP marker assays are performed using heat denatured DNA on non-denaturing DNA sequence gels. Special gels (e.g. mutative enhancement gels) have been developed to enhance the recovery of single strand conformational polymorphisms caused by INDELs, SNPs or SSRs.

Q. 27. What are advantages of SSCPs ?

Ans. Advantages of SSCPs as follows :

1. It is a simple PCR based assay.
2. Many SSCPs markers are multi allelic and highly polymorphic.
3. Most SSCPs are co-dominant and locus specific.
4. SSCP assay does not require special equipment.
5. The startup cost is low.
6. The SSCP marker assays can be performed using very small DNA samples (10-50 ng per individual).
7. The SSCP markers can be easily shared between Laboratories.
8. The SSCP gels can be silver stained (non-radioactivity).
9. The complexity of PCR products can be assessed and individual fragments can be isolated and sequences.

Q. 28. What are disadvantages of SSCPs ?

Ans. Disadvantages of SSCPs are as follows :

1. The development of SSCP markers is Labour intensive.
2. The development cost of SSCP markers is high.
3. The SSCP marker analysis can not be automated.

Q. 29. What are single nucleotide polymorphisms (SNPs) ?

Ans. The variations (differences) which are found at a single nucleotide position are referred to as single nucleotide polymorphisms or SNPs. This type of DNA polymorphism results due to substitution, deletion or insension. It occurs approximately every 1.3 Kb (Cooper et al. 1985 ; kwok et. at., 1996). Most polymorphisms of this type have only two alleles and also called biallelic loci. The SNP is the most common class of DNA polymorphism which found both in natural lines and after induced mutagenesis. Main features of SNPs are presented below:

1. SNP markers are highly polymorphic and mostly biallelic.
2. SNP markers provide ultra-high thorough put for genotyping.
3. SNP markers are locus specific.
4. SNP markers are excellent long term investment.
5. SNPs result due to substitution deletion or insersion.
6. SNP markers can be used to pinpoint functional polymorphisms.
7. Small amount of DNA is required for SNP markers.
8. Several SNP assay techniques have been developed.

Q. 30. What are advantages of SNPs ?

Ans. Advantages of SNPs are as follows :

1. SNPs are useful in gene mapping.
2. SNPs help in detection of mutations at molecular level.
3. SNPs are useful in positional cloning of a mutant locus.
4. SNPs are useful in identification of disease causing genes.

Q. 31. What are disadvantages of SNPs ?

Ans. Disadvantages of SNPs are as follows :

1. Most SNPs are biallelic and thus less informative than SSRs.
2. Multiplexting is not possible for all loci.
3. Some SNP assay techniques are costly.
4. In preparing genetic maps, three times more SNP markers are needed than SSR markers.
5. Development of SNP markers is labour oriented.

Q. 32 What are uses of SNPs ?

Ans. Uses of SNPs are given below :

1. SNPs are being used in generating genetic maps. SNPs have been used in preparing human genetic maps.
2. In plants, SNPs have not been extensively used so far.

Q. 33. What is heteroduplex Analysis (HA) ?

Ans. In refers to DNA polymorphisms produced by separating homoduplex from heteroduplex. DNA using non-denaturing gel electrophoresis (White et al. 1992) or partially denaturing high performance liquid chromatography (Wang et al. 1999; cho et al. 1999). The heteroduplexes are produced due to single base mismatches between genotypes. The presence of heteroduplexes indicates presence of DNA polymorphism.

Q. 34. What are advantages of HA ?

Ans. Advantages of HA are given below :

1. It is powerful technique for the discovery of single nucleotide polymorphism (SNP).
2. Very small quantity of DNA samples (10-50 ng per individual) is required to perform heteroduplex analysis.
3. Most of the heteroduplex markers are co-dominant and locus specific.
4. The heteroduplex analysis can be performed using automated high performance liquid chromatography.
5. The HA markers can be easily shared between laboratories.

Q. 35 What are disadvantages of HA ?

Ans. Disadvantages of HA are as follows :

1. The heteroduplex analysis reuires special equipment.
2. One protocol may not be sufficient for heteroduplex analysis of different targets using HPLC.

Q. 36. What are uses of HA ?

Ans. Uses of HA are as follows :

1. Heteroduplex analysis has been mostly used in human genetics to screen disease genes for DNA polymorphism. ,
2. In plant breeding, it can be used for detection of pathogens which are in latent stage and thus will be useful in selection of disease free plants.
3. It is useful in the discovery of Single Nucleotide Polymorphism.

Q. 37. What are Expressed Sequencetags (ESTs) ?

Ans. Expressed Sequence Tags (ESTs) are small pieces of DNA and their location on the chromosome and sequence are known. The term Expressed, Sequence Tags was first used by Venter and his colleagues in 1991. Main points related to ESTs are given below:

1. ESTs are short DNA sequences, usually 200-500 nucleotides long.
2. ESTs are a type of Sequenced Tagged Site (STS).

3. ESTs are created from CDNA, a single strand of DNA which has been copied from an mRNA molecule. In other words, an EST is a strand of DNA consisting only of the exons. Exon is the base pair sequence of a gene that is expressed.

Q. 38. What are advantages of ESTs ?

Ans. Advantages on ESTs are as follows :

1. ESTs are useful in discovering new genes particularly genes that are involved in genetic diseases.
2. It is a rapid and inexpensive technique of locating a gene.
3. ESTs provide information about gene expression. In other words, they can be used for tissue specific gene expression.
4. ESTs generate a large number of sequences quickly.
5. Through ESTs more information can be obtained from QTL mapping.
6. ESTs can be used to locate disease causing genes.

Q. 39. What are limitations of ESTs ?

Ans. Limitations of ESTs are given below :

1. ESTs have lack of primer specificity.
2. Development of ESTs is time consuming and labour oriented.
3. The error is higher than other conventional approaches.
4. It is difficult to obtain large transcripts (>.6 kb).

Q. 40. What are applications of ESTS ?

Ans. Uses of ESTs are as follows :

1. ESTs are commonly used to map genes of known function. In other words, they can be used for constructing genome maps.
2. ESTs are used for phylogenetic studies.
3. ESTs can be used to produce DNA arrays, a powerful tool for the analysis of the whole expression pattern of a tissue.

Q. 41. What is Sequence Tagged Sites (STSs) ?

Ans. It genomics, a sequence tagged site (STSs) is a short DNA sequence that has a single copy in a genome and whose location and base sequence are known. Main points related to STSs are given below:

1. STSs are usually 200-500 base pair long.
2. STS occurs only once in the genome. In other words, STSs have single copy in a genome.
3. STSs are detected by Polymerase Chain Reaction (PCR) in the presence of all other genomic sequences.
4. STSs are derived from CDNAs.

Q. 42. What are advantages of STS markers ?

Ans. Advantages of STSs are as follows :

1. STSs serve as landmarks in developing physical map of a genome. In other words, they are useful in physical mapping of genes .
2. STSs permit use of data from different Laboratories for gene mapping.
3. It is rapid and more specific technique than DNA hybridization techniques.
4. It can be automated resulting in performing large number of tests.
5. It is highly sensitive means has highly degree of accuracy.

Q. 43. What are limitations of STS ?

Ans. Limitations of STS are as follows :

1. Development of STS is a difficult task.
2. It is a time consuming and labour oriented technique.
3. Development of STS required high technical skill.

Q. 44. What are applications of STS ?

Ans. Applications of STS are given in below :

1. STS is the most powerful physical mapping technique.
2. It can be used to identify any locus on the chromosome.
3. STSs are used as standard markers to find out gene in any region of the genome.
4. It is used for constructing detailed maps of large genomes.

Q. 45. What are applications of DNA markers in crop improvement ?

Ans. Molecular markers or DNA markers have several useful applications in crop improvement programmes. As stated above, now several types of DNA markers are available. The choice of a type of DNA marker depends on the user's objectives. In plant breeding, molecular markers are useful in four principal ways, viz., (1) assessment of genetic diversity. (2) gene mapping (3) marker assisted selection, and (4) in crop evolution.

26

MARKER ASSISTED SELECTION

Q. 1. What is DNA marker ?

Ans. Easily identifiable traits are referred to as markers or marker traits. The markers are of three types viz. *(i)* morphological markers (shape, size, colour, surface of plants), *(ii)* biochemical markers (variations in protein structure) and *(iii)* molecular markers or DNA markers (variations in DNA fragments). The DNA markers or molecular markers have been defined in various ways by different workers. Some definitions of DNA markers are given below:

1. A molecular marker or DNA marker refers to a unique sequence of nucleotides found on a strand of DNA. The DNA markers are also known as genetic markers.
2. A marker may be defined as a distinctive stretch of DNA located close to the gene of interest. It may occur within the gene itself, in another nearby gene or in the non-coding DNA between genes.

Q. 2 What are main features of DNA markers ?

Ans. Molecular markers or DNA markers are tags which can be used to Identify specific genes and locate them in relation to other genes. The common features of DNA markers are briefly presented below:

(a) The marker is located very close to the gene of interest.
(b) The marker is almost always inherited along with the gene because it is linked to the gene.
(c) The marker is highly recognizable or identifiable.
(d) The marker acts as a red flag or tag indicating the presence of the gene of interest.
(e) DNA markers are useful or used in marker assisted selection (MAS). See later.
(f) DNA markers can also be used to identify corresponding gene in a distant plant species or even in animal or microbial species.

Q. 3. What is marker assisted selection ?

Ans. Marker Assisted Selection (MAS) refers to indirect selection for a desired plant phenotype based on the banding pattern of linked molecular (DNA) markers. MAS is based on the concept that it is possible to infer the presence of a gene from the presence of a marker which is tightly linked to the gene of interest. If the marker and the gene are located far apart then the possibility of their transmission together to the progeny individuals will be reduced due to double cross over recombination events.

Q. 4. What are main features of marker assisted selection ?

Ans. The main features of MAS are briefly presented below:

1. Other Terms Used

Marker assisted selection (MAS) is also termed as marker aided selection and marker assisted breeding (MAB). It differs from gene assisted selection (GAS) which refers to the selection which is based on QTLs (quantitative trait locus or loci).

2. Pre-requisites

There are two prerequisites for marker assisted selection. These are *(i)* a tight linkage between molecular marker and gene of Interest, and *(ii)* high heritability of the gene of interest.

3. Application

MAS is applicable for genetic improvement of plants as well as animals. In plants, it is equally applicable in both self pollinated and cross pollinated species.

4. Markers Used

MAS makes use of various types of molecular markers. The most commonly used molecular markers include amplified fragment length polymorphisms (AFLP), restriction fragment length polymorphisms (RFLP), random amplified polymorphic DNA (RAPD), simple sequence repeats (SSR) or micro satellites, single nucleotide polymorphisms (SNP). The use of molecular markers differs from species to species also.

5. Efficiency

The relative efficiency of MAS is greatest for characters with low heritability if a large fraction of the additive genetic variance is associated with the marker loci. In other words, MAS is useful when the heritability of the trait is low. Moreover, MAS is more efficient than purely phenotypic selection in quite large populations. It has been found by some workers that MAS may become less efficient than phenotypic selection in the long term. This is because the rate of fixation of unfavourable alleles at QTLs with small effects is higher under MAS than under phenotypic selection. It may be a consequence of the strong selection applied to QTLs with large effects under MAS in early generations. However, such problem comes after a long period.

6. Accuracy

Molecular markers have very .high accuracy. They are not affected by environmental conditions. MAS is a new breeding tool which is available to make more accurate and useful selections in breeding populations. MAS allows heritable traits to be linked to the DNA which is responsible for controlling that trait.

7. Speed of Progress

MAS is a rapid method of crop improvement. For example, in conventional breeding when we transfer a recessive character through backcross, one selfing is required after every backcross for identification of recessive character. MAS permits identification of recessive alleles even in heterozygous condition and thus speeds up the progress of crop improvement work.

8. Traits Improved

MAS can be used for improvement of both oligogenic and polygenic traits.. In the past, MAS has been mostly used for the genetic improvement of oligogenic traits and a little progress has been made with polygenic traits.

9. Material Developed

MAS leads to development of non-transgenic genotypes or cultivars. In other words, MAS is used for development of non-transgenic cultivars. The transgenic cultivars face public resistance. On the other hand, cultivars developed by MAS are acceptable by consumers.

10. Cost

MAS is very costly as compared to phenotypic selection. In MAS, the costly items include equipments, consumables, infrastructure, labour and DNA extraction process. MAS requires sophisticated and well equipped laboratory.

Q. 5. Give a comparison of MAS and phenotypic selection.

Ans. There are several differences between marker assisted selection and phenotypic selection. A brief comparison of these two methods is presented in Table 26.1

Table 26.1 Comparison of MAS and Phenotypic Selection

Particulars	Marker Assisted Selection	Phenotypic Selection
1. Markers used	Molecular Markers	Morphological Markers
2. Laboratory Required	Sophisticated Laboratory	Simple Laboratory
3. Technical Skill required	Very high	Medium
4. Accuracy	Very high	Medium to high
5. Time required to release a variety	Very short (3-5 years)	Very long (10-15 years)
6 Cost involved	Very high	Low to medium

7. Environmental effect	No effect	Very high
8. Effect of gene interaction	No effect	Very high
9. Mapping of QTL	Possible	Not possible
10. Screening of economic traits at seedling stage	Possible	Not Possible

Q. 6. What are the steps involved in marker assisted selection ?

Ans. In the marker aided selection, RFLP markers are widely used for genetic improvement of crop plants for various economic characters. The marker aided selection consists of five important steps, *viz.* (I) selection of parents, (2) development of breeding population, (3) Isolation of DNA from each plant, (4) scoring RFLPs, and (5) correlation with morphological traits. These are briefly discussed below:

1. Selection of Parents

Selection of suitable parents is an important step in marker aided selection. The parents should be such so that we can get usable level of polymorphism (variation) in the RFLP markers. In other words, parents with contrasting characters or diversgent origin should be chosen. This will help in identification of DNA of both the parents and also their segments in *F* 2 generation in various recombinations. For selection of parents, we have to screen germplasm and select parents with distinct DNA. The parents that are used for MAS should be pure (homozygous). In self-pollinated species, plants are usually homozygous. In cross-pollinated species, inbred lines are used as parents.

2. Development of Breeding Populations

This is the second important step for application of marker aided selection. The selected parents are crossed to obtain *F* 1 plants. *F* 1 plants between two purelines or inbred lines are homogeneous (alike phenotypically) but are heterozygous for all the RFLPs of two parents involved in the *F* 1. The *F* 2 progeny is required for the study of segregation pattern of RFLPs. Generally 50-100 *F* 2 plants are sufficient for the study of segregation of RFLP markers.

3. Isolation of DNA

The third important step is isolation of DNA from breeding population. The main advantage of MAS is that DNA can be isolated even from the seedlings and we need not to wait for flowering or seed development stage. The DNA is isolated from each plant of *F* 2 population. Standard procedures are available for DNA isolation. The isolated DNA is digested with specific restriction enzyme to obtain fragments of DNA. The DNA fragments of different sizes are separated by subjecting the digested DNA to agarose gel electrophoresis. The gel is stained with ethidium bromide and the variation in

DNA fragments can be viewed in the ultraviolet light. The DNA of chloroplasts, when digested with specific enzyme, produces about 40 fragments of different sizes. The nuclear DNA of higher plants, when digested with specific restriction enzymes, produces millions of fragments in a continuous range of sizes. It is a tedious job to identify individual DNA fragment in such cases.

4. Scoring RFLPs

The polymorphism in RFLPs between the parents and their involvement in the recombinants in F2 population is determined by using DNA probes. The labelled probes are used to find out the fragments having similarity .The probe will hybridize only with those segments which are complementary in nature. Generally 32P is used for radioactive labelling of DNA probe. Now non-radioactive probe labelling techniques are also available. In this way RFLPs are determined.

5. Correlation with Morphological Traits

The DNA marker (say RFLPs) are correlated with morphological markers and the indirect selection through molecular markers is confirmed. Once the correlation of molecular markers is established with morphological markers, MAS can be effectively used for genetic improvement of various economic traits.

Q. 7. What are applications of MAS in crop improvement ?

Ans. In crop improvement programmes MAS can be used in various ways. In other words, MAS has several useful applications in plant breeding. Important applications of MAS in plant breeding are briefly presented below:

1. MAS is very effective, efficient and rapid method of transferring resistance to biotic and abiotic stresses in crop plants.
2. It is useful in gene pyramiding for disease and insect resistance.
3. It is being used for transfer of male sterility and photoperiod insensitivity into cultivated genotypes from different sources.
4. MAS is being used for improvement of quality characters in different crops such as for protein quality in maize, fatty acid (linolenic acid) content in soybean and storage quality in vegetables and fruit crops.
5. MAS can be successfully used for transferring desirable transgene (such as Bt gene) from one cultivar to another.
6. MAS is very effective in introgression of desirable genes from wild into cultivated genotypes.
7. MAS is equally effective in genetic improvement of plants and animals.
8. MAS is useful in genetic improvement of tree species where fruiting takes very long time (say 20 years) because for application of phenotypic selection we have to wait for such a long time.

9. MAS has wide application for genetic improvement of oligogenic traits as compared to polygenic traits.

Q. 8 What are practical achievements of MAS ?

Ans. MAS has been used for genetic improvement of different field crops such as maize, barkey, rice, wheat, sorghum, soybean, chick pea, pea, sunflower, tomato, potato and some fruit crops for various economic characters. MAS has been mainly used for developing disease resistant cultivars in different crops (Table 17.2). Some notable examples of the use of MAS are given below.

1. Rice

In rice MAS has been successfully used for developing cultivars resistant to 'bacterial blight and blast. For bacterial blight resistance four genes *(Xa4, Xa5, Xal3* and *Xa21)* have been pyramided using STS (sequence tagged site) markers. The pyramided lines showed higher level of resistance to bacterial blight pathogen. In Indonesia, two bacterial blight resistant varieties of rice viz. Amgke and Conde have been released through MAS. For blast resistance, three genes (Pi1, Piz5 and Pita) have been pyramided in a susceptible rice variety Co39 using RFLP and PCR based markers.

2. Maize

In maize, normal lines have been converted into quality protein maize (QPM) lines through MAS using opaque 2 recessive allele. This work has been done at CIMMYT. (International Centre for Wheat and Maize Improvement, Mexico. Three SSR markers (ume 1066, phi 1057 and phi 112) present within opaque 2 gene have been used for this purpose. The MAS used for conversion of normal maize lines into QPM is simple, rapid and accurate.

3. Soybean

In soybean cyst nematodes pose serious problems and most of the varieties are susceptible to this parasite. The resistant gene (rhg I) is available in soybean, nematode resistant lines have been developed through MAS using SSR marker (Sat 309).

Table 26.2 Characters being used for MAS in different Crops.

Crop.	Disease Resistance	Insect Resistance	Other Traits
Rice	Bacterial blight Blast Rice Tungro Virus Gall midge	Brown hopper Green leaf hopper	Submergence tolerance Salt tolerance Temperature sensitive male sterility,

			Photoperiod sensitivity semidwarf stature, Shattering resistance, Grain Aroma, Amylose content
Maize	Northern Corn blight	–	Cytoplasmic male sterility QPM Days to pollen shed
Wheat	Leaf rust Powdery mildew, Loose smut	Hessian fly	Cyst nematode, Root lesion nematode, Earliness, vernalization requirement, Coleoptile colour, FlourColour
Sorghum	Head smut	–	Fertility restoration
Sorghum	Cyst nematoide Mosaic virus	–	Linolenic acid content
Pea	Fusarium wilt Powdery mildew	–	Nodulation ability
Tomato	Yellow leaf curl virus, Black mold, Bacterial wilt	–	–
Potato	Cyst nematoide, Potato Virus,x Late blight	–	–
Sunflower	Downy mildew	–	–
Chick pea	–	–	Double podding
Barley Mungbean Powdery mildew	Stripe just	–	–

Q. 9. What are the characters that have been improved using MAS ?

Ans. MAS has been used for genetic improvement of various characters in different crops. Important characters which have been improved through MAS in different crops include disease resistance, insect resistance, salinity resistance, shattering resistance. It has also been used for transfer of various characters such as male sterility , photoperiod insensitivity , earliness; and improvement of protein contents in some crop plants (Table 26.2)

Q. 10. What are the DNA markers which are commonly used in MAS ?

Ans. In MAS various types of DNA markers have been used in different crop plants. Molecular markers which have been widely used in MAS in different crops include Restriction Fragment Length Polymorphisms (RFLP), random amplified polymorphic DNA (RAPD) and simple sequence repeats (SSR) or microsatelite. Other markers which have been used in some crops include amplified fragment length polymorphism (AFLP), sequence tagged site (STS), expressed sequence tags (EST) and SCAR markers. Single nucleotide polymorphisms (SNP) are also used, SNPs have been identified for all the major cereal crops.

Q. 11. Explain use of MAS in forage and fruit crops.

Ans. MAS is also being used for genetic improvement of forage and fruit crops. Among fruit crops MAS has been extensively used In pomegranate, apple and pear. In fruit crops MAS is based on RFLP, RAPD, SSR and AFLP markers. In these fruit crops MAS is being used for higher fruit production , better keeping quality for storage and disease resistance.

Q. 12. Describe use of MAS in vegetable crops.

Ans. In vegetable crops MAS is being used in tomato and potato based on RFLP, RAPD and AFLP markers mainly for disease resistance. MAS has been found useful for genetic improvement of tree crops such as coconut and rubber.

Q. 13. What are advantages of MAS ?

Ans. MAS has several advantages over phenotypic selection and other breeding techniques. Some Important advantages of MAS are briefly discussed below:

1. Accuracy. The accuracy of MAS, as stated earlier, is very high because molecular are not affected by environmental conditions. It is very effective even with the (characters having low heritability.

2. Rapid Method. MAS is a rapid method of crop improvement. It takes 3-5 years for developing a new cultivar against 10-15 years taken by the conventional method of breeding.

3. Non-transgenic Product. MAS leads to development of non-transgenic cultivars which are acceptable to every body. In other words, it does not involve transgene. Hence there is no question of gene silencing.

4. Identification of Recessive alleles. MAS permits identification of recessive alleles even in heterozygous condition and thus speeds up the progress of crop improvement programmes. In other words, it is equally effective for the genetic improvement of recessive characters. ,

5. Early Detection of Traits. MAS permits early detection of traits that are expressed late in the life of plant. For example characters such as grain or fruit quality, flower colour, male sterility , photo period sensitivity that express late in the life of a plant can be screened in the seedling stage. In other words, DNA tested at seedling stage can throw light about the trait which are expressed later on.

6. Screening of Difficult Traits. MAS permits screening of difficult traits that are extremely difficult, expensive and time consuming to score phenotypically. For example, screening for traits such as root morphology and resistance to biotic (insects and diseases) and abiotic stresses (drought, salinity, heat, frast etc.) is very easy through MAS.

7. Gene Pyramiding. MAS is very effective method in accumulating multiple genes for resistance to specific pathogens and pests within the same cultivar. This process is called gene pyramiding. Marker assisted backcrossing is routinely applied in breeding programmes for gene introgression. MAS can provide an effective and efficient breeding tool for detecting, tracking, retaining, combining and pyramiding genes for disease resistance.

8. Small Sample for Testing. MAS requires only a small amount of plant tissue for DNA testing. In other words, MAS can be carried out with small breeding populations. Moreover, MAS can be applied at any stage of plant growth.

9. Permits QTL Mapping. MAS permits mapping or tagging of quantitative trait loci (QTL) which is not possible by conventional method.

10. Highly Reproducible. The MAS is based on DNA fingerprinting technique and the results of DNA fingerprinting pattern are highly reliable and reproducible.

Q. 14. What are limitations of MAS ?

Ans. MAS has several advantages as discussed. However, it has some limitations or drawbacks which are briefly presented below:

1. MAS is a costly method. It requires well equipped laboratory viz. expensive equipments, glasswares and chemicals.
2. MAS requires well trained manpower for handling of sophisticated equipments, isolation of DNA molecule and study of DNA markers.
3. The detection of various linked DNA markers (AFLP, RFLP, RAPD, SSR, SNP etc.) is a difficult, laborious and time consuming task.
4. MAS sometimes involves use of radioactive isotopes in labelling of DNA, which may lead to serious health hazards. This is a major disadvantage of RFLP based markers. The PCR based markers are safe in this regard.

5. It has been reported that MAS may become less efficient than phenotypic selection in the long term.
6. The use of MAS is more difficult for QTL because they have minor cumulative effects and are greatly influenced by environmental conditions and genetic background.

Q. 15. What are the future prospects of MAS ?

Ans. It is believed that MAS is a potential tool for crop improvement and hence it should be an integral part of plant breeding techniques. It has been reported that MAS has been used mainly by developed countries because of high cost of the technology in terms of infrastructure, equipments, chemicals and glasswares. The high cost of MAS technology can not be afforded by many developing countries. To make this technology usable by all developing and underdeveloped world, the following ways and means may be helpful in the spread of this technology.

1. Costly technology like MAS should be supported by consultative group on International Agricultural Research (CGIAR) that promotes collaborative research and training on global basis. It will help in rapid spread of technology (MAS) in developing countries.
2. The FAO and Rockfeller Foundation may play key role in making MAS technology available to resource poor countries.
3. The private Industries should support MAS technology for the benefit of common man and in the global interest.
4. To take advantage of such technology, there is need for International collaboration of research institutes engaged in MAS programmes.
5. There is need for collaborative approach between public and private organizations to provide access to MAS technology in developing countries.
6. International organizations such as FAO and CGIAR should organize training programme on regional level for capacity building in MAS technology .
7. The marker aided selection can be used as a tool in the crop improvement programme. It can speed up the progress of breeding programmes. It can cut short the time required for development of new varieties. However, this technique cannot be used as a substitute for conventional breeding approaches. Marker aided selection has its own limitations.

SECTION-IV
STRESS RESISTANCE

27

BREEDING FOR DISEASE AND INSECT RESISTANCE

Q. 1. What is disease ?

Ans. In crop plants, various disorders caused by pathogens are called desease.

Q. 2. What is pathogen ?

Ans. Various disease causing organisms such as fungi, bacteria, viruses and mycoplasm refer to pathogen.

Q. 3. What is pathogenecity ?

Ans. Ability of a pathogen to attack a host is called pathogenecity.

Q. 4. What is host ?

Ans. The plant which is attacked by a disease, insect or parasitic weed is called host.

Q. 5. What is parasite ?

Ans. An organism or virus which lives upon or within another living organism is called parasite.

Q. 6. What is pest ?

Ans. Any animal or higher plant which parasitizes crop plants *i.e.* insects, mematodes, birds and parasitic needs refer to pest.

Q. 7. What do you mean by immune ?

Ans. Completety or fully resistant plants are known as immune.

Q. 8. What is virulent ?

Ans. A race of a pathogen which is capable of attacking a host with specific resistance is called virulent race.

Q. 9. What is avirulent ?

Ans. A race of a pathogen which is unable to attack a host having specific resistance is known as avirulent race.

Q. 10. What is physiological race ?

Ans. Strain of a pathogen differing in pathogenecity is known as physiological race; also called pathotype. Races of a pathogen are of two types, viz. *virulent* and *avirulent.*

Q. 11. What is biotype ?

Ans. Variant or new strain of an insect is called biotype.

Q. 12. What is epidemic ?

Ans. Wide spread uncontrolled incidence of a disease is known as epidemic.

Q. 13. What is genetic resistance ?

Ans. Ability of some genotypes to give higher yield of good quality than other varieties/genotypes at the same initial level of disease or insect infestation under similar environmental conditions. Genetic resistance is of two types, viz. *vertical* and *horizontal.*

Q. 14. What is vertical resistance ?

Ans. Resistance of a host to the particular (specific) race of a pathogen is called vertical resistance. It is also known as major gene resistance, oligogenic resistance, race specific resistance and qualitative resistance.

Q. 15. What is horizontal resistance ?

Ans. Resistance of a host to all the prevalent races of a pathogen is called horizontal resistance. It is also known as minor gene resistance, polygenic resistance, nonspecific resistance, general resistance and quantitative resistance.

Q. 16. Who coined terms vertical and horizontal resistance ?

Ans. Terms vertical resistance and horizontal resistance were first used by Van der Plank in 1963.

Q. 17. What is durable resistance ?

Ans. The long lasting resistance is known as durable resistance. It may be vertical or horizontal.

Q. 18. What is seedling resistance ?

Ans. The resistance which is exhibited by young seedlings is known as seedling resistance.

Q. 19. What is adult resistance ?

Ans. The resistance which is exhibited by adult plants is termed as adult resistance.

Q. 20. What is field resistance ?

Ans. The resistance which gives an effective control of a parasite under field conditions is known as field resistance.

Q. 21. What do you mean by major genes ?

Ans. Genes with distinct individual effect which can be easily identified are known as major genes.

Q. 22. What are minor genes ?

Ans. Genes with small effect on a character which can not be easily identified are called minor genes.

Q. 23. What is hypersensitivity ?

Ans. A host pathogen interaction which leads to death of infested tissues is called hypersensitivity. It is an important mechanism of disease resistance.

Q. 24. What is difference between yield breeding and biotic resistance breeding ?

Ans. In productivity breeding, there is interaction between genotype and environment. In resistance breeding for biotic factors, there is triangular interaction of host-pest environment.

Q. 25. What are advantages of genetic resistance ?

Ans. Main advantages of genetic resistance are given below :

(i) It reduces environmental pollution and health hazards caused by pesticides.

(ii) It permits multiplication of natural enemies of insect pests.

(iii) It also reduces cost of crop production.

Q. 26. What is gene for gene hypothesis ?

Ans. This hypothesis states that for each gene controlling resistance in the host, there is a corresponding gene controlling pathogenicity in the pathogen. It is also known as Flor hypothesis.

Q. 27. Who developed the concept of gene for gene hypothesis ?

Ans. The concept of gene for gene hypothesis was developed by Flor in 1956 working with linseed crop.

Q. 28. What are mechanisms of disease resistance ?

Ans. There are three mechanisms of disease resistance as given below.

(i) Resistance to establishment of the pathogen in the host tissue.

(ii) Resistance to growth and development of pathogen already established in the host tissue.

(iii) Ability of a host to perform well despite the establishment of the pathogen in the host tissues.

Q. 29. What are mechanisms of insect resistance ?

Ans. There are four mechanisms of insect resistance as follows.

(i) Nonpreference, or non-acceptance.

(ii) Antibiosis

(iii) Tolerance and

(iv) Eacape or avoidance.

Q. 30. What do you mean by non-preference ?

Ans. Non-preference refers to various features of host plant which make the host undesirable or unattractive to insects for food, shelter or reproduction. It is also known as non-acceptance or antixenosis.

Q. 31. Which plant characters are associated with non-preference ?

Ans. Following plant characters are associated with non- preference :

(i) Colour of leaves
(ii) Light penetration
(iii) Hairiness of stem & leaf
(iv) Leaf angle
(v) Odour
(vi) Taste

Q. 32. What are the characters associated with non-preference to bollworm in cotton ?

Ans. In cotton, following characters are non-preference to boll worms.

(*i*) Red plant body
(*ii*) Smooth leaves
(*iii*) Okra leaf
(*iv*) Open canopy
(*v*) Nectarilessness
(*vi*) Fregobract.
(*vii*) Thick and hard boll rind
(*viii*) Long pedicel.

Q. 33. What is non-preference to stem sawfly in wheat ?

Ans. In wheat, solid stem is non-preference for stem sawfly.

Q. 34. Which traits are non-preference to brown plant hopper in rice ?

Ans. In rice, low aspargine, red pericarp and purple stigma are non-preference to brown plant hopper.

Q. 35. What is non-preference for jassids in cotton ?

Ans. In cotton, hairiness of leaves is non-preference to jassids.

Q. 36. What is non-preference for pea aphid in pea ?

Ans. In pea, yellow green leaves are non-preference to pea aphid.

Q. 37. What is antibiosis ?

Ans. Antibiosis refers to adverse effects of the host on feeding, development and reproduction of insect pests. Resistant plants retard the growth and rate of reproduction of insect pests. In some cases, antibiosis even leads to death of an insect. Antibiosis may involve morphological, physiological and biochemical features of the host plant.

Q. 38. What are the characters associated with antibiosis to bollworm in cotton ?

Ans. In cotton, high levels of gossypol, tannins, heliocides and silica contents are associated with antibiosis to bollworm.

Q. 39. What is associated with antibiosis to green bugs in wheat ?

Ans. In wheat, high benzyl alcohol is associated with antibiosis to green bugs.

Q. 40. What acts as antibiosis to cereal leaf beetle in barley ?

Ans. In barley, waxiness of leaves acts as antibiosis to cereal leaf beetle.

Q. 41. What acts as antibiosis to stem borer in rice ?

Ans. In rice, high silica content acts as antibiosis to rice stem borer.

Q. 42. What acts as antibiosis to stem borer in maize ?

Ans. In maize, high aspartic acid acts as antibiosis to maize stem borer.

Q. 43. What acts as antibiosis to European corn borer in maize ?

Ans. In maize, high DIMBOA in leaves acts as antibiosis to European corn borer.

Q. 44. What acts as antibiosis to spotted aphid and pea aphid in alfalfa ?

Ans. In alfalfa, high saponin content acts as antibiosis to spotted aphid and pea aphid.

Q. 45. What acts as antibiosis to cabbage aphid in Brassica ?

Ans. In Brassica, low sinigrin content acts as antibiosis to cabbage aphid.

Q. 46. What is tolerance to insects ?

Ans. Tolerance refers to the ability of a variety to produce greater yield than susceptible variety at the same level of insect attack. Tolerant cultivars have greater recovery of damaged parts than susceptible ones.

Q. 47. What are plant features associated with insect tolerance ?

Ans. The following plant characters are associated with insect tolerance :

(i) Speedy recovery of damaged parts.

(ii) Healthy leaf growth.

(iii) Superior plant vigour.

(iv) Potential for flower compensation.

Q. 48. What is insect avoidance or escape ?

Ans. Avoidance refers to escape of a variety from insect attack either due to earliness or its cultivation in the season where insect population is very low. For example, early maturing cotton varieties escape pink boll worm infestation which occurs late in the season.

Q. 49. Who first gave mechanisms of insect resistance ?

Ans. Mechanisms of insect resistance were first proposed by Painter in 1951. He gave three mechanisms of insect resistance in crop plants, viz. non-preference, antibiosis and tolerance.

Q. 50. What are basis of insect resistance ?

Ans. There are following three important bases of insect resistance.

(i) Morphological factors. It includes various morphological traits such as hairiness, colour, thickness and toughness of tissues.

(ii) Physiological factors. It includes osmatic concentration of cell sap and leaf exudates.

(iii) Biochemical factors. It includes various types of chemical compounds such as high gossypol, tannins, heliocides and silica contents for bollworms in cotton.

Q. 51. What are sources of insect resistance in crop plants ?

Ans. In crop plants, there are five important sources of insect resistances as follows :

(i) Cultivated varieties. (ii) Germplasm collections. (iii) Wild species and wild relatives. (iv) Induced mutations (v) Micro-organisms such as Bt.

Q. 52. What are factors affecting durability of insect resistance ?

Ans. The durability of insect resistance depends an the following factors.

(i) Formation of new biotypes of an insect.
(ii) The mode of inheritance of insect resistance.
(iii) Morphological and biochemical characters associated with insect resistance.

Q. 53. What breeding methods are used for developing insect resistant cultivars ?

Ans. The following breeding methods are used for developing insect resistant cultivars.

(i) Plant introduction.
(ii) Pure line selection and mass selection.
(iii) Backcross, bulk and Pedigree breeding methods.
(iv) Mutation breeding, and
(v) Transgenic breeding.

Q. 54. How much yield losses are caused by insect in crop plants ?

Ans. In crop plants, about 14% yield losses are caused annually by insect pests (Cramer, 1967).

Q. 55. How much annual yield losses are caused by diseases in crop plants ?

Ans. In crop plants, about 20-30% annual yield losses are caused by various diseases.

Q. 56. What is gene pyramiding ?

Ans. Incorporation of two or more major genes in a variety for specific resistance to a disease is called gene pyramiding.

Q. 57. What is difference between gene pyramiding and multilines ?

Ans. In multiline approach, each major gene is incorporated into separate genotype and then resistant genotypes are mixed together. In gene pyramiding approach, all major genes are incorportated in a single cultivar.

Q. 58. What are advantages of gene pyramiding ?

Ans. Varieties which are developed by gene pyramiding approach have durable resistance. New cultivars are developed with a combination of 2, 3 or more major genes.

Q. 59. In which crop gene pyramiding approach has been used ?

Ans. Gene pyramiding approach has been used in oats against crown rust in Canada.

Q. 60. What is gene deployment ?

Ans. Planned use of major genes in development of disease resistant cultivars for various geographical areas is called gene deployment. This approach helps in preventing diseases and maintaining the diversity for major genes in different geographical regions.

Q. 61. How vertical resistance can be exploited in plant breeding ?

Ans. In plant breeding, vertical resistance can be exploited in following ways.

(i) By developing multilines.

(ii) By developing cultivars with individual major gene.

(iii) By gene pyramiding and

(iv) By gene deployment.

Q. 62. What is Verticellium tolerant variety of upland cotton ?

Ans. In upland cotton, variety MCU 5 is tolerant to Verticillium wilt.

Q. 63. What is yellow mosaic resistant variety of okra ?

Ans. In okra, variety Parbhani Kranti is resistant to yellow mosaic virus.

Q. 64. What is stemsawfly resistant variety of wheat ?

Ans.f In wheat, variety 'Rescue' with solid stem is resistant to stem sawfly. It was released in Canada in 1946.

Q. 65. What is green bug resistant variety of barley ?

Ans. In barley, variety "Will" is resistant to greenbugs. It was released in USA.

Q. 66. What are resistant varieties of alfalfa to spotted alfalfa aphid ?

Ans. In alfalfa, varieties Cody, Moapa and Zia are resistant to alfalfa spotted aphid. These varieties were released in USA.

Q. 67. What is stem eelworm resistant variety of oat ?

Ans. In oat, variety Greta is resistant to stem eelworm. It was developed in Belgium.

Q. 68. What are jassids resistant varieties of cotton ?

Ans. In upland cotton, varieties B 1007, Khandwa 2, SRT 1, DHY 286 and PKV 081 are resistant to jassids.

Q. 69. What are whitefly resistant/tolerant varieties of cotton ?

Ans. In upland cotton, varieties Kanchana, Supriya and LK 861 are resistant to whitefly.

Q. 70. Give name of leaf curl resistant hybrid of cotton.

Ans. Intra-lursutum cotton hybrid LHH 144 is resistant to leaf curl virus. It was developed from Punjab Agricultural University, Ludhiana.

Q. 71. Who first proposed use of multilines for control of diseases ?

Ans. The use of multilines for controlling diseases in crop plants was first proposed/suggested by Jensen in 1952.

Q. 72. What is parasitic weed of sorghum in India ?

Ans. In India, *Striga* is the parasitic weed of sorghum. Striga is found in Maharashtra state.

Q. 73. What is the Striga resistant variety of Sorghum ?

Ans. The variety SAR 1 of *Sorghum* is resistant to *Striga*. It was released from ICRISAT for cultivation in those areas of Maharashtra where *Striga* is a problem.

Q. 74. What is serious parasitic weed of sugarbeet in Turkey ?

Ans.f In Turkey, *Cuscuta* is a serious parasitic weed of sugarbeet.

Q. 75. What is serious parasitic weed of *Sorghum* in Africa ?

Ans. In Africa, *Orobanche* is the serious parasitic weed of *Sorghum*.

Q. 76. Where have been developed *Orobanche* resistant varieties of chickpea ?

Ans. Orobanche resistant varieties of chickpea have been developed at ICARDA in Syria.

Q. 77. Where have been developed *Orobanche* resistant varieties of sunflower ?

Ans. Orobanche resistant varieties of sunflower have been developed in Russia.

Q. 78. What are damages caused by Insects ?

Ans. Insect attack leads to following damages.

(i) Stunting–reduction in plant growth.
(ii) Damage of vegetative and reproductive parts.
(iii) Premature defoliation.
(iv) Wilting of plants.

Q. 79. What are biological methods of insect control ?

Ans. Biological methods of insect control are given below :

(i) Use of resistant varieties.
(ii) Use of botanical pesticides such as Neem, Ipomea, Datura etc.
(iii) Use of predators and parasites.

28

BREEDING FOR DROUGHT AND SALINITY RESISTANCE

Q. 1. What is stress ?

Ans. Stress refers to adverse conditions for crop growth and production imposed by either biotic or abiotic factors.

Q. 2. What are biotic factors ?

Ans. Living organisms such as diseases, insects and parasitic weeds which cause stress conditions for crop growth and production are called biotic factors.

Q. 3. What is biotic stress ?

Ans. Biotic stress refers to adverse conditions for crop growth and production imposed by biotic factors such as diseases, insects and parasitic weeds.

Q. 4. What are abiotic factors ?

Ans. Non-living factors such as drought, salinity, alkalinity, acidity, cold, frost, heat etc. which adversely affect crop growth and production are called abiotic factors.

Q. 5. What is abiotic stress ?

Ans. Abiotic stress refers to adverse conditions for crop growth and production caused by abiotic factors. It is also known as environmental stress.

Q. 6. What is drought ?

Ans. Condition of soil moisture deficiency or water scarcity is known as drought.

Q. 7. What are mechanisms of drought resistance ?

Ans. There are four mechanisms of drought resistance as follows.

(i) Drought escape
(iii) Drought avoidance
(iii) Drought tolerance
(iv) Drought resistance

Q. 8. What is drought escape ?

Ans. Ability of crop plants to complete their life cycle before the start of moisture stress in the field is known as drought escape. Early maturing varieties generally escape from late season drought.

Q. 9. What is drought avoidance ?

Ans. Ability of plants to maintain a favourable internal water balance under moisture stress is referred to as drought avoidance.

Q. 10. What is drought tolerance ?

Ans. Ability of crop plants to grow, develop and reproduce normally under moisture deficit conditions is called drought tolerance. In other words, it refers to survival of crop plants under water deficit conditions without injury.

Q. 11. What is drought resistance ?

Ans. Ability of crop plants to give higher yield under moisture stress conditions is known as drought resistance. It refers to good yielding ability under moisture deficit conditions.

Q. 12. What is moisture stress ?

Ans. In the soil, moisture deficit conditions which make survival of the plants difficult or have adverse effects on the plant growth and reproduction is referred to as moisture stress.

Q. 13. What is drought hardening ?

Ans. Improvement in drought tolerance ability of a genotype through various seed and seedling treatments is called drought hardening.

Q. 14. What is salt tolerance ?

Ans. Ability of crop plants to prevent, reduce or overcome the injurious effects of soluble salts present in the root zone is called salt tolerance.

Q. 15. What is hydroponics ?

Ans. Hydroponics refers to growing of plants in the nutrient solution.

Q. 16. What is lysimeter ?

Ans. Cemented micro plots of various sizes which are used to under take study on roots and salt tolerance are called lysimeters.

Q. 17. What is porometer ?

Ans. An instrument which is used for measuring stomatal aperture is called porometer.

Q. 18. What is psychrometer ?

Ans. An instrument which is used to measure tissue water potential in plants is known as psychrometer.

Q. 19. What is tensiometer ?

Ans. An instrument which is used for measuring soil moisture is known as tensiometer.

Q. 20. What are xerophytic plants ?

Ans. Plant species showing normal growth and development under moisture stress conditions are called xerophytic plants.

Q. 21. What are mesophytic plants ?

Ans. Those plant species which grow well under average or normal moisture conditions are known as mesophytes.

Q. 22. What are hydrophytic plants ?

Ans. Those plant species which exhibit normal growth and development under excessive moisture conditions are called hydrophytes. Such plants can withdstand water stagnation in the field.

Q. 23. What are oxylophytes ?

Ans. Plant species which exhibit normal growth and development under acidic soil (soil acidity) are called oxylophytes.

Q. 24. What are halophytes ?

Ans. Plant species which exhibit normal growth and development under salinity conditions are termed halophytes.

Q. 25. What are glycophytes ?

Ans. Plant species which are sensitive to high salt concentrations (can not tolerate high salt concentrations) are called glycophytes.

Q. 26. What are lithophytes ?

Ans. Plants which can grow on hills are called lithophytes.

Q. 27. What are sociophytes ?

Ans. Plants which exhibit normal growth and development under shade conditions are called sociophytes.

Q. 28. Which areas are more pone to drought ?

Ans. Arid and semi-arid areas are more prone to drought conditions. About 36% of the land area constitutes arid and semi-arid zones.

Q. 29. What are effects of drought ?

Ans. In crop plants, drought leads to reduction in both quality and yield. It has adverse effects an plant growth and development. Drought damages chloroplasts and lowers output of the photosynthetic apparatus. Drought leads to increase in proline, abscisic acid and ethylene in the leaves.

Q. 30. What are morphological traits associated with drought ?

Ans. Following morphological characters of crop plants are associated with drought resistance.

(i) Earliness.

(ii) Lesser number of stomata per unit area on leaves.
(iii) Thick and small leaves.
(iv) Leaf and stem waxiness.
(v) Deep root system.
(vi) Determinate growth habit.

Q. 31. What are physiological characters associated with drought ?

Ans. Following physiological characters are associated with drought resistance.

(i) High photosynthetic rate.
(ii) Low transpiration rate.
(iii) High osmotic concentration.

Q. 32. What are sources of drought resistance ?

Ans. In crop plants, following are main sources of drought resistance.

(i) Cultivated varieties.
(ii) Germplasm collections.
(iii) Wild relatives and wild species.
(iv) Induced mutations.

Q. 33. What are measurements of drought resistance ?

Ans. Main ways of measuring drought tolerance/resistance are given below :

(i) Higher leaf water retention under drought conditions.
(ii) High rate of photosynthesis under drought conditions.
(iii) Superior yield performance under drought conditions.
(iv) Deep root system.

Q. 34. What are indicators of drought ?

Ans. Indications of drought are given below :

(i) Rolling of leaves.
(ii) Increase in proline level in leaves.
(iii) Soil moisture deficit.

Q. 35. How drought tolerance is measured in rice ?

Ans. In rice, the force required to pull out the seedlings from paddy soil is used as a measure of drought tolerance.

Q. 36. What are drought tolerant varieties of sugarcane ?

Ans. Drought tolerant varieties of sugarcane are COS 510, BO 1, CO 1148, CO 449, CO 997, CO 1158, CO 62175 and CO 6907.

Q. 37. What are drought tolerant varieties of wheat ?

Ans. In bread wheat, drought tolerant varieties include NP 4, NP 720, PB 9-D, C 217, C 281, K 46, RS 31-1, AO 88, AO 113 and AO 115.

Q. 38. What are drought tolerant varieties of durum wheat ?

Ans. In *durum* wheat, varieties Motia, Gulab, Jay and Arnej are resistant to drought.

Q. 39. What are drought tolerant varieties of Cowpea ?

Ans. In Cowpea, varieties IGFRI 437 and IGFRI 450 are resistant to drought.

Q. 40. Which is drought tolerant variety of rice ?

Ans. In rice, variety Akashi is drought tolerant.

Q. 41. What is drought tolerant variety of barley ?

Ans. In barley, variety Karan-280 is tolerant to drought.

Q. 42. What are drought tolerant varieties of groundnut ?

Ans. In groundnut, varieties Jyoti and HD 3-30 are drought tolerant.

Q. 43. What are drought tolerant varieties of sunflower ?

Ans. In sunflower, varieties EC 68414 and EC 68415 are tolerant to drought.

Q. 44. What is drought tolerant variety of pearlmillet ?

Ans. In pearlmillet, variety BJ 104 is tolerant to drought.

Q. 45. What are salinity tolerant varieties of sugarcane ?

Ans. In sugarcane, varieties CO 453, CO 997, CO 62399, CO 7717, CO 1148, COS 767 and BO 91 are tolerant to soil salinity.

Q. 46. What is salinity tolerant variety of okra ?

Ans. In okra, variety Pusa Sawani is salinity tolerant.

Q. 47. What are salinity tolerant varieties of barley ?

Ans. In barley, varieties Ratana, RS 6, Karan 18, Karan 19 and Karan 92 are tolerant to salinity.

Q. 48. Give name of one salinity tolerant variety of rice.

Ans. Variety Mohan of rice is tolerant to salinity.

Q. 49. What are drought tolerant varieties of cotton ?

Ans. In cotton, varieties MA 9 and MCU 10 are tolerant to drought.

Q. 50. What are drought tolerant varieties of onion ?

Ans. In onion, varieties Hisar 2 and Punjab Selection are tolerant to salinity.

Q. 51. What is full name of Soil Salinity Institute of India ?

Ans. In India, research work on soil salinity is carried out at the Central Soil Salinity Research Institute, Karnal and State Agricultural Universities.

Q. 52. What are differences between drought avoidance and drought tolerance ?

Ans. Main differences between drought avoidance and drought tolerance are presented below in Table 28.1.

TABLE 28.1 Differences between drought avoidance and drought tolerance.

Drought avoidance	Drought tolerance
1. Plants maintain favourable tissues water content.	Plants do not maintain favourable tissue water content.
2. It reduces photosynthesis and increases root development.	Tolerant genotypes exhibit better seed germination, seedling growth and photosynthesis.
3. In cereals, it operates during vegetative phase.	In cereals, it operates during reproductive phase.
4. Plants can not withstand low tissue water content.	Plants can withstand low tissue water content.
5. It involves various morpholo- gical and anatomical features of plant which reduce water loss through transpiration.	It generally involves those characters which support for better photosynthesis under drought conditions.

29

BREEDING STRATEGIES FOR CLIMATE CHANGE

Q. 1. What is climate change ?

Ans. Climate change is any long-term significant change in the "average weather" of a region or the earth as a whole. Average weather may include average temperature, precipitation and wind patterns. It involves changes in the variability or average state of the atmosphere over durations ranging from decades to millions of years.

Q.2. What are main features of climate change ?

Ans. Main points related to climate change are listed below:

(i) Climate change has continued throughout the entire history of Earth.

(ii) These changes can be caused by dynamic processes on Earth, external forces including variations in sunlight intensity, and more recently by human activities.

(iii) Climate change is a gradual process which takes very long time. (decades to millions of years)

(iv) In recent usage, especially in the context of environmental policy, the term "climate change" usually refers to changes in modern climate.

(v) Climate includes the average temperature, amount of precipitation, days of sunlight, etc.

(vi) It is believed beyond doubt that human activity is contributing to the current rapid changes in the world's climate.

Q. 3. What are the factors associated with climate change ?

Ans. Climate changes reflect variations within the Earth's atmosphere, processes in other parts of the Earth such as oceans and ice caps, and the effects of human activity .The external factors that can shape climate are often called climate forcing and include such processes as variations in solar radiation, the Earth's orbit, and greenhouse gas concentrations.

Weather is the day-to-day state of the atmosphere. On the other hand, *climate* refers to the average state of weather and is fairly stable and predictable. Climate includes the average temperature, amount of precipitation, days of sunlight, and other variables that might be measured at any given site.

Climate change is the result of several factors such as (i) glaciation, (ii) ocean variability , (iii) Co2 concentration, (iv) plate tectonics, (v) solar variations, (vi) orbital variations, (vii) volcanism, (viii) fossil fuel, (ix) cement manufacture, (x) land use (irrigation and deforestation), (xi) live stock etc. Detailed discussion of all these factors is beyond the scope of the present discussion. According to a 2006 United Nations report, **Livestock's Long Shadow,** livestock is responsible for 18% of the world's greenhouse gas emissions as measured in CO_2 equivalents. All the above factors singly or in combination significantly contribute to climate change.

The sun is the source of over 99% of the heat energy in the climate system. Less than 1% of the energy is provided by the gravitational pull of the Moon (manifested as tidal power), in addition to geothermal energy provided by the hot inner core of the Earth. The energy output of the sun, which is converted to heat at the Earth's surface, is an integral part of shaping the Earth's climate.

Q. 4. What are the effects of climate change ?

Ans. The climate change has two types of effects, viz. direct effects and indirect effects. The direct effects of climate change include effect on (i) temperature, (ii) rainfall, (iii) sunlight, (iv) CO_2 concentration, etc. As a result of change in these aspects in a region, there are indirect effects or change in the (i) pest scenario, (ii) disease situation, (iii) water availability, (v) day-length, {vi)' biodiversity and (vii) cropping pattern.

Q. 5. What is the effect of climate change on temperature ?

Ans. There is an increase in temperature which will lead to drought/water deficit conditions. It will promote the incidence of certain types of insects diseases and weeds that were not prevailing earlier in a particular region. Extreme temperature expands the desert area. Higher temperatures are expected to improve or retard seed germination, plant growth and/ or plant development, depending on the relative sensitivity or tolerance of crop genotypes.

Differences in climatic conditions, such as increases in temperature, will have a direct impact on local insect vector populations. Reproduction, physical activity and metabolic rates would all be increased. These processes would act together to increase the efficiency of vectors and consequently virus spread. In future, the use of insecticides will become restricted and methods to control vectors will require constant innovation.

The severe climate change leads to death or sometimes extinction of some plants and animal species. For example, in the UK during the drought year of

2006 significant numbers of trees died or showed dieback on light sandy soils. In Australia, since the early $9O_s$, tens of thousands of flying foxes (Pteropus) have died as a direct result of extreme heal. Water deficit summers can lead to more periods of drought, potentially affecting many species of animal and plant.

Q. 6. What are the effects of climate change on Rainfall Pattern ?

Ans. Climate change predictions are for increased variability of rainfall in many regions worldwide, resulting in greater fluctuations in soil water regime. The climate change may lead to increase or decrease of rainfall over the average value of rainfall in a particular region. The increase in the average rainfall may lead to flood and water stagnation conditions, whereas decrease in average rainfall may cause drought or water deficit conditions. The presence of many of the pathogens in the environment correlates with seasonal temperature and their spread is aided by rainfall.

Q. 7. What are the effects of climate change on days of sunlight ?

Ans. The change in the average value of sunlight days in a particular region will have effect on the thermo and photoperiod requirements of the plant species grown in that region. As a result, new set of crop varieties will have to be bred for such situations. Only thermo and photo insensitive varieties can be successful under such situations.

Q. 8. What is the effect of climate changeon CO2 ?

Ans. The global warming and climate change lead to rise in atmospheric CO_2 concentration which will have a positive effect on productivity, though in a crop genotype-dependent manner .

Q. 9. What are the effects of climate change on biodiversity ?

Ans. Climate change and biodiversity interact in many important ways. In some cases, life cycles of many wild plants and animals are closely linked to the passing of the seasons. If two species are interdependent (e.g. a wild flower and its pollinating insect), climatic changes can disturb their synchronization. For example, if one has a cycle dependent on day length and the other on temperature or precipitation. In principle, at least, this could lead to extinctions or changes in the distribution and abundance of species.

Several organizations such as Wildlife Trust, World Wide Fund for Nature, Birdlife International and the Audubon Society are actively monitoring and researching the effects of climate change on biodiversity and advance policies in areas such as landscape scale conservation to promote adaptation to climate change are being framed.

Q. 10. What breeding problems are expected to emerge due to climate change ?

Ans. Gradual change in the climate over a long period will lead to several new problems such as appearance of new insects, new diseases, new weeds, high temperature, drought conditions, change in rainfall etc. As a result of climate change, the following problems emerge.

(i) Water scarcity or excess depending upon the change in the rainfall pattern.

(ii) Drought if there is low rainfall and increase in temperature.

(iii) Change in response to thermo and photoperiods depending upon the change in sunlight and day length.

(iv) Appearance of new insects, diseases and weeds in the changed climate.

(v) Change in the adaptation of cultivars in the areas of climate change.

(vi) Change in the cropping pattern in the region of climate change.

Table 29.1 Climate change and its effects

SN.	Climate Change	Anticipated Effects
1.	Rise in Temperature	Drought, desert expension, insects, diseases, crop adaptation, etc
2.	Rainfall Pattern	More Rainfall: Flood situation, water stagnation, cropping pattern. Low Rainfall: Water deficit, drought
3.	Sunlight	Thermo and photo period requirement
4.	Wind velocity	Lodging of crop cultivars

Such types of changes will require new set of crop cultivars for successful crop production. To keep pace with the climate change, breeders have to make continuous concerted efforts. The climate change needs to be converted from a difficulty into an opportunity .

Q. 11. What are the sources of resistance to climate change ?

Ans. In crop plants, there are five important sources which can be used for developing suitable cultivars for cultivation in the region where clirnate change is taking place. The sources include (i) germplasm collection, (ii) cultivated varieties, (iii) wild species and relatives of crop plants, (iv) induced mutations, and (v) transgenes. These are briefly discussed below:

1. Germplasm Collections: Germplasm Collections are the potential sources that can be effectively used in developing suitable crop cultivars for areas of climate change. The new cultivar has be developed, to solve the specific problem or a set of problems, using appropriate breeding techniques.

2. Cultivated Varieties: Crop cultivars can also be developed using obsolete or current crop varieties. Again the cultivar has to be developed to combat the problem of specific disease or insect or weed, or drought, or water stagnation that has emerged as a result of climate change. In the past, cultivated crop cultivars have been effectively used for developing varieties resistant to above mentioned changed situations.

3. Wild Species and Relatives: Wild Species and Relatives of cultivated plants are potential sources of developing crop cultivars suitable for climate change. However, use of these poses several problems such as cross in compatibility , hybrid in viability and hybrid sterility. Hence, such sources are rarely used.

4. Induced Mutations: Induced mutations are potential sources of developing crop cultivars suitable for climate change. In the past, induced mutations have been successfully used for developing varieties resistant to diseases and insects in several crop species.

5. Transgenic Technology: Transgenic plants can be used to combat the problems arising due to climate change. Crop cultivars resistant to various insects and diseases can be developed through the use of transgenic technology.

Q. 12. What breeding approaches should be adopted for developing varieties suitable for climate change ?

Ans. Traditional breeding approaches will continue to develop cultivars suitable for climate change. These approaches include introduction, selection, hybridization and mutations. In developing suitable crop cultivars for climate change due importance should be given to: (i) development of MAR lines, (ii) genetic enhancement, (iii) transgenic technology , (iv) durable resistance, and (v) crop adaptation.

Q. 13. What is the role of MAR lines in developing varieties suitable for climate change ?

Ans. Emphasis should be given to develop crop cultivars with multiple adversities resistance. This may include resistance to insects, diseases, drought and excessive moisture. This includes both breeding for multiple insects resistance (MIR) and breeding for multiple disease resistance (MDR).

Q. 14. Describe the role of genetic enhancement and transgenic breeding in developing varieties suitable for climate change ?

Ans. Genetic enhancement can be used in developing novel genetic diversity that can be used in breeding crop cultivars suitable for climate change. Transgenic breeding can be successfully used for developing crop cultivar resistant to diseases, insects and drought conditions arising due to climate change.

Q. 15. Explain the role of durable resistance and crop adaptation in developing varieties suitable for climate change ?

Ans. Plant breeders should make use of long lasting resistance (durable resistance) for developing insect and disease resistant varieties. Multi-lines can be used for achieving such goal. Maintaining food production under conditions of climate and environmental change will require the breeding of new crop varieties better adapted to these conditions. Heterogeneous crops also exhibit greater stability across environments. The best sources of pest and pathogen resistance can be further enhanced by deployment in such mixed crop stands.

For climate change, crop cultivars with wide adaptation should be selected based on multi-location under diversified environmental conditions for 3-years. In other words, crop adaptation should be given due importance in developing varieties suitable for climate change. Breeding strategies for different situations of climate change are presented in Table 29.2.

Table 29.2. Situation of Climate Change and Breeding Strategies

S. No.	Situation of Climate Change	Breeding Strategies
1.	Low rainfall in paddy growing region.	(i) Development of upland Paddy cultivars that can be grown with irrigation without standing water. (ii) Shifting of paddy with crops with low water requirement.
2.	Very high rainfall in paddy growing areas.	Development of floating varieties of paddy.
3.	Rise in temperature leading to drought situations.	Development of drought and heat resistant crop cultivars.
4.	Very high rainfall in pulses growing areas.	Shifting of pluse crops with paddy and sugarcane.
5.	Sudden appearance of new insects and diseases due to climate change.	Development of crop cultivars resistant to new insects and diseases.
6.	Moisture deficit in wheat growing areas.	Development of drought resistant cultivars of wheat.
7.	Low intensity of light and lesser sunlight days.	Development of thermo and photo insensitive crop cultivars.
8.	Delayed on set of Monsoon.	Development of crop varieties suitable for late sowing.
9.	Increase in wind velocity.	Development of crop cultivars resistant to lodging.

Q. 16. How would you screen the material for resistance to climate change ?

Ans. The screening of breeding material can be carried out looking to the new problem arising as a result of climate change. The new problems may include heat, drought, excessive moisture, new insects, new diseases and new weeds. The standard procedures available can be adopted for screening of material to tackle these problems. The material can be screened under both natural and controlled conditions to select genotypes suitable for a particular situation. Under artificial conditions the screening of breeding material should be carried out under controlled conditions representing climate change.

Q. 17. What should be the future breeding strategies for developing crop cultivars suitable for climate change ?

Ans. The climate change is occurring at the global level. As a result, India is facing a climate change of considerable magnitude. It is envisaged that as the twenty-first century progresses, there will be:

(i) rise in temperature ranging from 2 to 6 degree Celsius. The level of warming will be higher in the northern part of India than in the southern parts;

(ii) rapid increase in night temperatures compared to day temperatures;

(iii) increase in rainfall (15 to 40%) over all the states, especially those in the western and central-west parts of India, excepting Punjab, Haryana, Delhi, Rajasthan and Tamil Nadu.

In view of the climate change in Indi, the future plant breeding efforts need to be directed towards the following thrust areas.

1. The new crop varieties should be tolerant to drought and heat that are caused due to increasse in temperature. Crop breeding programs to develop temperature and drought-tolerant high yielding cultivars of the identified crops should be initiated urgently, so that the desired kinds of crop cultivars are available when effects of climate change are noticed.
2. The new varieties, especially grown during winter season should have characteristics of early flowering, photo- and thermo-insensitivity , early maturity and high productivity.
3. The plant genetic resources, especially land races from those areas that have similarities with climate change can be used to start breeding programs for developing varieties suitable for climate change.
4. The new crop cultivars for climate change can be developed using a combination of conventional breeding approaches, marker assisted

selection, induced mutations and transgenic-breeding. Crop based coordinated programs need to be launched to develop early-maturing, high-yielding and temperature and drought tolerant varieties as early as possible.

5. The desirable genotypes for climate change can also be selected in the breeding populations of some ongoing research programs. There will be need for identification of areas where the climate change conditions already exist or resemble. In such areas, large segregating breeding populations can be screened for selection of desirable genotypes.

Q. 18. Cite some examples of practical achievements of breeding for climate change.

Ans. In the past, several cultivars in different crop species have been developed with resistance to various insects, diseases and drought. Conventional breeding approaches along with modern crop improvement techniques will be rewarding to solve various problems that arise in future as a result of climate change. Progress has been made in the genetic dissection of flowering time, inflorescence architecture, temperature and drought tolerance in certain model plant systems and by comparative genomics in crop plants. Recently, the Indian Agricultural Research Institute, New Delhi has released an early-maturing wheat variety suitable for late planting.

SECTION-V
BREEDING FOR SPECIFIC PURPOSES

IDEOTYPE BREEDING

Q. 1. What is ideotype ?

Ans. A biological mode which is expected to perform or behave in a predictable manner within a defined environment is known as ideotype. It is also known as ideal plant type or model plant type.

Q. 2. Who developed the concept of plant type ?

Ans. The concept of plant type was first introduced in rice breeding by Jennings in 1964.

Q. 3. Who coined the term ideotype ?

Ans. The term ideotype was coined by Donald in 1968 working with wheat crop.

Q. 4. What is crop ideotype ?

Ans. A plant model which is expected to yield greater quantity of grains, fibre, oil or other useful product when developed as a cultivar is called crop ideotype.

Q. 5. Who developed the concept of crop ideotype ?

Ans. The concept of crop ideotype was developed by Donald 1968 on wheat.

Q. 6. What is ideotype breeding ?

Ans. A method of crop improvement which is used to enhance genetic yield potential through genetic manipulation of individual plant character is termed ideotype breeding.

Q. 7. What is harvest index ?

Ans. The ratio of economic yield to the biological yield or the ratio of economic produce to total biomass is called harvest index.

Q. 8. Who coined the term harvest index ?

Ans. The term harvest index was first used by Donald in 1962.

Q. 9. What is biological yield ?

Ans. Total dry matter production per plant is called biological yield or biomass.

Q. 10. What is economic yield ?

Ans. The dry weight of useful plant part such as grain yield in cereals, pulses and oilseeds and seed cotton yield in cotton is referred to as economic yield.

Q. 11. What is determinate plant type ?

Ans. The plant type in which growth of apical bud is ceased after a definite period is known as determinate plant type. Usually such plants produce only one flush of flowers such as early genotypes of pigeonpea, greengram, blackgram, cowpea etc.

Q. 12. What are indeterminate plants ?

Ans. Plant genotypes in which apical bud grows continuously are called interminate plants. Such genotypes can produce second flush of flowers such as long duration genotypes of cotton and pigeonpea.

Q. 13. What is compact plant type ?

Ans. Plant types which have erect stature and have very less plant spread are known as compact plant types ; also called erect plant types. Such plant types are suitable for closer spacings.

Q. 14. What is robust plant type ?

Ans. Plant genotypes which have vigorous plant growth are called robust plant types. Such genotypes are generally spreading types.

Q. 15. What is dwarf plant type ?

Ans. Plant genotype having dwarf stature is called dwarf plant type.

Q. 16. What is the source of dwarfing gene in wheat ?

Ans. In wheat, Norin 10 is the important source of dwarfing gene.

Q. 17. What is the source of dwarfness in rice ?

Ans. In rice, Dee-Geo-Woogen is the important source of dwarfing gene.

Q. 18. What is tall plant type ?

Ans. Plant genotypes having tall stature are called tall plant types. Such genotypes are generally prone to lodging.

Q. 19. What is pyramidal plant type ?

Ans. In dicots, genotypes having pyramidal shape such as in cotton refer to pyramidal plant type.

Q. 20. What are erect leaves ?

Ans. In cereals, leaves having upright position are called erect leaves. Such leaves have acute angle with the stem and also have high CO_2 fixation efficiency.

Q. 21. What are drooping leaves ?

Ans. In cereals, leaves having downward bending position are known as drooping leaves. Such leaves have right or obtuse angle with the stem and are considered physiologically less efficient.

Q. 22. What are tillers ?

Ans. In cereals, the replica of main stem originating from the base of a plant is called tiller.

Q. 23. What is culm ?

Ans. In cereals, the shoot length from first internode to the base of the spike is called culm.

Q. 24. Define physiologically efficient plants.

Ans. Plant genotypes with following characteristics are considered physiologically efficient.

(i) High CO_2 fixation efficiency.

(ii) High mineral absorption efficiency.

(iii) High response to nutrients.

(iv) Low photo respiration.

(v) Photo and thermo insensitivity.

Q. 25. What is physiological crop breeding ?

Ans. Genetic improvement of crop plants in relation to following physiological parameters is known as physiological crop breeding.

(i) CO_2 fixation efficiency. (ii) Nutrient uptake capacity. (iii) High harvest index. (iv) Reduction in photo-respiration and transpiration.

Q. 26. What is synchronous maturity ?

Ans. The maturity of all fruits of a plant at the same time is known as synchronous maturity. Such trait is desirable for harvesting by machine.

Q. 27. What is idiotype ?

Ans. Morphological features of the chromosomes of a particular plant species refers to idiotype.

Q. 28. What are C_3 plants ?

Ans. Plant species producing three carbon compounds (phosphoglyceric acid) as the first stable substance during photosynthesis are called C_3 plants. C_3 plants are physiologically lesser efficient than C_4 plants.

Q. 29. What are examples of C_3 plants ?

Ans. Majority of crop plants are C_3. Examples of C_3 plants include rice, wheat, barley, pulses, oil seeds etc.

Q. 30. What are C_4 plants ?

Ans. Those plant species which produce four carbon compounds (oxaloacetic acid) as the first stable product of photosynthesis are called C_4 plants. Physiologically C_4 plants are more efficient than C_3 plants.

Q. 31. Cite examples of C_4 plants.

Ans. C_4 plants include Maize, Sorghum, Sugarcane and *Amaranthus.*

Q. 32. Who first proposed ideal plant type of maize ?

Ans. In maize, ideal plant type was first suggested by Mock and Pearce in 1975.

Q. 33. Who first suggested ideal plant type of six-rowed barley ?

Ans. In six-rowed barley, ideal plant type was first suggested by Rasmusson in 1987.

Q. 34. Who proposed ideal plant type of cotton for irrigated conditions ?

Ans. The ideal plant type of American and Desi cottons for irrigated conditions was proposed by Singh *et al* (1974).

Q. 35. Who proposed ideal plant type of cotton for rainfed conditions ?

Ans. For rainfed conditions, ideal plant type of upland and arboreum cottons was proposed by Singh and Narayanan in 1993.

Q. 36. Who proposed ideotype of Brassica ?

Ans. Ideotype of Brassica was proposed by Bhargava et al in 1984.

Q. 37. Who proposed ideotype of wheat for rainfed conditions ?

Ans. In India, ideotype of wheat for rainfed conditions was proposed by R.D. Asana.

Q. 38. Who proposed ideotype of *Brassica napus* for rainfed conditions ?

Ans. Ideotype of *Brassica napus* for rainfed conditions was proposed by Thurling in 1991.

Q. 39. What are main features of ideotype breeding ?

Ans. Main features of ideotype breeding are given below :

(i) Emphasis is given on individual trait which enhances the yield.
(ii) It includes morphological and physiological traits.
(iii) It exploits physiological variation.
(iv) It involves interdisciplining approach.
(v) It is a slow method of cultivar development.
(vi) Selection is focussed an yield enhancing traits.
(vii) Values of each character are decided in advance.

Q. 40. What are main features of wheat ideotype ?

Ans. Main features of wheat ideotype are listed below :

(i) short and strong stem.

(ii) Erect leaves.
(iii) Few small leaves.
(iv) Larger and erect ear.
(v) Presence of awns.
(vi) A single culm.

Q. 41. What are main features of rice ideotype ?

Ans. Main features of rice ideotype as given below :

(i) Semi dwarf stature.
(ii) High tillering capacity.
(iii) Short, erect, thick and highly angled leaves.

Q. 42. What are main features of maize ideotype ?

Ans. In maize, main features of ideal plant type are given below :

(i) Low tillers.
(ii) Large cobs.
(iii) Engled leaves.

Q. 43. What are the factors affecting ideotype ?

Ans. Ideotype is affected by following factors :

(i) The crop species. (ii) Type of cultivation or cultivation practices. (iii) Socio-economic conditions of farmers. (iv) Economic use. (v) Market requirements etc.

Q. 44. What steps are involved in ideotype breeding ?

Ans. Ideotype breeding consists of following steps :

(i) Development of conceptual model.
(ii) Selection of base material from germplasm.
(iii) Incorporation of desirable traits.
(iv) Selection of desirable plant type.

Q. 45. In which crops ideotype breeding has been rewarding ?

Ans. Ideotype breeding has been rewarding in cereals (wheat and rice) and millets (Sorghum and pearl millet).

Q. 46. What is the plant type suggested by M.S. Swaminathan ?

Ans. M.S. Swaminathan has listed the following desirable attributes of crop ideotypes with special reference to multiple cropping in the tropics and sub-tropics.

(i) Superior population performance.
(ii) High productivity per day.
(iii) High photosynthetic ability.
(iv) Low photorespiration.
(v) Photo and thermo insensitivity.
(vi) High response to nutrients.

(vii) High productivity per unit of water.
(viii) Multiple resistance to insects and diseases.
(ix) Better protein quantity and quality.
(x) Crop canopy that can retain and fix a maximum of CO_2.
(xi) Stability to mechanization.

Q. 47. What are advantages of ideotype breeding ?

Ans. It provides solution to several problems at a time like disease, insects and lodging resistance, maturity duration, yield and quality by combining desirable genes for these traits from different sources into a single genotype.

Q. 48. What are demerits of ideotype breeding ?

Ans. Ideotype breeding has some limitations or demerits. Incorporation of several morphological, physiological and resistance traits from different sources into a single genotype is a difficult task. Sometimes, combining of some characters is not possible due to tight linkage between desirable and undesirable characters. Presence of such linkage hinders the progress of ideotype breeding. Moreover, it is a slow method of cultivar development.

Q. 49. What are differences between traditional breeding and ideotype breeding ?

Ans. Main differences between traditional breeding and ideotype breeding are presented below in Table 30.1.

TABLE 30.1 Differences between traditional and ideotype breeding

Traditional Breeding	Ideotype Breeding
1. The main objective is defined before initiating the breeding work.	The conceptual theoretical model is prepared before initiation of breeding work.
2. Selection is focussed on yield and some other characters.	Selection is focussed on individual plant characters.
3. It usually includes various morphological and economic characters.	It includes various morphological, physiological and biochemical plant characters.
4. Value of each character is not fixed in advance.	Value of each trait is defined in advance.
5. This is a simple and rapid method of cultivar development.	This is a difficult and slow method of cultivar development.
6. The phenotype of a new variety is not specified in advance.	Phenotype of new variety to be developed is specified in advance.

31

BREEDING FOR QUALITY TRAITS

Q. 1. What is quality ?

Ans. Quality refers to suitability or fitness of an economic plant product to its end use. It is of different types. viz. market quality, industrial quality, nutritional quality, etc.

Q. 2. What is market quality ?

Ans. Fitness of a product for marketing is called market quality. In food and vegetable crops, it includes uniformity in shape, size, colour, texture, keeping quality etc.

Q. 3. What is industrial quality ?

Ans. Suitability of a plant product for use as a raw material in industries is called industrial quality. It includes :

(i) backing quality in wheat. (ii) matting quality in barley. (iii) crushing quality in sugarcane. (iv) canning quality in fruit crops.

Q. 4. What is nutritional quality ?

Ans. Suitability of a plant product for human and animal consumption is called nutritional quality. A quality product has balanced combination of carbohydrates, proteins and lipids.

Q. 5. What is quality breeding ?

Ans. Genetic improvement of crop plants in relation to various quality attributes is called quality breeding. Quality traits differ from crop to crop.

Q. 6. What is food ?

Ans. The substance which is used (eaten) to meet the requirement of body for energy, maintenance of health, growth and reproduction is called food.

Q. 7. What are nutrients ?

Ans. Various chemical components of food which provide nourishment to the body such as carbohydrates, proteins, fat, vitamins and minerals are called nutrients.

Q. 8. What is nutrition ?

Ans. The scientific study of food in relation to health of human and animals is called nutrition.

Q. 9. What is malnutrition ?

Ans. Adverse effect of imbalanced diet on human health is called malnutrition. It results from deficiency, excess or imbalance of nutrients in the diet.

Q. 10. What is under nutrition ?

Ans. Low intake of food is called under nutrition.

Q. 11. What are amino acids ?

Ans. Organic compounds which contain carboxyl (COOH) or acidic group and an amino or basic (NH_2) group are called amino acids. Amino acids are of two types, viz. essential and non-essential.

Q. 12. What are essential amino acids ?

Ans. Those amino acids which cannot be synthesized in human body and their requirement has to be met through dietary intake are called essential amino acids. These are given below :

Methionine	Tryptophan
Isoleucine	Valine
Leucine	Phenylalanine
Lysine	Histidine
Threonine	Arginine

Q. 13. What are non-essential amino acids ?

Ans. Those amino acids which can be synthesized by human body and they need not be supplied through diet are known as non-essential amino acids.

Q. 14. What are proteins ?

Ans. Organic compounds which are composed of amino acids are called proteins. In other words, proteins are polymers of amino acids.

Q. 15. What is vitamin ?

Ans. A vital substance which is required in very small quantity and is essential for proper growth and good health is called vitamin.

Q. 16. What are quality characters in wheat ?

Ans. In wheat, amber grain colour, medium to bold grain size, hard viterous texture, and lustrous appearance are important features for good market quality. High lycine content and good backing quality are essential for use in biscuit and bread making.

Q. 17. What are quality characters in barley ?

Ans. In malting barley, low protein content and high extract of soluble oligosaccharides after malting are desirable characters. Low protein produces less haze in beer and high oligosaccharides are essential for fermentation.

Q. 18. What are quality characters in rice ?

Ans. Important quality characters of rice are given below :

(i) White, fine and long slender grains.

(ii) Taste and fragrance.

(iii) Less breakage in milling.

(iv) More hulling recovery.

(v) Better cooking quality, and

(vi) High protein and lycine contents.

Q. 19. What are quality traits in maize ?

Ans. Main quality traits of maize are given below :

(i) Bold flint grains with attractive colour.

(ii) High lycine, sugar and oil contents.

Q. 20. What are quality traits in sorghum ?

Ans. In sorghum, white grains of attractive shape and size, high protein and lycine contents are main quality traits.

Q. 21. What are quality traits in Pearlmillet ?

Ans. In pearlmillet, bold lustrous and pearly amber coloured grains and high iron contents are desirable quality characters.

Q. 22. What are quality characters in pulses ?

Ans. In pulse crops, main quality characters are given below :

(i) Attractive shape, size and colour of grains.

(ii) High protein content.

(iii) High methionine and tryptophan contents.

(iv) Less flatulence.

Q. 23. What are quality traits in oilseeds ?

Ans. Main quality characters of oil seed crops are given below :

(i) Attractive shape, size and colour of seeds.

(ii) High oil content.

(iii) Absence of antinutritional factors.

(iv) High proportion of unsaturated fally acids.

Q. 24. What are quality characters in sugarcane ?

Ans. Main quality characters of sugarcane are as follows :

(i) Moderate hardness.

(ii) Long internodes.
(iii) Optimum fibre for milling.
(iv) High sucrose content.
(v) Good quality of juice.

Q. 25. What are quality characters in cotton ?

Ans. Quality characters of cotton are given below :

(i) Fibre length, strength, fineness, maturity and uniformity.
(ii) White colour of lint.

Q. 26. What are quality characters in tobacco ?

Ans. Main quality characters differ according to uses as follows :

(i) Thin leaves are preferred for cigar and pipe smoking.
(ii) Thick leaves are suitable for cigarettes.
(iii) High nicotin is suitable for bidi, hookah and chewing.
(iv) Low nicotin is preferred for cigarettes.
(v) High sugar content is preferred.

Q. 27. What are quality characters in potato ?

Ans. In potato, main quality characters are :

(i) Attractive shape, size and colour of tubers.
(ii) Good taste and thin skin.
(iii) Good keeping quality and cooking quality.
(iv) High starch content.

Q. 28. What are quality traits in vegetables ?

Ans. Main quality traits of vegetables are as follows :

(i) High vitamin and mineral contents.
(ii) Good taste and cooking quality.
(iii) Keeping quality.

Q. 29. What are quality traits in forage crops ?

Ans. Main quality traits of forage crops are given below :

(i) Greater nutritive value.
(ii) More polatability, and
(iii) Absence of toxic substances.

Q. 30. What are goals of breeding for improved nutritional quality ?

Ans. There are four major goals of breeding for improved nutritional quality as follows :

(i) High quantity and quality of protein.

(ii) High content and quality of oil.

(iii) High vitamin and mineral contents.

(iv) Low toxic substances.

Q. 31. What are water soluble vitamins ?

Ans. Vitamins of B and C groups are water soluble.

Q. 32. What are fat soluble vitamins ?

Ans. Fat soluble vitamins are A, D, E and K.

Q. 33. What are limiting amino acids in cereals ?

Ans. In cereals, limiting amino acids are lysine, threonine and sometimes tryptophan.

Q. 34. What are limiting amino acids in pulses ?

Ans. In pulses, limiting amino acids are methionine and tryptophan.

Q. 35. What is limiting amino acid in nuts and oil seeds ?

Ans. In nuts and oil seeds, lysine is the limiting amino acid.

Q. 36. What is limiting amino acid in vegetables ?

Ans. In vegetables, limiting amino acid is methionine.

Q. 37. What are sulphur containing amino acids ?

Ans. Sulphur containing amino acids are :

Tryptophan	Lysine
Threonine	Valine
Isolencine	Methionine

Q. 38. What are proteins found in cereals ?

Ans. In cereals, following four types of proteins are found :

(i) Albumins–soluble in water.

(ii) Globulins–soluble in saline solutions.

(iii) Prolamins–soluble in relatively strong alcohol.

(iv) Glutelins–soluble in dilute alkaline solutions.

Cereals contain about 70% prolamins and glutelins and 30% albumins and globulins.

Q. 39. What are proteins found in pulses ?

Ans. Pulses contain two types of protein viz albumin and globulin. Pulses contain about 80% globulin.

Q. 40. What are types of globulin found in pulses ?

Ans. In chickpea, globulin is of three types, viz. alpha, beta and gamma. Alpha globulin accounts for more than 80% of total protein.

Q. 41. What are types of globulin found in faba and pea ?

Ans. In faba and pea, globulin is of two types, viz. legumin and vicilin.

Q. 42. What are toxic substances found in legumes ?

Ans. In legumes, following toxic substances are found :

(i) Tripsin inhibitor : It is found in pigeonpea, french bean and soybean.
(ii) Haemagglutinins : It is found in french bean.
(iii) Lathyrogenes : It is found in Lathyrus sativus.
(iv) Glucocides : It is found in white clover.
(v) Goitrogens : It is found in soybean.
(vi) Antivitamin E factor : It is found in field pea.

Q. 43. What are important sources of nutritional quality ?

Ans. Important sources of nutritional quality are as follows :

(i) Cultivated varieties.
(ii) Germplasm collections.
(iii) Wild species.
(iv) Mutations etc.

Q. 44. What are sources of high lysine in maize ?

Ans. Sources of high lysine in maize are given below :

Opaque 2   Opaque 7
Flory 2   Brittle 2

All these are spontaneous mutations.

Q. 45. What source of high lysine in sorghum ?

Ans. In sorghum, there are three main sources of high lysine as follows :

(i) IS 11167 and IS 11758–Spontaneous mutants.
(ii) P 721–Induced mutant (DES).

Q. 46. What are sources of high lysine in barley ?

Ans. In barley, sources of high lysine are given below :

(i) Induced mutants–Notch 1, Notch 2 and Rio 1508.
(ii) Spontaneous mutant–Hiproly.

Q. 47. What is toxic substance in Lathyrus ?

Ans. In Lathyrus, toxic susbtance is Lathyrogen or neurotoxin which leads to paralysis of lower limbs in human.

Q. 48. What is toxic substance in pigeonpea and cowpea ?

Ans. In pigeon pea and cowpea, toxic substance is tripsin inhibitor.

Q. 49. What are toxic substances in soybean ?

Ans. In soybean, toxic substances are tripsin inhibitor and goitrogens.

Q. 50. What are toxic substances in rapeseed and mustard ?

Ans. In rapeseed and mustard, toxic substances are Erucic acid and Eicosenoic acid.

Q. 51. What is toxic substance in cotton ?

Ans. In cotton, gossypol, a phenotic compound, is the toxic substance.

Q. 52. What is toxic substance in sunflower ?

Ans. In sunflower, polyphenotics is toxic substance.

Q. 53. What are toxic substances in potato ?

Ans. In potato steroidal akaloids are toxic compounds.

Q. 54. What are toxic substances in cucurbits ?

Ans. In cucurbits, cucurbitacins are toxic substances.

Q. 55. What is toxic substance in brinzal ?

Ans. In brinzal, bitter principle is the toxic substance.

Q. 56. What is toxic substance in cassava ?

Ans. In cassava, the toxic substances in CN glucocide.

Q. 57. What are toxic substances in mango ?

Ans. In mango, resigns are toxic substances.

Q. 58. In which crop cynogenic glucocides are found ?

Ans. Cynogenic glucocides are found in sorghum, sudan grass and white clover.

Q. 59. What are toxic substances in alfalfa ?

Ans. In alfalfa, saponins and plant estrogens are toxic substances.

Q. 60. What is the toxic substance in sweet clover ?

Ans. In sweet clover, coumarin is the toxic substance.

Q. 61. What are toxic substances in safflower ?

Ans. In safflower, polyphenolics are toxic substances.

Q. 62. What are toxic substances in Lespedeza ?

Ans. In Lespedeza, tannins are toxic substances.

Q. 63. What are high lysine varieties of maize ?

Ans. In maize high lysine varieties are Protina, Shakti and Ratan. All these are composite varieties.

Q. 64. What are sources of high protein and lysine in wheat ?

Ans. In wheat, Atlas 66 and Indian line Naphal are important sources of high protein and lysine.

Q. 65. What are high sucrose varieties of sugarcane ?

Ans. In sugarcane, varieties CO 671, CO 6806, CO 7314, CO 7704, and CO 62174 have high sucrose content.

Q. 66. What are varieties of barley suitable for malting ?

Ans. In barley, varieties Karan 15, Karan 92 and Karan 280 are good for malting quality.

Q. 67. Which is the high protein and high oil variety of soybean ?

Ans. In soybean, variety Lee has high protein (43–45%) and high seed oil contents.

Q. 68. Which is high oil variety of rapeseed ?

Ans. In rapeseed, variety K 88 has high oil content. It contains 48.8% oil.

Q. 69. Which is low neurotoxin variety of Lathynes ?

Ans. In *Lathynes*, variety Pusa 24 has low neurotoxin. The level of neurotoxin is much below critical level.

Q. 70. What are maize varieties developed using opaque 2 ?

Ans. In maize three varieties, viz. Protina, Shakti and Ratan have been developed through *the use of opaque 2, a high lysine mutant.

Q. 71. What are limitations of breeding for quality traits ?

Ans. Main limitations of breeding for quality traits are given below :

(i) In cereals and pulses, there is negative association between grain yield and seed protein.

(ii) There is negative association between seed oil and seed protein.

(iii) Sometimes selection for high lysine leads to reduction in grain size as in barley.

(iv) Transfer of quality character from wild species is a difficult task and long-term programme.

(v) Such work requires cooperation of biochemist which some times becomes limiting factor.

32

BREEDING FOR MULTIPLE CROPPING SYSTEMS

Q. 1. What do you mean by multiple cropping ?

Ans. Multiple cropping refers to growing two or more crops on the same field in a year. It is of two types, viz. sequential cropping and intercropping.

Q. 2. What is sequential cropping ?

Ans. Cultivation of two or more crops in sequence (one after another) on the same field in a year is known as sequential cropping. It is of three types, viz. double cropping, triple cropping and rotoon cropping.

Q. 3. What is double cropping ?

Ans. Cultivation of two crops on the same field one after another in a year is called double cropping.

Q. 4. What is triple cropping ?

Ans. Cultivation of three crops one after another on the same field in a year is called triple cropping.

Q. 5. What do you mean by rotoon cropping ?

Ans. Cultivation of crop regrowth after harvest on the same field in a year is known as rotoon cropping. It is followed in Rice and Sorghum under irrigated conditions.

Q. 6. What is inter-cropping ?

Ans. Cultivation of two or more crops together on the same field in a year is called intercropping. It is also known as associated cropping or simultaneous polyculture.

Q. 7. What are different types of inter cropping ?

Ans. Inter cropping is of following four types :

(i) Mixed inter cropping.

(ii) Row inter cropping.

(iii) Strip inter cropping, and

(iv) Relay inter cropping.

Q. 8. What is mixed inter-cropping ?

Ans. Raising two or more crops by mixing their seeds or without distinct row arrangement is known as mixed inter-cropping, also called mixed cropping.

Q. 9. What is row inter-cropping ?

Ans. Raising two or more crops together but in separate rows is called row inter cropping.

Q. 10. What is strip inter-cropping ?

Ans. Raising two or more crops together in a field but in different distinct strips is known as strip inter cropping.

Q. 11. What is relay inter-cropping ?

Ans. Raising two or more crops in a field in a year in such a way that the second crop is planted before the harvest of first crop is called relay inter cropping. In relay inter cropping, there is some overlapping period of two crops.

Q. 12. What is cropping pattern ?

Ans. Yearly sequence of crops on a farm or given land is known as cropping pattern.

Q. 13. What is cropping system ?

Ans. Yearly sequence of crops on a farm and its interaction with farm resources is called cropping system.

Q. 14. What is sole cropping ?

Ans. Raising one crop alone in pure stand is known as sole cropping ; also called solid planting.

Q. 15. What is monoculture ?

Ans. Repetitive growing of the same sole crop on the same field in a year is called monoculture.

Q. 16. What is mixed farming ?

Ans. Growing crops in combination with animals such as dairy, poultry, pigary etc. is known as mixed farming.

Q. 17. What is cropping index ?

Ans. Number of crops grown on the same field in a year multipled by hundred is called cropping index.

Q. 18. What is allelopathy ?

Ans. Suppression of plant growth of one species by another species due to liberation of biologically active substance from the root or shoot is called allelopathy.

Q. 19. What are allelochemicals ?

Ans. Chemical substances which are liberated in allelopathy and inhibit the growth of another species growing together are known as allelochemicals.

Q. 20. What is main objective of crop improvement in multiple cropping ?

Ans. The main objective of crop improvement for multiple cropping system is to increase the total yield of the system over monocultures.

Q. 21. What are criteria which should be kept in mind while selecting crop species for multiple cropping system ?

Ans. Following points should be given due importance white selecting crop species for multiple cropping system.

(i) Both crops should not have similar rooting pattern. If one is shallow rooted, the other should be deep rooted so that both can absorb nutrients from different layers of soil.

(ii) Pulse crop should be included as one of the components.

(iii) Selected species should not have allelopathy.

(iv) There should be difference in the height of two species.

(v) Selected crops should not have common diseases and insects.

Q. 22. What are desirable plant traits for multiple cropping ?

Ans. The following plant traits are desirable for multiple cropping :

(i) Earliness : It permits inclusion of another crop in the system.

(ii) Uniform ripening because it permits harvest of crop in one picking in greengram, blackgram, maize, cotton etc.

(iii) Photo insensitivity : It permits sowing of crop at any date.

(iv) Rapid seedling growth : It helps in competition with weeds.

(v) Crops included in multiple cropping system should differ in root length.

(vi) Semidwarf genotypes with erect status are desirable in many multiple cropping systems.

(vii) Resistance to biotic and abiotic stresses.

(viii) Wider adaptability etc.

Q. 23. What are sources of desirable traits for multiple cropping systems ?

Ans. Following are important sources of desirable traits for multiple cropping systems.

(i) Cultivated varieties. (ii) Germplasm collections. (iii) Wild species of cultivated plants.

Q. 24. In which crops erect genotypes give higher yield at closer spacing than spreading types ?

Ans. In some pulse crops such as green gram, black gram, soybean and cowpea, erect genotypes give higher yield than spreading types at closer spacings.

Q. 25. What types of Sorghum and cereal cultivars are expected to give better yield at higher plant densities ?

Ans. In sorghum and cereals, cultivars with low tillering capacity are expected to give higher yields at higher plant densities.

Q. 26. What type of maize cultivars are expected to give higher yield at low plant densities ?

Ans. In maize, cultivars with prolific bearing capacity are expected to give higher yields at low plant densities.

Q. 27. In which crops, cultivars with high tillering capacity are expected to perform better at low plant densities ?

Ans. In sorghum, sugarcane, wheat, rice, barley and triticale, cultivars with high tillering capacity are expected to perform better at low plant densities.

Q. 28. What is the significance of earliness in multiple cropping system ?

Ans. In multiple cropping system, earliness permits more than one planting of the same species or addition of another crop species in the crop rotation.

Q. 29. What is significance of uniform ripening of a cultivar ?

Ans. Uniform ripening is a desirable character which permits harvest of the crop at one time.

Q. 30. What is significance of rapid seedling growth ?

Ans. Rapid growth of seedlings leads to partial suppression of weed growth due to high competitive ability.

Q. 31. Which of the four types of inter-cropping is widely used ?

Ans. Among four types of inter-cropping, row inter-cropping is the most popular.

Q. 32. What points should be kept in mind for selection of crop for sequential cropping ?

Ans. Following points should be kept in mind in selecting crops for sequential cropping.

(i) Both grain and fodder legumes should find a place in the rotation.

(ii) Both crops should not have common diseases and insects.

(iii) Both crops should differ in root length so that they can absorb nutrients from different layers. It will help in maintaining soil fertility.

Q. 33. What are advantages of multiple cropping ?

Ans. Main advantages of multiple cropping are given below :

(i) It provides diversity of diets by planting different food crops together or in sequence.
(ii) It provides employment to rural labourers throughout the year.
(iii) It leads to stability of production.
(iv) It permits best possible use of land resources.
(v) Inter cropping and mixed cropping act as barriers in the spread of diseases and insects.
(vi) It reduces the risk of crop failure.

Q. 34. What are factors determining cropping systems ?

Ans. Cropping systems are determined by following factors.

(i) Crop adaptation. (ii) Market demand. (iii) Socio-economic conditions. (iv) Preference of diet. (v) Taste etc.

Q. 35. What type of cultivars were in use in traditional cropping systems ?

Ans. In traditional cropping systems, primitive cultivars or land races were being used.

Q. 36. What are main features of land races ?

Ans. Main features of land races are given below.

(i) They are low yielding than improved cultivars.
(ii) They possess high genetic diversity.
(iii) They have high competitive ability with weeds.
(iv) They have broad genetic base and wide adaptability.
(v) They have high degree of resistance to biotic and abiotic stresses.

Q. 37. What are main demerits of land races ?

Ans. Main demerits of land races are given below :

(i) They are low yielding.
(ii) The produce of land races is less uniform and less attractive.
(iii) Land races are generally late maturing.

Q. 38. What are breeding methods used for development of suitable genotypes for multiple cropping systems ?

Ans. The following breeding methods are used for developing suitable genotypes for multiple cropping system.

(i) Plant introduction.
(ii) Selection (pure line, mass selection, progeny selection depending upon mode of pollination of the crop.)
(iii) Hybridization (Pedigree, bulk & backcross methods)
(iv) Heterosis, composite and synthetic breeding in cross pollinated species.

(v) Mutation breeding in self pollinated and asexually propogated species.

Q. 39. What is the cropping index of multiple cropping ?

Ans. The cropping index of multiple cropping system is always 200% or more.

Q. 40. What is example of ratoon cropping ?

Ans. Rice–Rice regrowth is an example of ratoon cropping.

Q. 41. In which crops ratoon cropping is used ?

Ans. Ratoon cropping is used in Rice and Sorghum under irrigated conditions.

Q. 42. Give an example of double sequential cropping.

Ans. Cotton–Wheat is an example of double sequential cropping.

Q. 43. Cite an example of sequential triple cropping.

Ans. Mung–Maize–Wheat is an example of triple sequential cropping.

Q. 44. Give an example of mixed inter-cropping.

Ans. Wheat + Mustard is an example of mixed inter-cropping.

Q. 45. Give an example of monoculture.

Ans. Rice–Rice–Rice is an example of monoculture.

Q. 46. Give an example of mixed farming.

Ans. Farms crops + dairy is are example of mixed farming.

Q. 47. In which area Rice-Wheat-Greengram sequence is popular ?

Ans. The Rice–Wheat–Greengram sequence is popular in North- West India.

Q. 48. In which area Rice-Wheat-Jute sequence is popular ?

Ans. The Rice–Wheat–Jute sequence is popular in Assam and West Bengal.

Q. 49. In which region Pigeonpea + blackgram (1 : 1) row inter cropping is popular ?

Ans. The Pigeonpea + blackgram (1 : 1) row inter-cropping is popular in Varanasi region.

Q. 50. Why late maturing cotton varieties are preferred in Cotton + Groundnut intercropping ?

Ans. In Cotton + Groundnut inter-cropping, late maturing cotton cultivars are preferred over early maturing cultivars, because late cultivars produce most of the cotton after the harvest of groundnut.

Q. 51. What type of bean cultivar gives higher yield in Maize-Bean intercropping with simutaneous sowing ?

Ans. In Maize-Bean inter-cropping with simultaneous sowing, a climber cultivar of bean gives higher yield than a bushy type cultivar.

33

ORGANIC PLANT BREEDING

Q. 1. What is organic plant breeding ?

Ans. Looking to the adverse effects of chemicals (insecticides, pesticides, fungicides, weedicides, chemical fertilizers) on human health, animal health, soil health and environment, organic plant breeding is gaining increasing significance, the world over. Organic plant breeding refers to development of new crop cultivars without the use of any chemical.

Q. 2. What is organic seed ?

Ans. The seed that has ben multiplied under organic conditions at least for one generation is referred to as organic seed. It may be developed by organic or nonorganic breeding programme. The organic seed is relatively common.

Q. 3. What are organic varieties ?

Ans. Those cultivars that originate from organic plant breeding techniques are known as organic varieties. Organic varieties are limited in number.

Q. 4. What are main features of organic breeding ?

Ans.

1. In the organic breeding, chemicals are not used neither in the development of varieties nor in their testing.
2. Those breeding methods which do not use any chemical are used in developing organic varieties. In other words, breeding techniques that are free from use of chemicals are permitted for organic plant breeding.
3. The entire evaluation of breeding material is carried out under organic conditions.
4. In the organic plant breeding higher priority is given to quality of the product than yield. Emphasis is laid on developing varieties for better human health. Such varieties should contain some secondary metabolites in addition to high protein content.
5. In the organic plant reing programmes much emphasis is given in maintaining enough diversity in the varieties to provlde them high buffering capacity to environmental changes.

6. Patenting of breeding material is not permitted because it will put restriction on free exchange of breeding material among breeders and farmers.
7. Organic varieties differ are produced under certified organic plant breeding programmes.

Q. 5. What are differences between organic variety and organic seed ?

Ans. Organic varieties differ from organic seed. The main differences between them are given in Table 33.1

Table 33.1 Differences between organic varieties and organic seed.

Organic Varieties	Organic Seed
I. Originate from organic plant breeding.	I. Originate from organic or non-organic plant breeding.
2. Tested always under organic conditions.	2. Multiplied at least for one generation under organic condition.
3. Number of such varieties is limited.	3. Such seed is very common.

Q. 6. What are objectives of organic plant breeding ?

Ans. The breeding objectives of organic plant breeding are some what different from those of non-organic plant breeding. However, some objectives are common in both the cases. Important objectives are briefly discussed below:

1. **Competitiveness with weeds.** It is considered important for organic varieties where weedicides are not used. In non-organic breeding it is not so important because weeds can be suppressed through use of weedicides.
2. **Plant Stature.** Generally, varieties for organic use tend to be taller than those developed for non-organic production.
3. In organic plant breeding higher priority is given to quality than productivity or yield.
4. In organic plant breeding higher emphasis is given to maintain biodiversity in the variety.
5. Higher degree of resistance to biotic stresses because insecticides/pesticides are not for controlling insects and diseases.

Q. 7. What are common objectives of organic and non-organic plant breeding ?

Ans. There are some common objectives for organic and non-organic varieties which are given below:

1. Earliness so that a variety can fit well into multiple cropping system.

2. Resistance to abiotic stresses such as drought, soil salinity, metal toxicity, heat, frost, excessive moisture.
3. Resistance to lodging in cereals and sugarcane.
4. Non shattering habit in legumes.
5. Wider adaptability and stability in production.
6. Freedom from toxic substances.
7. High photosynthetic and nutrient uptake efficiency.
8. Photo and thermo insensitivity.
9. Amenability to machine harvesting.

Besides these common objectives, non-organic plant breeding gives emphasis on yield and uniformity of varieties.

Q. 8. What are the breeding methods used for developing organic cultivars ?

Ans. The main objective of organic plant breeding is to develop crop cultivars that enhance productivity and adaptability under organic farming. All plant breeding techniques are not applicable for organic plant breeding. The breeding methods that are suitable for organic breeding are briefly discussed below:

1. **Plant introduction.** Any exotic variety developed without the use of any chemical. It can sed as a variety after thorough testing under organic conditions.
2. **Selection.** It includes pure line selection, mass selection, progeny selection, clonal selection etc.
3. **Hybridization.** It refers to combination breeding. It includes backcross, pedigree method, bulk breeding method and single seed descent.
4. **Biotechnology.** The marker assisted selection can be permitted if GMOs and radiations are not involved in marker production. The meristem culture can be permitted particularly because of its key role in virus elimination.
5. Population improvement approaches such as recurrent selection, biparental mating, diallel selective mating and disruptive mating can be used for developing superior inbred lines that can be utilized in hybridization programmes. .
6. The F 1 hybrids are permitted provided the F_1 offspring are fertile.

Q. 9. What techniques are not permitted in organic plant breeding ?

Ans. Those methods that involve use of chemicals or radiations or donot produce fertile plants are not permitted. Such methods include :

1. Genetic modification or genetically modified organisms.
2. Cytoplasmic male sterility based hybrids without restorer genes.

3. Somatic hybridization or protoplast fusion.
4. Radiated mentor pollen for mutation induction.
5. Mutation induction with radiations or chemical substances.

There are some techniques that are suitable for organic plant breeding but need further proof. These techniques are given below:

1. **For Variation Induction.** It includes embryo culture, ovary culture and in vitro pollination.
2. **For selection.** It includes in vitro selection and DNA diagnostic methods.
3. **For Maintenance.** It includes another culture, micro-propagation and somatic embryogenesis.
4. **Substances.** Some chemical substances such as silver thiosulphate, silver nitrate, growth stimulants and colchicine need further proof for permission of use in organic plant breeding.

Q. 10. Give a comparison of organic breeding and organic farming.

Ans. There is difference between organic breeding and organic farming. The former refers to development of new crop cultivars without the use of any chemical. Varieties developed in this way are used for organic cultivation. The organic farming, on the other hand, refers to crop cultivation without the use of any chemical (insecticides, fungicides, herbicides, growth regulators, defoliants, inorganic fertilizers etc.). The organic produce is duly certified by recognized certifying organization. The organic cultivation is also known as clean cultivation, natural cultivation, green cultivation or eco-friendly cultivation. The organic produce is of two types, viz., organic certified A and organic certified B. These are defined below:

1. Organic A. The crop which is grown after three years without the use of prohibited chemicals is called organic certified A.

2. Organic B. The crop which is produced in the first and second year without use of chemicals is referred to as organic B.

Q. 11. What chemicals are prohibited in organic plant breeding ?

Ans. In organic breeding as well as organic farming use of chemicals is prohibited. Various agrochemicals which are prohibited include inorganic fertilizers, insecticides, fungicides, herbicides, growth regulators, defoliants. Seed treatment with fungicides or insecticides is not permitted. The treatment with irradiations (X-rays, gamma-rays, etc.) is also not permitted.

Q. 12. What materials are allowed in organic plant breeding ?

Ans. Various materials which are permitted include organic manures (FYM, compost, green manure, cakes, etc.). The fish meal, leather meal, the cotton seed meal are permitted to maintain soil fertility. Gypsum is allowed to correct soil salinity. For control of pests, genetic resistance is used. Application of botanical pesticides (Neem, Datura, Ipomea, tobacco, chillies) and bioagents (parasites and predators) is permitted.

Q. 13. Who can take up organic farming ?

Ans. The organic farming (cultivation) is to be carried out by registered growers only. The registered producers have to be recognized by certifying agency. The producer has to sign an agreement to abide by the rules of certifying agency. The fields of organic crop are inspected by the inspectors of certifying agency during the crop period to verify that only permissible practices have been followed by the producer. The certifying agency has right to reject any field for certification, if the producer has not followed instructions of the inspectors strictly. The certifying agency may charge fee for inspection.

Q. 14. Give a comparison of organic and non-organic breeding.

Ans. The organic plant breeding differs from conventional or non-organic breeding in several aspects as given in Table 33.2

Table 33.2 Differences between organic and non-organic breeding

Particulars	Organic Breeding	Non-organic Breeding
1. Breeding methods used	Those methods which donot involve radiations and chemicals	All breeding methods
2. Major emphasis is on	Quality	Yield
3. Competitiveness with weeds	High	Low
4. Biodiversity	High	Low
5. Uniformity of produce	Low	High
6. Adaptation	High (wide)	Narrow
7. Genetic base	Broad	Narrow
8. Use of chemicals	Not permitted	Permitted
9. Use of radiations	Not permitted	Permitted
10. Insect control	Genetic resistance, use of botanicals and bio-agents	Mainly by the use of insecticides
11. Disease control	By Genetic Resistance	By fungicides and pesticides

12. Weed control	By mechanical means	By weedicides and mechanical means
13. Nutrient supply	By organic manures	By mainly inorganic fertilizers
14. Type of cultivation	Sustainable	Non-sustainable
15. Cost of ciltivation	Low	High
16. Environmental pollution	Nil	Very high

Q. 15. What are advantages of organic plant breeding ?

Ans. There are several advantages of organic plant breeding. Some important advantages are given below

1. It is used for developing cultivars and hybrids suitable for organic farming. The use of organic varieties helps in reducing the cost of cultivation by prohibiting use of various agrochemicals.
2. Use of organic varieties is ecofriendly. It permits multiplication of natural enemies (parasites and predators) of harmful insects resulting in effective biological control.
3. In organic plant breeding more emphasis is given on improvement of quality rather than yield. The organic varieties of food grains, vegetables and fruits would be more suitable for human health because of their better quality.
4. In organic plant breeding use of chemicals is prohibited. As a result there is no environmental pollution. This has favourable effect on soil health, human health and animal health. The insecticides will not reach the food chain. The beneficial soil microflora will increase and ground water and rain water will be free from contamination of chemicals.
5. Use of orgainc varieties will lead to sustainable agriculture. It will not deplete the resources. In other words it will promote judicious use of resources.
6. It promotes use of biological inputs such as organic manures (FYM, Compost, green manure, cakes etc.), botanical pesticides (Neem, Datura, Ipomea, tobacco, calatropies, chillies etc.) which do not have any adverse effects on the ecosystem. Rather use of organic manures improves the physico-chemical properties of the soil.

Q. 16. What are disadvantages of organic breeding ?

Ans. In organic breeding, use of several breeding techniques and some important chemicals is banned or prohibited which will have adverse effects in the progress of plant breeding. Some such examples are given.

1. Induced mutations play important role in developing new varieties and creating extra variability for selection. Banning use of this technique will have adverse effect on progress of plant breeding.
2. Protoplast fusion permits gene transfer between any two species related or unrelated. It has overcome the barriers of gene transfer between species and genera. Banning of this technique will act as a barrier in somatic hybridization.
3. Genetically modified organisms are used to solve those problems that can not be resolved by conventional breeding methods. The ban on the use of GMO will have adverse effects in solving such problems.
4. Tissue culture techniques such as embryo culture, ovary culture, anther culture etc. lead to extra variation which is used for selection of suitable material. Bann on these techniques will prolong development of organic varieties.
5. The CMS based hybrids without restorer genes are used in sugar-beets, cabbage and some forage crops. Bann on the use of this technique will adversely effect breeding of these crops.
6. The well known chemical colchicine is used for making interspecific hybrids fertile through chromosome doubling. Banning the use of colchicine will have adverse effects on interspecific hybridization.
7. Banning above useful techniques/chemical will prolong the period of developing organic varieties.
8. The breeding efficiency will go down and breeding will be more expensive.
9. Restriction on the exchange of material between traditional and organic breeding will have adverse effects in the progress of breeding.

34

PARTICIPATORY PLANT BREEDING

Q. 1. What participatory plant breeding ?

Ans. Plant breeding programmes involving scientists, farmers and others such as consumers, marketers, processors and policy markers refer to Participatory Plant Breeding (PPB). In other words, participatory plant breeding is a breeding programme in collaboration with breeders, farmers, consumers, industry etc.

Q. 2. What are the main features of participatory plant breeding ?

Ans. The main features of participatory plant breeding are presented below:

1. PPB in volves breeders, farmers, consumers and processors in the development of new crop cultivars.
2. Participatory Plant Breeding is also called as Collaborative Plant Breeding (CPB), Farmer Participatory Breeding (FPB) and Participatory Crop Improvement (PCI). All these terms are interchangeably used.
3. The participatory crop improvement is subdivided into two areas, viz., Participatory Varietal Selection (PVS) and PPB. The PVS refers to selection within stable (homozygous) lines or populations involving farmers. The PPB deals exclusively with segregating material.
4. In PPB, the testing of advanced generation breeding material is carried out at both research station and farmers fields. The crossing is done on research station where F_1 and F_2 generations are also grown. From F_3 onwards, the material is tested both at Research Farm and farmers fields.
5. At farmers fields, the yield testing of segregating material begins three years after making a cross i.e. from F_3 onwards. The material is tested in three different trials as follows.
   (a) **Farmers Initial Trials (FIT).** These trials are conducted as unreplicated with large plot size (12-15 m2). The number of FIT in each village depends on number of farmers willing to participate in the trial.

(b) **Farmers Advanced Trials (FAT).** The breeding materials selected from the-FIT are yield tested in FAT using local check. The plot size is 45 m2 to produce enough seed to plant the selected entries on larger plots in the third stage. The number of FAT in each village depends on the number of farmers willing to participate in the trial. In each village same set of entries is tested. The FAT are conducted in a variety of soils and managements.

(c) **Farmers Elite Trials.** The entries selected from the FAT are then tested in these trials with a plot size twice as large as the FAT (90 m^2). These trials are conducted using local check which may vary from village to village.

6. The best entry thus identified is released as a variety for specific areas by the variety release committee.
7. The users have a research role in all major stages of breeding and selection process. The farmers can give their input at any stage. Thus users become co-researchers.
8. Farmers and other users help in (*i*) setting over all goals, (*ii*) determining specific breeding priorities. They can make crosses, screen germplasm entries in the pre-adaptive phase of research, take charge of adaptive testing and lead seed multiplication and distribution process.
9. PPB is very much useful in developing crop cultivars for unfavourable conditions and remote areas.

Q. 3. What are types of participatory plant breeding ?

Ans. The participatory plant breeding is of two types, viz. (*i*) formal led PPB and (*ii*) farmer led PPB. These are discussed below:

1. Formal Led PPB

The breeding programmes initiated by research stations and later joined by farmers refer to formal led PPB. Researchers or specifically breeders run such programmes and invite farmer participation in formal research. The expenditure of conducting research is born by the research organization. The breeding objectives are also decided by researchers.

2. Farmer Led PPB

When the breeding is initiated by farmers and scientists have to support them in selection and seed maintenance, it is called farmer led PPB. In this, farmers decide their breeding objectives in consultation with researchers. The entire expenditure of conducting experiments is born by the farmers. The farmer led PPB has the objective to provide varieties or populations suitable for specific local environment.

Q. 4. What are the stages of participation ?

Ans. There are five stages of plant breeding. PPB may incorporate farmers input at various stages. These stages are given below:

1. Setting breeding objectives. Breeding objectives are decided for solving problems of common interest of the farmers.

2. Generating variation through Crossing. The new variability is created by crossing between selected parental lines.

3. Selecting in Segregating Populations. The selection is made in segregating populations to develop new varieties for specific situations.

4. Variety testing and characterization. The material selected from segregating populations is tested for yield performance and other economic characters in different trials and the best entry is identified for release.

5. Popularization and Seed Production. The new variety is popularised by conducting demonstrations. The seed multiplication is carried out to make available seed of new variety to the farmers of the region for which variety has been released.

Farmers input can be incorporated in any of the above five stages of breeding. It is not possible in traditional breeding programmes.

Q. 5. What are objectives of participatory plant breeding ?

Ans. The objectives or goals of PPB are decided jointly by researchers and farmers. The major goals of PPB are given below:

1. Enhancing biodiversity and germplasm conservation.
2. Increasing production and profitability through development and enhanced adoption of improved (suitable) varieties.
3. Providing benefits to a specific type of user or addressing the needs of a broad range of users.
4. Building farmers skills to enhance farmer selection and seed production efforts.

The participatory plant breeding programmes should be implemented under following situations.

1. Areas that are not dedicated to large scale crop production such as tribal areas and remote areas where traditional breeding schemes are not in operation.
2. Marginal crop production areas, where environments are highly variable so that selection based an G x E interaction is not the best selection strategy.
3. Areas where agriculture is risk prone, complex and require low input. For such situation heterogeneous breeding populations have to be developed.

4. Areas where crop end uses are diverse and/or locally unique.
5. Areas, where important major crops exist but are not the focus of formal plant breeding efforts.

Q. 6. Give a comparison of participatory varietal selection and PPB.

Ans. Comparison of farmer participatory varietal selection (PVS) and PPB is given in Table 34.1

Table 34.1 Comparison of PVS and PPB

Particulars	Participatory Varietal Selection	Participatory Plant Breeding
1. Intra-cultivar variability	No effect	Increases
2. Number of cultivars	There is smaller increse	There is large increase
3. Impact in increase in cultivar number over large areas	Uniform impact	More Impact in Participatory Villages.
4. Breeding Methods used	Mass selection	Hybridization techniques

Q. 7. What are defferences between participatory plant breeding and traditional plant breeding ?

Ans. Participatory plant breeding and traditional breeding differ from each other in several aspects (Table 34.2)

Table 34.2 Comparison of Participatory Plant Breeding and Conventional Plant Breeding

Particulars	Participatory Plant Breeding	Conventional Plant Breeding
1. Other terms used	Collaborative Plant breeding Farmer Participatory Breeding Participatory Crop Improvement	Classical plant breeding Traditional plant Breeding Formal plant breeding
2. Involvement of farmers	Yes	No
3. Incorporation of farmer input	Permitted in all stages	Not permitted
4. Testing of material	At research station and farmers fields	At research station only
5. Adaptation	Local adaptation	Wide adaptation

6. Biodiversity	Promoted	Reduced
7. Genetic base of varieties developed	Broad	Narrow
8. Varietal homogeneity	Low	High
9. Use of land races	More	Less
10. Germplasm evaluation	For local adaptation	For wide adaptation
11. Varieties Release Process	First adoption by farmers then release	First release then adoption by farmers
12. Varieties for low input	Developed	Rarely developed
13. Breeding Methods	Usually mass selection and bulk breeding	All breeding techniques
14. Development of Pure Line cultivars	Not taken up	Taken up

Q. 8. What are advantages of PPB ?

Ans. PPB has several advantages. Important advantages of PPB are briefly presented below:

1. PPB is most appropriate for developing cultivars for low input and high stress conditions.
2. PPB permits incorporation of farmers input at all the stages of plant breeding.
3. Farmers are able to decide which varieties better suit their needs and conditions.
4. PPB gears up the speed of developing new varieties for the farmers.
5. It leads to rapid adoption of the new varieties by the farmers. In PPB the new varieties are first adopted by the farmers and then released. In traditional breeding, first the variety is released and then adopted by the farmers.
6. The seed of new varieties is multiplied in sufficient quantity before release so that the seed can be made available to the farmers.
7. Varieties developed by PPB have greater biodiversity that can provide protection from biotic and abiotic stresses.
8. In PPB, varieties are released only after monitoring their acceptance by farmers through an initial adaptation. The seed of these trusted varieties is then multiplied.

9. In PPB, users have a research role in all major stages of breeding and selection.
10. Plant breeders can direct their research according to needs of the specific group of farmers (women, men, rich, poor). Socio-economic conditions of different people require different cultivars/hybrids.
11. The on farm research ensures that varieties will perform well under farmers fields (real life conditions). Moreover, on farm testing can be easily managed by breeders and farmers.
12. PPB is useful in increasing knowledge and skills of farmer. As a result farmers can participate more actively in such programmes.

Q. 9. What is the role of farmers in PPB ?

Ans. Farmers play very important role in PPB. They play key role in management, input and information supply and skill building. These are briefly discussed below:

1. Management Role

Farmers provide technical and social leadership in PPB. Farmers help in deciding what type of varieties are to be developed to suit specific environment and farmers need. For example, they help in developing suitable varieties for low input conditions (remote areas) and moisture deficit conditions. Farmers help in selection of representative sites for testing of breeding material on farmers fields. Farmers also provide information about varietal preferences, desirable traits and plant types required for cultivation in a participation area.

2. Input Supply Role

In PPB, farmers play imporlant role in making arrangement of various inputs for conducting plant breeding trials at their fields. They help in providing following inputs :

(i) Farmers provide representative sites (land) for testing of breeding material.

(ii) They help in arranging labour required for land preparation, sowing, intercultural .operations and harvesting of experimental material.

(iii) Farmers help in providing fertilizers, manures and insecticides at the village level.

(iv) They provide land races for use in further breeding work.

(v) They provide their input during selection process.

3. Skill building Role

Farmers also play key role in skill building. The breeders provide training to a group of farmers in the adopted village. Then farmers give this training to other farmars in the same village or adjacent villages and thus help in skill building.

Q. 10. What are the degrees of participation in PPB ?

Ans. In practice, three degrees of participation, viz.,

(i) consultative, *(ii)* collaborative and *(iii)* collegial are found. These are explained below:

***(i)* Consultative.** When information is obtained from farmers and other clients to start participatory plant breeding programmes, it refers to consultative degree of participation. In the global PPB programmes, the most frequently observed degree of participation is consultative. It is required in setting breeding goals.

***(ii)* Collaborative.** Collaborative participation means that there is task sharing between researchers and breeders as defined by the formal research programme. This is second most frequent type of participation observed in plant breeding programmes.

***(iii)* Collegial.** It means that researchers support a farmer initiated, farmer managed programme. Such programme is directly accountable to farmers and other client groups with a stake in the results of germplasm development.

SECTION-VI
SEED RELATED TOPICS

35

SEED PRODUCTION TECHNIQUES

Q. 1. What is seed ?

Ans. In broad sense, any plant part which is used for commercial multiplication of a crop is called seed. In strict sense, the product of fertilized orvule which consists of embryo, seed coat and cotyledon(s) is known as seed.

Q. 2. What is variety ?

Ans. A genotype released for commercial cultivation either by State Variety Release Committee (SVRC) or Central Variety Release Committee (CVRC) is called variety. A variety is notified by the Government of India.

Q. 3. What is improved seed ?

Ans. Seed of a released, notified and popular variety produced by scientific method is called improved seed.

Q. 4. What are advantages of improved seed ?

Ans. Improved seed results in :

(i) Better germination.
(ii) Vigorous seedling growth.
(iii) Higher crop stand.
(iv) Better quality of produce, and
(v) Higher crop yield.

Q. 5. What are types of improved seed ?

Ans. Improved seed is of following five types :

(i) Nucleus seed
(ii) Breeder seed
(iii) Foundation seed.
(iv) Registered seed
(v) Certified seed

Q. 6. What is nucleus seed ?

Ans. Initial seed of an improved variety which is always limited in quantity and produced by the originating plant breeder is called nucleus seed. Main features of nucleus seed are given below :

(i) It is produced at the research farms under the direct supervision of originating plant breeder.

(ii) It is genetically and physically cent per cent pure.

(iii) Certification is not required for nucleus seed.

(iv) It is used for production of breeder seed.

Q. 7. What is breeder seed ?

Ans. Progeny of nucleus seed or breeder seed produced under strict supervision of original plant breeder is known as breeder seed. Main features of breeder seed are given below :

(i) It is produced in isolation from other varieties.

(ii) It is genetically and physically pure.

(iii) Certification for breeder seed is not required. The breeder seed plot is inspected by the monitoring team consisting of original/sponsored breeder and one representative each from National Seeds Corporation and State Seed Certification Agency.

(iv) It is used for production of foundation seeds.

Q. 8. What is foundation seed ?

Ans. Progeny of breeder seed produced by National Seeds Corporation/ State Seeds Corporation is called foundation seed. Main features of foundation seed are given below :

(i) It is produced by NSC/SSC under the supervision of trained staff.

(ii) It is genetically pure and has physical purity of 100%.

(iii) Certification is done by State Seed Certification Agency.

(iv) It is used for the production of certified seed.

Q. 9. What is registered seed ?

Ans. It is the progeny of either foundation seed or registered seed. In India, this category of seed is omitted and certified seed is produced directly from the foundation seed. Main features of registered seed are given below :

(i) It is produced at the farms of progressive farmers under the supervision of National Seeds Corporation.

(ii) It is geneticallty pure.

(iii) It has physical purity of 98%.

(iv) Certification is done by State Seed Certification Agency.

(v) It is used for production of certified seeds.

Q. 10. What is certified seed ?

Ans. It is the progeny of either foundation or registered or certified seed. Its main features are as follows :

(i) It is produced on the fields of progressive farmers under the strict supervision of State Seed Certification Agency.

(ii) It is genetically pure.

(iii) It has physical purity of 98%.

(iv) Certification is done by State Seed Certification Agency.

(v) It is distributed/sold to the farmers for commercial crop production.

Q. 11. What is seed certification ?

Ans. A legal system which ensures production of high quality seed in terms of genetic purity and germination is called seed certification. The basic requirements of seed certification are given below :

(i) An improved variety

(ii) Genetic purity

(iii) Physical purity

(iv) High germination.

Q. 12. What is genetic purity ?

Ans. It refers to absence of seeds of other variety of the same crop as well as of other crops. The genetic purity is determined by grow out test. It differs from crop to crop. Generally, more contamination is allowed in cross pollinated species and hybrids than in self-pollinated species.

Q. 13. What is physical purity ?

Ans. The freedom from inert matter and defective seeds is known as physical purity. It varies from 95% to 98% in different crops.

Q. 14. What is inert matter ?

Ans. Inert matter includes non-living materials like sand, pebbles, soil particles, straw etc.

Q. 15. What are defective seeds ?

Ans. Seeds which are broken, disease infested, insect damaged, undeveloped and unfit for germination are called defective seeds.

Q. 16. What is germination ?

Ans. Emergency of normal seedlings from the seeds under ideal conditions of light, temperature, moisture, oxygen and nutrients is called germination.

Q. 17. What are components of seed based on germination ?

Ans. Based on germination, seed is divided into following five components :

(i) *Normal seedlings* : It includes seedlings with normal growth of root and shoot.

(ii) *Abnormal seedlings* : Seedlings which are unable to develop into normal plants.

(iii) *Hard seeds* : Seeds which do not absorb water.

(iv) *Fresh ungerminated seeds* : Viable seeds which can absorb water but do not germinate and remain fresh in the germination test.

(v) *Dead seeds* : Inviable seeds.

Q. 18. What is seed testing ?

Ans. Process of evaluation of seed in terms of purity and germination is called seed testing.

Q. 19. What is roguing ?

Ans. The process of removal of off types (phenotypically different) plants from the field of an improved variety to avoid contamination is known as roguing

Q. 20. What is isolation distance ?

Ans. Separation of the field of a variety from that of another variety of the same crop to prescribed standard distance to avoid contamination is called isolation distance. It differs from crop to crop. Generally, more isolation distance is required for cross pollinated species than self pollinated species.

Q. 21. What is growout test ?

Ans. The test which is conducted by growing the seed in the field for determining genetic purity is known as growout test.

Q. 22. What is seed viability test ?

Ans. The test which is conducted in the laboratory for determining the percentage of living seeds is called seed viability test.

Q. 23. What is seed viability ?

Ans. The living state of seeds is called seed viability.

Q. 24. What is seed health test ?

Ans. The test which is conducted to examine the presence or absence of pathogen in the seed is called seed health test.

Q. 25. What is true seed ?

Ans. The product of fertilized ovule which consists of embryo, seed coat and cotyledon(s) is called true seed.

Q. 26. What are off types ?

Ans. Phenotypically different plants which are found in the field of an improved variety are called off types.

Q. 27. What is seed science ?

Ans. A science which deals with various aspects of seeds such as development, structure, production, testing etc. is called seed science.

Q. 28. What is seed technology ?

Ans. A science which deals with production, processing, testing, packaging, storage and marketing of crop seeds is known as seed technology.

Q. 29. What is seed processing ?

Ans. Various operations related to seed such as drying, cleaning, grading, packaging refer to seed processing.

Q. 30. What is composite seed sample ?

Ans. A seed sample which is true representative of a seed lot is called composite seed sample.

Q. 31. What is working seed sample ?

Ans. A small seed sample which is drawn from the composite seed sample for seed testing purpose is called working seed sample.

Q. 32. What is seed production technology ?

Ans. A branch of plant breeding which deals with principles and methods of improved seed production is known as seed production technology.

Q. 33. What steps are involved in quality seed production ?

Ans. Quality seed production consists of following steps :

(i) Proper isolation distance.
(ii) Roguing at appropriate stages of crop growth.
(iii) Use of standard package of practices.
(iv) Adoption of recommended plant protection measures.

Q. 34. What is the objective of field inspection ?

Ans. In the seed crop, field inspections are made to examine :

(i) Isolation distance. (ii) Proportion of off types. (iii) Objectionable weeds. (iv) Incidence of diseases and insects, and (v) General crop health.

Q. 35. What is varietal deterioration ?

Ans. Permanent reduction either in the genetic or agronomic value of a variety is called varietal deterioration. Varietal purity can be maintained by keeping proper isolation distance and removing of off types.

Q. 36. What are causes of varietal deterioration ?

Ans. There are several causes of deterioration of a variety which can be divided into following two groups :

(*i*) Genetic causes : It includes :
(*a*) Natural out crossing.
(*b*) Spontaneous mutation.
(*c*) Residual variability at the time of release of a variety, and
(*d*) Adapitive changes.

(*ii*) *Non-genetic causes* : It includes :
(*a*) Mechanical admixture.
(*b*) Disease infestation, and
(*c*) Adverse agro-climatic conditions.

Q. 37. What are seed producing organizations in India ?

Ans. Following organizations are engaged in seed production in India :

(i) National Seeds Corporation (NSC)
(ii) State Seeds Corporaticns (SSCs)
(iii) State Seed Certification Agencies (SSCAs)
(iv) Private Seed Companies.

Q. 38. When was NSC established ?

Ans. The NSC was established in March 1963. However, it started functioning in July 1963. It is located in the Pusa Campus in New Delhi. It was established to take up the work of quality seed production and promote seed industry in the country.

Q. 39. What are functions of NSC ?

Ans. Main functions of NSC are given below :

(i) Production of quality seed (breeder, foundation, certified) of selected crops.
(ii) Interstate seed marketing of national varieties.
(iii) Impart short term training about seed production work.
(iv) It coordinates the work of certified seed production with State Seed Corporations.
(v) It provides consultancy about designing, procurement and installation of seed processing plants.
(vi) Takes up seed certification work in those States where State Seed Certification Agencies have not been established.
(vii) Also deals with import and export of seeds.

Q. 40. When was National Seed Project launched ?

Ans. The NSP was launched in 1976 to establish State Seed Corporations. Main functions of NSP are as follows :

(i) Establishment of New State Seed Corporations.
(ii) Strengthening of existing State Seed Corporations.
(iii) Strengthening facilities for breeder, foundation and certified seed production, processing and storage facilities at selected research Institute of ICAR and SAUs.

Q. 41. In which states SSCs have been established ?

Ans. State Seed Corporations have been established in the following states :

Uttar Pradesh	Punjab
Andhra Pradesh	Orissa
Bihar	Rajasthan
Haryana	Karnataka
Maharashtra	

Q. 42. What are States where State Seed Certification Agencies are available ?

Ans. State Seed Certification Agencies are available in following states :

Uttar Pradesh	Orissa	J & K
Andhra Pradesh	Rajasthan	Kerala
Bihar	Tamil Nadu	Madhya Pradesh
Haryana	Himachal Pradesh	Assam
Punjab	Delhi	
Maharashtra	West Bengal	
Karnataka	Gujarat	

Q. 43. What are various types of seed tests ?

Ans. Seed testing included following tests :

(i) Genetic purity test
(ii) Physical purity test
(iii) Germination test
(iv) Seed moisture test
(v) Seed viability test
(vi) Seed health test.

Q. 44. How will you test seed viability ?

Ans. Seed viability is tested with tetrazolium test. The seeds are first soaked in fresh water for 24 hours. Then seeds are split longitudinally into two haves and are dipped into 1% aqueous solution of tetrazolium chloride in the petridishes for 4 hours. Then these seeds are taken out and washed with fresh water and examined for colour. Only viable seeds take the stain and dead seeds remain colourless.

Q. 45. What is objectionable weed in wheat ?

Ans. In wheat, convolvulus arvensis (Hiran Khuri) is objectionable weed.

Q. 46. What is objectionable weed in rapeseed and mustard ?

Ans. In rapeseed and mustard, Argemone maxicana is objectionable weed.

Q. 47. What is objectionable weed in rice ?

Ans. In rice, wild rice is objectionable weed.

Q. 48. What is objectionable weed in Okra ?

Ans. In okra, wild okra is objectionable weed.

Q. 49. What is objectionable weed in Lucern ?

Ans. In lucern, *Cuscuta* spp is objectionable weed.

Q. 50. What is objectional weed in Methi ?

Ans. In Methi, Melilotus spp is objectionable weed.

Q. 51. What is full form of ISTA ?

Ans. ISTA–refers to International Seed Testing Association.

Q. 52. Compare UPOV acts of 1978 and 1991.

Ans. Comparison of UPOV Acts 1978 and 1991 is presented below in Table 35.1 :

TABLE 35.1 Brief comparison of UPOV Acts of 1978 and 1991

Particulars	UPOV Act, 1978	UPOV Act, 1991
1. Criteria for Protection 2. Authorization of Rights 3. Rights provided to the breeder 4. Breeders Privilege 5. Farmers Privilege 6. Period of Protection (*a*) Field crops (*b*) Forest, Fruit and ornametal trees, shrubs and vines,	1. Distinctness 2. Uniformity 3. Stability Admissible Commercial production, marketing and export of his variety Provided Provided Fifteen years Eighteen years,	1. Novelty (newness) 2. Distinctness 3. Uniformity 4. Stability Admissible Same as in UPOV Act,, 1978 Not provided Not provided Twenty years Twenty five years

36

MAINTENANCE BREEDING

(VARIETAL MAINTENANCE)

Q.1. What is maintenance breeding ?

Ans. A branch/area of plant breeding which deals with principles and methods of breeder seed production is referred to as maintenance breeding. In other words, it deals with ways and means of maintaining genetic and physical purity of released and notified varieties and parents of hybrid. It is also referred to as varietal maintenance technology.

Q.2. What are main features of maintenance breeding ?

Ans. Thus main points related to maintenance breeding are listed below.

1. It deals with principles and methods of pure seed production.
2. It deals with continuous breeder seed production of released and notified varieties. In other words, it leads to fresh breeder seed production every year.
3. It also undertakes breeder seed production of the parental lines of released hybrids.
4. Genetic purity, physical purity and germination are the main points taken into account.
5. The seed health is also taken into account.
6. The breeder and foundation seed is used as base material for starting maintenance breeding programme.
7. The main objective of maintenance breeding is to enhance the life span of released and notified varieties and commercial hybrids.
8. Its entire focus is on the planting seed quality of a crop species rather than milling seed quality.
9. It prevents varietal deterioration

Q.3. What are genetic causes of varietal deterioration ?

Ans. Genetic causes include *(i)* cross pollination, *(ii)* mutations, *(iii)* residual variability, and *(iv)* adaptive changes. These are briefly discussed below:

(i) Natural cross pollination or outcrossing with other genotypes leads to genetic deterioration of a variety in subsequent cycles of seed multiplication.

(ii) Spontaneous mutations after crossing with the original variety lead to its deterioration in subsequent years.

(iii) Sometimes, residual variability remains in a variety at the time of its release. Such variability leads to deterioration of a variety in few years.

(iv) Sometimes shift in gene frequency either due to natural selection or artificial selection leads to varietal deterioration.

Q.4. What are non-genetic causes of varietal deterioration ?

Ans. Non-genetic causes of varietal deterioration include (i) mechanical mixture, *(ii)* disease infestation, and *(iii)* adverse agroclimatic conditions. These are briefly discussed below:

(i) Mixing of different varieties may take place during thrashing, ginning, sowing with seed drills. Planting of such mixed seed leads to varietal deterioration due to multiplication and natural outcrossing.

(ii) Development of new virulent races of a pathogen may lead to deterioration of a variety. The variety may become susceptible to such races and deteriorate.

(iii) Adverse agro-climatic conditions such as flood, drought, soil salinity etc. may lead to deterioration of a variety.

Q.5. What are important activities of plant breeding ?

Ans. Plant breeding has three important activities, viz., *(i)* varietal development, *(ii)* seed multiplication, and *(iii)* varietal maintenance. These are defined below:

(i) Varietal Development. It makes use of various plant breeding methods (selection, hybridization, mutation and genetic engineering) to develop new plant varieties.

(ii) Seed Multiplication. It deals with multiplication/production of successive generations of various categories of seeds such as nucleus seed, breeder seed, foundation seed, registered seed and certified seed. The seed multiplication/increase is regulated by seed certification agencies and breeders.

(iii) Varietal Maintenance. It deals with continuous fresh supply (annually) of breeder seed which used to start another cycle of seed multiplication. It makes use of various selection/maintenance procedures.

Q.6. What are types of maintenance procedures ?

Ans. Maintenance procedures are of two types, viz., *(i)* selection procedures and *(ii)* non-selection procedure. The second approach deals with maintenance of breeder seed in cold storage. Merits and demerits of these methods are given below:

1. Selection Procedures

(i) It involves continuous extensive field testing.

(ii) The material is exposed to outcrossing, spontaneous mutations and change in gene frequencies.

2. Storage Methods

(i) Field testing is not required.

(ii) The seed material is not exposed to outcrossing, mutations, diseases, insects, mechanical mixtures, change in gene frequency caused due to natural selection and selection exerted by breeders.

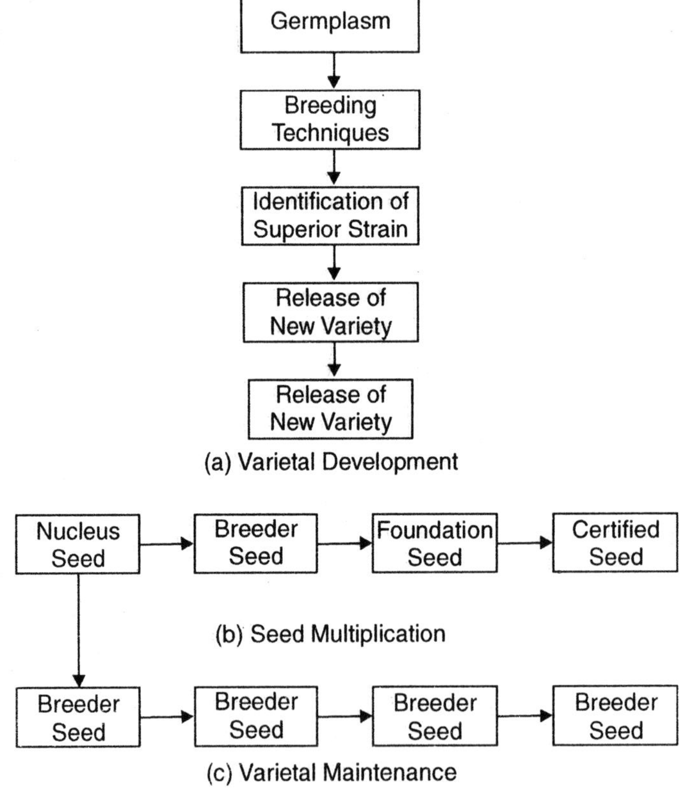

Fig. 36.1 Plant Breeding Activities

Q.7. What breeding approaches are used for varietal maintenance ?

Ans. In crop plants various breeding methods/approaches are used for maintaining the genetic purity of cultivars and hybrids. In crop plants four methods, viz. *(i)* positive mass selection, *(ii)* negative mass selection, *(iii)* model bulk selection, and *(iv)* plant to row selection used for maintaining varietal purity.

Q.8. Describe the role of positive mass selection in varietal maintenance ?

Ans. Selection of desirable plants (true to type) from the field of a variety and mixing the seed of individual selected plants to grow next generation is referred to as positive mass selection. This is also known as bulk of large number of true to type plants. This method is widely used for varietal purification in various field crops. In cotton, this method was first used by Menbane in Texas, USA. Now this method is widely used for maintenance of varietal purity in cotton. Merits and demerits of this method are listed below.

Merits

(i) It is a simple method of varietal maintenance.

(ii) It is used for purification of improved varieties.

(iii) It can also be used for purification of parental lines of hybrids.

(iv) It retains diversity in the variety.

Demerits

(i) Selection is based on phenotype which may not always be an indication of superior genotypes.

(ii) The progeny test is not carried out.

Q.9. What is the role of Negative Mass Selection in varietal maintenance ?

Ans. When only undesirable off type plants are removed from the field of a variety and rest are allowed to grow further, it is known as negative mass selection. It is also known as roguing of off type plants. It is widely used for maintaining varietal purity in various field crops. In cotton, it has been found to be a satisfactory maintenance procedure. This is a simple method. Selection is based on phenotype and progeny test is not conducted. The main advantages of this method are presented below.

Merits

1. It is a simple method of varietal maintenance and is widely used for maintenance of varietal purity in different crops.
2. The maintenance procedure is easy and requires little skill.
3. The number of discarded plants are very few in proportion to plants retained.
4. It can be used even for maintenance of inbred lines used for developing hybrids.
5. The genetic diversity of the variety is not affected.
6. It is used both in self pollinated and cross pollinated species.

Demerits

1. The progeny test is not conducted.
2. Sometimes, off type plants may escape which may lead to contamination of the variety in later cycles of seed multiplication.
3. It requires careful examination of offtype plants.

Q. 10. Give comparison of positive and negative mass selections.

Ans. A Brief comparison of positive mass selection and Negative mass selection is presented below:

Table 36.1 Comparison of positive and negative mass selections

S.No.	Particulars	Positive Mass Selection	Negative Mass Selection
1.	Selection Procedure	Selection of True to the type plants	Removal of off type plant
2.	Basis of Selection	Phenotype	Phenotype
3.	Progeny test	No conducted	Not conducted
4.	Applicable to	Both self and cross pollinated species	Pollinated species
5.	Genetic diversity	Retained	Retained
6.	Category of seed used	Nucleus/breeder seed	Nucleus/breeder seed
7.	Plants retained and bulked	Selected plants only	All plants after roguing.

Q.11. Describe role of Model Bulk Selection in maintenance breeding.

Ans. This method was first used by Manning in 1955. According to this method large number of single plants (300-500) which are productive and true to the type are selected in a variety. Then these plants are examined under laboratory conditions for three parameters, viz. fibre length, lint per seed and seed weight. Some additional characters which were not used by Manning (1955) such as lint index and ginning out turn can also be included. Those plants which conform to the mean of a vaiety for above three parameters are selected and their seed is bulked to grow new generation. Rest of the plants are discarded. This method is widely used for maintenance breeding. Merits and demerits of this method are given below.

Merits

(i) This is a simple method of varietal purification.
(ii) This is more scientific than earlier two methods.
(iii) It is based on both field and laboratory observations.
(iv) It used in cotton and some other crops.

Demerits

(i) Large population has to be grown for selection.
(ii) Selection is based on phenotype.
(iii) It requires more labour, space and time.

Q.12. Describe role of Plants to Row method in maintenance breeding.

Ans. This method of maintenance breeding is based on progeny test. It is also known as progeny row selection. In this method large number of single plants (500-1000) true to the type are selected from the parent variety. The selected plants are screened for fibre length and lint percentage in the laboratory. Plants conforming to the norms of a variety are selected and rest are discarded. In the next crop season, progeny rows of selected plants are grown. Those progenies which fulfill the norms of the variety are bulked and rest are discarded. Several modifications of this scheme have been suggested by various researchers as follows.

1. Some researchers are of the opinion that progeny test should be conducted atleast for one year to assess the performance of the variety.
2. Others are of the opinion that one year progeny test is not sufficient. Progeny test should be conducted for two, three or more seasons. When several years progeny testing is carried out, it is referred to as a pedigree system or pedigree method of selection. The merits and demerits of plant to row method are given below:

Merits

(i) The selection is based on breeding value *i.e.* progeny performance.

(ii) It is more scientific method than earlier three methods.

(iii) It involves both field and laboratory evaluation.

(iv) Progeny tests are based on large samples which provide more precession.

Demerits

(i) It is more laborious and time consuming method.

(ii) It requires more experimental area for evaluation and testing.

(iii) It requires two or more crop seasons for testing.

Q.13. What are advantages of maintenance breeding ?

Ans. There are several advantages of maintenance breeding which are briefly presented below :

1. It helps in purification of improved cultivars and parental lines of hybrids.
2. It prevents cultivars from genetic deterioration and thus prolongs life of a variety.
3. It is useful in studying the efficacy of various maintenance procedures.
4. It helps in quality seed production which in turn leads to higher crop yields.

Q.14. What are limitations of maintenance breeding ?

Ans. There are some limitations of maintenance breeding which are briefly presented below.

1. Some of the maintenance procedures require lots of experimental material for evaluation.
2. Large number of single plants have to be evaluated in terms of agronomic performance and quality parameters. Hence only limited number of cultivars can be handled at a time.
3. Maintenance procedures are used for varietal purification. Hence chances of evolving new varieties through such procedures are rare.
4. Progeny row method require more time (2-3 crop seasons) for evaluation of purity of a variety .
5. Most of the testing procedures are based on phenotypic performance only.
6. In positive mass selection major portion of the plants are rejected which adds to the expenditure.

37

INDIAN SEEDS ACT 2004

Q. 1. What is Indian Seed Act 2004 ?

Ans. The Indian Seeds Bill 2004 hereinafter called Indian Seeds Act 2004 was approved by the government of India in 2004 and came into force with effect from January 2005. The main objective of the Seeds Bill, 2004 is to ensure availability of quality seeds to farmers. It replaces the Seeds Act, 1966.

Q. 2. What the scope of Indian seeds Act 2004 ?

Ans. This Act has wider coverage than seeds act of 1966. It includes agriculture, horticulture, forestry, plantation crops, medicinal and aromatic plants. The seeds of all these crops are covered by the new act.

Q. 3. What category of seeds permitted by the seeds Act 2004 ?

Ans. This act permits sale of certified seed only. The name of variety, physical purity and germination percentage have to be indicated on the seed container or bag.

Q. 4. Is license compulsory ?

Ans. The seed dealers, sellers and growers should have license which is compulsory.

Q. 5. Is registration compulsory ?

Ans. Registration for all kinds and varieties of seed is compulsory. No producer can grow seeds unless he is registered. It is compulsory for all cultivated varieties of field crops, vegetable crops, fruit crops, medicinal and aromatic plants, forest species and plantation crops. Every seed producer and dealer, and horticulture nursery has to be registered with the state government.

Q. 6. What are the Criteria for Registration ?

Ans. The released and notified varieties can be registered for seed multiplication. There are four criteria for protection of varieties under PPV & FR Act. These are novelty, distinctiveness, uniformity and stability.

Q. 7. What is the period of registration ?

Ans. The Indian Seeds Act 2004 has adopted UPOV Act 1978. According to this Act, the period of protection is 15 years for annual and biennial crops and 18 years for perennial plants such as trees and vines.

Q. 8. How the Certification is done ?

Ans. The Act permits self certification of seeds by accredited agencies and allows the central government to recognize certification by foreign seed certification agencies. Seed producers are permitted to self certify the performance of their seed under certain conditions. The seed companies need to provide the results of multiplication trials before registration.

Q. 9. What is the procedure of claiming Compensation ?

Ans. The disputes about the quality of seeds have to be settled in the consumer court. Any loss in the production due to poor seed quality is claimed in the consumer court.

Q. 10. What are the Farmers' Privilege in the seeds Act 2004 ?

Ans. The Seeds Act 2004 allows farmers' privilege. The Bill does not restrict the farmer's right to use or sell his farm seeds and planting material, provided he does not sell them under a brand name. All seeds and planting material sold by farmers will have to conform to the minimum standards of germination, physical purity and genetic purity applicable to registered seeds.

Q. 11. What are the Penalties for violation of the Act 2004 ?

Ans. Any person who contravenes any provisions of the Act or imports, sells or stocks seeds deemed to be misbranded or not registered, can be punishable by a fine between Rs 5,000 and Rs 25,000. The penalty for giving false information is a prison term up to six months and/or a fine up to Rs 50,000. In case of companies, the person(s) in charge of conduct of the business of the company will be held accountable: .

Q.12. What are the technologies which have been Excluded in the seeds Act 2004 ?

Ans. Those techniques which are harmful to animals and the environment such as terminator technology and traitor technology are excluded and not allowed by the seed act.

Q.13. What Research Exemption have been provided in the seeds Act 2004 ?

Ans. This act allows use of protected material by breeders for development of new varieties.

Q.14. What is the structure of Central seed Committee in the seeds Act 2004 ?

Ans. In the Seed Act 2004, the central seed committee consists of Chairman (Secretary DARE), Agricultural Commissioner, DDG horticulture (ICAR), Joint Secretary Seeds(DAC), representatives from DBT, Ministry of Environment and Forest, Secretary Agriculture (from five States), Director (SSCA) from one State and MD-SSC from one State. In addition, two representatives each of farmers and seed industry are nominated.

Q.15. What are the differences between Seeds Act 1966 and Seeds Act 2004 ?

There are several differences between seeds act 1966 and seeds Act 2004 which are presented in Table 37.1

Table 37.1 Differences between Seed Act 1966 and Seed Act 2004

S.No.	Particulars	Seeds Act 1966	Seeds Act 2004
1.	Crop covered	Agriculture and horticulture	Agriculture, horticulture, forestry, plantation crops, medicinal and aromatic plants.
2.	Registration of transgenic varieties	No provision	Special provision
3.	Registration with PPVFR authority	Not required	Required
4.	Period of protection	Not defined	Defined
5.	Penalty for violation of Act	Rs. 100-1000/ and 6 months imprisonment	Rs. 500-50,000/ and six months imprisonment.
6.	Self Certification	Not permitted	Permitted
7.	Seed multiplication of land Races	Permitted	Not permitted
8.	Representatives in central seed committee	From all States	From five States only
9.	Involvement of Private Sector	No	Yes

Q.16. What are the similarities between Seeds Act 1966 and Seeds Act 2004 ?

Ans. There are several similarities between seeds act 1966 and seeds Act 2004 which are presented in Table 37.2.

TABLE 37.2 Similarities between Seed Act 1966 and Seed Act 2004

S.No.	Particulars	Seeds Act 1996	Seeds Act 2004
1.	Category of seed permitted for sale	Certified Seed	Certified Seed
2.	License	Compulsory	Compulsory
3.	Certification	Compulsory	Compulsory
4.	Farmers' exemptions	Available	Available
5.	Purity and germination standards	Adopted	Adopted

6.	Compensation of loss by use of certified seed	Through consumer court	Through consumer court
7.	Requirements for seed multiplication	Released and notified varieties	All seeds for sale must be registered.

Q.17. What are the differences between Seeds Act 2004 and PPV & FR Act 2001 ?

Ans. There are several differences between Seed Act 2004 and Protection of Plant Variety and Farmers' Rights Act 2001 which are presented in Table 37.3

Table 37.3 Differences between Seeds Act 2004 and Protection of Plant Varieties and Farmers' Rights Act 2001

SN	Particulars	Seeds Act 2004	PPV & FR Act 2001
1.	Record Maintained	National Register of Seeds	National register of plant varieties
2.	Compensation	Through consumer court	Through PPV & FR Authority
3.	Disclosure of variety's parentage	Not required	Required
4.	Variety's performance to be declared by	Seed producer, distributor or vendor	Breeder of the variety
5.	Penalty for violation of the Act.	Rs. 5000-50,000/	Rs. 50,000-20 lakh
6.	Imprisonment for violation of Act	Up to six months	From 2-3 years

Q. 18. What are the similarities between Seeds Act 2004 and PPV & FR Act 2001 ?

Ans. There are several similarities between Seed Act 2004 and Protection of Plant Variety and Farmers' Rights Act 2001 which are presented in Table 37.4 .

Table 37.4 Similarities between Seeds Act 2004 and Protection of Plant Varieties and Farmers' Rights Act 2001

SN	Particulars	Seeds Act 2004	PPV & FR Act 2001
1.	Registration	Compulsory	Compulsory
2.	Period of registration	15 years for annual & biennial crops & 18 years for perennials.	15 years for annual & biennial crops, 18 years for perennials.
3.	Renewal of registration	Permitted for similar period.	Permitted for similar period.
4.	Farmer Rights	Allowed	Allowed

5.	Plant breeders' Rights	Available	Available
6.	Disclosure of variety's performance	Compulsory	Compulsory
7.	Sale of branded seeds by farmers	Not permitted	Not permitted

Q.19. What are the Merits of Seed Act 2004 ?

Ans. There are several merits of the seeds Act 2004 which are briefly discussed as follows.

1. Wider Coverage. This act has wider coverage (agriculture, horticulture, forestry, plantation crops, medicinal and aromatic plants) than Seed Act of 1966.

2. Easy Certification. This act has made the process of certification very simple means allowed self certification. But it should not be misused.

3. Bans Harmful Technologies. The act did not permit registration of harmful technologies such as terminator technology and traitor technology .

4. Permits Farmers' Rights. It permits farmers' rights. The farmers need not to register their varieties.

5. Ensures Quality Seed. This act ensures availability of good quality seeds to farmers by imposing heavy penalties on the sale of spurious seeds.

6. Registration for long Duration. This act provides registration for 15 years for annual and biennial crops and 18 years for perennial crops which is sufficiently long period. Moreover, the registration is renewable for another term of similar duration.

7. Central Seed Committee. In the Seed Act 2004, the constitution of central seed committee is better than that of Seed Act 1966.

8. Promotes Seed Industry. The new seed act will lead to fast development of seed industry due to involvement of representatives of multinational seed companies or industries in the central seed committee.

9. Increase in Productivity. The new act will help in increasing crop productivity due to availability of better quality seeds.

10. The present seed replacement rate is 20-25 %. The seed act will enhance seed replacement rate of various crops.

11. The seed act will boost the export of seeds and encourage import of useful germplasm for use in crop improvement programmes.

12. It will encourage investment in seed research and development.

Q. 20. What are the Demerits of Seed Act 2004 ?

Ans. There are some demerits of the seeds Act 2004 which are briefly discussed as follows.

1. Although farmers are exempt from registering their seed varieties, the seeds have to conform to standards prescribed for commercial seeds. Farmers may find it difficult to adhere to the standards required of commercially sold seeds, because the seed saved and exchanged by farmers constitute above 80% of the seeds planted.
2. It is not clear whether the compensation would include the value of the crop or only the cost of the seed.
3. It is not clear whether a seed producer may sell seed which is registered by a different producer .
4. The disclosure of parentage is not essential. It may lead to a situation where seeds can be registered without disclosing the parentage or origin of the seed.
5. The Act does not have the provision of benefit sharing unlike the Convention on Biological Diversity and the PPVFR Act.
6. The Seed Act does not provide for a mechanism to trace back a packet of seed to the dealer, processor and producer. This would make it difficult to trace back a defective lot, and rectify any deficiencies in the supply chain.
7. Seed producers would be permitted to self-certify the performance of their seeds under certain conditions. This opens up the possibility of false declaration by seed companies. To prevent this, only government agencies should be allowed to conduct these trials and grant certification.
8. Every horticultural nursery has to be registered with the state government and has to maintain records of layout plan, source of every planting material etc. Such nurseries in the unorganized sector may find it difficult to adhere to these conditions.
9. The Seed Inspector has the power to enter and search as well as break open container or break open doors. without a warrant. This is against the normal procedure.
10. The new seeds act will encourage cultivation of new varieties. Thus old varieties and land races will be replaced by new varieties on vast areas resulting in significant reduction in biodiversity. Reduction in biodiversity will invite danger of uniformity and lead to narrow genetic base and narrow adaptation.
11. It will encourage multinational seed companies which may lead to monopoly of such companies. Seed companies may hike seed prices resulting in exploitation of farmers.

38

PROTECTION OF PLANT VARIETIES AND FARMERS' RIGHTS ACT 2001

Q. 1. What is Protection of Plant Varieties and farmers' Rights Act 2001 ?

Ans. The Protection of Plant Varieties and Farmers' Rights Act 2001 was approved by the government of India in 2004 but came into force with effect from January, 2005. The main objective of this act is to encourage development of new plant varieties and promote development of seed industry in India.

Q. 2. What are the objectives of Protection of Plant Varieties and farmers' Rights Act 2001 ?

Ans. There are four main objectives of enforcing the Protection of Plant varieties & Farmers' Rights Act 2001 as given below.

(i) To establish an effective system for protection of plant varieties, farmers' rights and plant breeders' rights.

(ii) To recognize and protect farmers' rights for their contribution in conserving, improving and making available plant genetic resources for development of new plant varieties.

(iii) To stimulate investment for research and development both public and private sector for development of new plant varieties.

(iv) To promote the growth of seed industry in the country to ensure availability of high quality seeds to the farmers.

Q. 3. What is the Legislative Structure of PPV & FR Act 2001 ?

Ans. The Authority consists of a chair person called as authority and 15 persons as members. The members of the Authority include (i) Agricultural Commissioner- GOI, (ii) DDG Crop Sciences-ICAR, (iii) Joint Secretary Seeds-GOI, (iv) Horticulture Commissioner-GOI, (v) Director, NBPGR, (vi) one member each (not below the rank of Joint Secretary) from DBT, (vii) Ministry of Environment and forests-GOI, (viii) Ministry of Law Justice and Company Affairs-GOI; (IX) One representative each (nominat by central government from farmers, (x) tribal organization, (xi) seed industry, (xii) an agricultural university, (xiii) and women organization; and (xiv) two nominated representative of state government.

Q. 4. What is the Headquarters of PPV & FR Act Authority ?

Ans. The headquarters of Protection of Plant Varieties and Farmers' Rights Authority is located in Pusa Campus (National Agricultural Science Complex), New Delhi.

Q. 5. What are the Plant Material Protected by PPV & FR Act 2001 ?

Ans. Various types 'of crop cultivars can be protected under PPV & FR Act. The following breeding material can be protected.

(i) All new varieties of self pollinated species such pure line and multi line varieties.

(ii) New varieties of cross pollinated species such as open pollinated variety, synthetics and composites.

(iii) New varieties of vegetatively/clonally propagated species.

(iv) Single, three-way and double cross hybrids.

(v) Inbred parental lines of hybrid varieties.

(vi) Trees, fruits, ornamental plants and vines, etc..

Q. 6. What are the requirements for Registration of new variety under PPV & FR Act 2001 ?

Ans. The PPV & FR Act Authority establishes a Plant Varieties Registry, which maintains a National Register of Plant Varieties. Specifies details under which a variety may be registered such as a complete passport data of the parental lines from which a variety has been derived have to be furnished by the breeder. A new variety will be registered under the act if it conforms to the criteria of novelty , distinctiveness, uniformity and stability.

(i) Novelty. It refers to newness of a variety. A variety which has not been grown for more than one year prior the application for registration can be protected.

(ii) Distinctiveness. The variety should be clearly distinguishable fro previously available varieties at-least for one characteristics.

(iii) Uniformity. It refers to same type of population of a variety. In cross pollinated species, the range of variation should be within the tolerable limit.

(iv) Stability. It means that the variety should give stable performance after repeated reproduction or propagation. Moreover; a variety should have a designation.

For registration of a variety , an application has to be filed with the Registrar, PPV & FR Act. No fee is charged if a variety is registered by a farmer.

Q. 7. What the duration of Protection under PPV & FR Act 2001 ?

Ans. The PPV & FR Authority has accepted UPOV Act 1978. According to this

Act, the period of protection is 15 years for annual and biennial crops and 18 years for perennial plants such as trees and vines. The registration is renewable after its expiry for a similar term of duration.

Q. 8. What Farmers' Rights are provided by PPV & FR Act 2001 ?

Ans. A farmer is entitled to save, use, sow, resow, exchange, share or sell his farm produce including seed of a variety protected under the Act in the same manner before this Act came into force. He cannot sell branded seed of a variety protected under the Act.

Q. 9. What Plant Breeders' Rights have been provided by PPV & FR Act 2001 ?

Ans. A certificate of registration for a variety issued by this act shall confer an exclusive right on the breeder or his successor, his agent or licensee, to produce, sell, market, distribute, import or export seed of his variety .The breeder has right to authorize and other person for production and marketing of his variety .

Q. 10. What are the technologies which have been Excluded in PPV & FR Act 2001 ?

Ans. Those techniques which are harmful to animals and the environment such as terminator technology and traitor technology are excluded and not allowed by the Protection of Plant Varieties and farmers' Rights Act 2001.

Q. 11. How the Compensation is claimed if a variety does not perform as declared ?

Ans. If a breeder of a propagating material of a variety registered under the Act sells his product to a farmer, he has to disclose the expected performance under given conditions. If the propagating material fails to perform, the farmer can claim compensation in the prescribed manner before the Protection of Plant Varieties and Farmers' Rights Authority.

Q. 12. What are the Penalties for violation of the PPV & FR Act 2001 ?

Ans. Penalty for applying false denomination to a variety is imprisonment up to two years and/ or a fine between Rs 50,000 and Rs. five lakh. Penalty for falsely representing a variety as registered is imprisonment up to three years and/ or a fine between Rs. one lakh and Rs. five lakh or both. Penalty for subsequent offence

Q. 13. What are the similarities between Seed Act 2004 and PPV & FR Act 2001?

Ans. There are some similarities between Seed Act 2004 and Protection of Plant Variety and Farmers' Rights Act 2001 which are presented in Table 38.1

Table 38.1 Similarities between Seeds Act 2004 and Protection of Plant Varieties and Farmers' Rights Act 2001

SN	Particulars	Seeds Act 2004	PPV & FR Act 2001
1.	Registration	Compulsory	Compulsory

2.	Period of registration	15 years for annual & biennial crops, 18 years for perennials.	15 years for annual & biennial crops, 18 years for perennials.
3.	Renewal of registration	Permitted for similar Period	Permitted for similar Period
4.	Farmers Rights	Allowed	Allowed
5.	Plant breeders' Rights	Available	Available
6.	Disclosure of variety's performance	Compulsory	Compulsory
7.	Sale of branded seeds by farmers	Not permitted	Not permitted
8.	Involvement of Private Sector	Yes	Yes

Q.14. What are the differences between Seed Act 2004 and PPV & FR Act 2001?

Ans. There are several differences between Seed Act 2004 and Protection of' Plant Variety and Farmers' Rights Act 2001 which are presented in Table 38.2

Table 38.2 Differences between Seeds Act 2004 and Protection of Plant Varieties and Farmers' Rights Act 2001

SN	Particulars	Seed Act 2004	PPV & FR Act 2001
1.	Record maintained	National Register of Seeds	National register of plant varieties
2.	Compensation	Through consumer court	Through PPV & FR Authority
3.	Disclosure of variety's parentage	Not required	Required
4.	Variety's performance to be declared by	Seed producer, distributor or vendor	Breeder of the variety
5.	Penalty for violation of the Act	Rs. 5000-50,000/	Rs. 50,000-20 lakh
6.	Imprisonment for violation of the Act	Up to six months	From 2-3 years

Q.15. What are the Merits of PPV & FR Act 2001 ?

Ans. There are several advantage of Protection of Plant varieties and Farmers' Rights Act 2001 which are briefly presented below.

1. It provides an effective system for protection of plant varieties, farmers' rights and plant breeders' rights.
2. It recognizes the contribution of farmers in conserving, improving and making available plant genetic resources for development of new plant varieties.

3. It provides rights to farmers to save, use, sow, resow, exchange, share or sell his farm produce including seed of a variety protected by PPV & FR Act.
4. It confers an exclusive right on the breeder or his successor, his agent or licensee, to produce, sell, market, distribute, import or export seed of his variety. Thus it gives recognition to breeders for their contribution and encourage them to breed still better varieties.
5. It encourages investment in plant breeding research and development both by public and private sectors for development of superior varieties. It will lead to fast development of seed industry and improvement in seed quality due to competitions among breeders.
6. It promotes the growth of seed industry in the country to ensure availability of high quality seeds to the farmers.
7. It involves representatives of farmers, women and seed industry in its legislative structure who will look to the interest of these sectors.
8. The Act has provision for compensation. If the seed gives poor performance than the declared by the breeder, farmer can claim compensation in the prescribed manner before the Protection of Plant Varieties and Famlers' Rights Authority .
9. The Act has provision for rewarding farmers for their role in conserving plant genetic resources. Such expenditure can be met out from the gene fund available with the PPV & FR Act Authority.
10. The act has provision of benefit sharing. The farmers can get share from the benefits arising from cultivation of farmers varieties and knowledge.
11. The act has provision to impose penalties to curb unhealthy practices.

Q.16. What are the Demerits of PPV & FR Act 2001 ?

Ans. There are some demerits of Protection of Plant varieties and Farmers' Rights Act 2001 which are briefly presented below.

1. The PPV& FR Act will promote cultivation of new high yielding varieties which are highly uniform. This will result in reduction in genetic diversity leading to narrow genetic base and danger of uniformity .
2. This will encourage monopoly of some breeders or seed companies for genetic material with special traits.
3. The holders of plant breeders rights will produce less seed than market demand and hike price to earn more profit. The high price can not be afforded by small and marginal farmers. Thus all farmers can not take advantage of new varieties.
4. It will inhibit free exchange of seed material and encourage unlawful practice.

39

PLANT BREEDERS RIGHTS

Q. 1. What are Plant Breeders' Rights ?

Ans. The Plant Breeders' Rights programme was first established in 1987 under the Plant Variety Rights Act (PVRA), which was succeeded by the current Plant Breeders' Rights Act 1994. (PBRA).Plant breeders' rights (PBR), also known as plant variety rights (PVR), are intellectual property rights granted to the breeder of a new variety of plant. Plant Breeders Rights are granted to novel plant varieties that are distinctive, uniform, and stable (e.g, cultivars breed true-to-type for desired traits). The legal protection of a new plant variety is granted to the breeder or his successor. The effect of PBR is that prior authorization is required before the material can be used for commercial purposes.

Q. 2. What is the procedure for Registration of Plant Breeders' Rights ?

Ans. Registration is essential to get legal rights. There is no protection without registration. The registration consists of following steps:

(i) Filing application in the prescribed form,

(ii) Payment of prescribed processing fee along with application. The registration fee varies from country to country. In USA, after registration annual renewal fee (300 Dollars) is payable.

(iii) Examination of the application by the experts, and

(iv) Issue of protection certificate, if application if found to meet the desired requirements.

(v) Normally the registration takes about 2.5 years for most of the species takes 2.5 years.

Q. 3. What is the Duration for protection of Plant Breeders' Rights ?

Ans. The period of protection varies with plant species. The protection is provided for a definite period. The duration of protection for different plant species is as follows':

(i) **For Trees and Vines:** Initially for nine years and maximum for 18 years.

(ii) **For extant varieties:** Initially six years and maximum for 15 years fro the date registration of the variety .

(iii) **In other cases:** Initially six years and maximum for 15 years fro the date registration of the variety .

Thus breeders' rights allow plant breeders to get benefit of their variety for a limited period. When the PER period is over, the variety reverts to the public domain and is available to every body. The limited duration of Plant Breeders' Rights ensure a balance between private and public interest.

Q. 4. What is the Validity of PBRs ?

Ans. The protection of the Plant Breeders' Rights is valid only in the country where it has been registered. The protection in other countries can be obtained by filling separate application in each country .

Q. 5. What are the Matters Covered by Plant Breeders' Rights ?

Ans. The protection right can be granted for varieties of all botanical genera and species. The variety should have a designation (name) as per the rule of International Code of Nomenclature.

Q. 6. What are the Requirements for protection of Plant Breeders' Rights ?

Ans. There are four basic requirements for protection of a variety under PBR, viz., (i) novelty , (ii) distinctiveness, (iii) uniformity, and (iv) stability. See later.

Q. 7. Is it possible to Transfer Plant Breeders' Rights ?

Ans. An authorized plant breeder has right to authorize other interested persons for commercial production and marketing of his variety on royalty basis.

Q. 8. Who has Control over the Plant Breeders' Rights ?

Ans. The authorized plant breeder has right to prevent other from commercial production and marketing of his variety without permission.

Q. 9. When does the Plant Breeders' Rights become effective ?

Ans. The Plant Breeders' Rights come into force immediately after registration of the variety .

Q. 10. To whom plant breeders rights are granted ?

Ans. BPR can be granted only to breeder or legal owner of a new variety. ,

Q. 11. What is protected by plant breeders' rights ?

Ans. The plant breeders 'rights protect the variety but not the standard breeding procedures that are used to develop new variety .

Q. 12. What is the purpose of plant breeders' rights ?

Ans. The main purpose of PBR is to encourage breeders to develop better plant varieties. The PBRs provide monopoly of new varieties to breeders for a limited period.

Q. 13. Can the same variety be patented and protected by plant breeders' rights ?

Ans. Yes, the same variety can be patented and protected by PBR. The gene technology or gene sequences can be patented and the end product that is the plant variety can be protected under PBR.

Q. 14. What are the Requirements for protection of plant breeders' rights ?

Ans. The PBR cannot be granted to any variety. There are four basic requirements for protection of a variety under Plant Breeders' Rights. These are (i) novelty, (ii) distinctiveness, (iii) uniformity , and (iv) stability.

Q. 15. What do you mean by Novelty ?

Ans. It refers to newness of a variety .The variety should be new and should not have been commercially cultivated for more than one year before granting protection under PBR. In other words, there should not be a prior sale for more than one year.

Q. 16. What is distinctiveness ?

Ans. The new variety must be distinguishable in one or more morphological, quality or other character from previously available varieties. When a variety differs atleast in one characteristic from varieties of common knowledge, it is known as distinct.

Q. 17. Define Uniformity.

Ans. The variety should be uniform. In other words, all plants in a variety should look alike.

Q. 18. What is stability ?

Ans. The should give stable performance in different generations. All these attributes are determined by grow out test.

Q. 19. What are the Rights Provided to the breeder of a new variety ?

Ans. The *Plant Breeders' Rights Act* provides plant breeders the exclusive right to produce and sell new plant varieties which they have developed. In other words, it provides exclusive rights to the breeder for commercial production and marketing of his variety. The important Plant Breeders' Rights are as follows:

(i) Rights for commercial Seed production: PBR provide legal right to the breeder of a variety for large scale seed production. This he can do either on his own farm or on the farms of authorized farmers on payment basis.

(ii) Rights for Marketing: The breeder or owner of a variety has exclusive rights to regulate marketing of his variety .

(iii) Rights to export and import: The breeder or owner of a variety has full rights to regulate export and import of his or her variety.

(iv) Rights of Authorization: The breeder or owner of a protected variety has rights to authorize other interested persons for commercial production, and marketing, export and import of his variety . However, prior authorization of the breeder or owner of a variety is required for such purpose.

(v) Rights to prevent Infringement: The breeder or owner of a variety has rights to prevent others from unauthorized commercial production and marketing of his variety .

Q. 20. What is Infringement of plant breeders' rights ?

Ans. Unauthorized production and marketing of a registered variety by other person amounts to infringement. The owner has the right to take legal action against the infringer and claim damages. The PBR Authority may initiate legal action against a person who is involved in the infringement of a protected plant variety. The Authority can recover both damages and profits from such person. The Plant Breeders' Act (PBRA) provides for heavy penalties against jnfringement of the breeders' right. In USA it 55,00 Dollars for individuals and 275000 Dollars for companies.

Q. 21. What are exemptions of Plant Breeders' Rights Act ?

Ans. The Plant Breeders' Rights Act provides two exemptions *viz.,* (i) breeders' exemptions and (ii) farmers' exemptions.

Q. 22. What do you mean by Breeders' Exemptions ?

Ans. The legal rights that are provided to plant breeders to use protected material for further research refer to breeders' exemptions. Breeders exemptions are also called research exemptions or breeders' privilege. The UPOY Act 1978 provides Breeders' exemptions. However, the Act-1991 has curtailed & breeders' Exemptions.

Q. 23. What are Farmers' Exemptions ?

Ans. The legal rights that are provided to farmers to save, use, exchange, share or sell his farm produce of a protected variety are known as farmers' exemptions. Farmers' exemptions are also called Farmers' Rights or Farmers privilege. Here the sale is restricted to non-commercial sale. The UPOV Act 1978 provided farmers Exemptions, however, these exemptions were curtailed by UPOV Act 1991. Because of these reasons, UPOV Act is not accepted by many countries.

Q. 24. What are Advantages of plant breeders' rights ?

Ans. There are several advantages of plant breeders' rights which are listed below.

1. Breeders get benefit of their variety.
2. PBR help in faster development of seed industry.
3. PBR lead to improvement in quality because of competition.

4. PBR are useful in procurement of good material on payment basis.
5. PBR help in enrichment of genetic resources.

Q. 25. What are Disadvantages of plant breeders rights ?

Ans. There are some disadvantages of plant breeders' rights which are listed below

1. It will promote monopoly.
2. It will encourage unhealthy practices.
3. It may lead to increase in prices.
4. There will be reduction in genetic variability.
5. There will be compulsion to purchase fresh seed every year.

40

FARMER'S RIGHTS

Q.1. What are Farmers' Rights ?

Ans. Farmers, rights refer to the rights arising from the past, present and future contributions of farmers in conserving, improving and making available plant or animal genetic resources, particularly those in the centres of origin/diversity. In other words, the legal rights provided to farmers to save, use, sow, replant, exchange, share or sell his farm produce including seed of a variety protected under Plant Variety Protection Act refer to Farmers Rights. The purpose of these rights is to "ensure full benefits to farmers and support the continuation of their contributions.

The FAO Conference held in Rome from 11-29 November, 1989 endorsed the concept of Farmers Rights with a view to:

(i) ensuring global recognition of the need for conservation and the availability of sufficient funds for these purposes;

(ii) assisting farmers and farming communities throughout the world, especially those in areas of original diversity of plant genetic resources, in the protection and conservation of their PGR and of the natural biosphere; and,

(iii) allowing the full participation of farmers, their communities and countries in the benefits derived, at present and in the future, from the improved use of PGR.

Q.2 Is Registration required for a farmers' Variety ?

Ans. A farmer who has bred or developed a new variety shall be entitled for registration and protection of his variety like a breeder.

Q.3. What is the procedure of registration of farmers' variety ?

Ans. The registration process consists of the following steps:

(i) Filing of application. The application is to be filed in the prescribed form and submitted in the office of the registrar of Plant Variety Protection and farmers' Rights Act. The application should contain all the desired documents and information.

(ii) Examination of the application. The application is examined by experts in the registrar office. If found correct, it is advertised for opposition. Three months time is given for filing opposition.

(iii) Registration. If there is no opposition, and the registrar and Authority are satisfied, the variety is registered and the registration certificate is issued to the breeder, farmer or other owner of the variety.

Q.4. What are the Requirements for registration of farmers' variety ?

Ans. The farmers; variety shall be entitled for registration if it meets the following requirements.

(i) It should have a denomination (name) assigned by the applicant

(ii) An affidavit that the variety does not contain terminator gene.

(iii) A complete passport data of parental lines from which the variety has been derived.

(iv) The contribution of any farmer, village community, institution or organization in the development of the variety .

(v) The source of genetic material used in developing the variety and the parental material used in developing new variety was lawfully acquired.

(vi) A brief description of the variety about its novelty, distinctiveness, uniformity and stability.

(vii) A proof of the right to make the application.

Q.5. How much fee is charged from farmers for registration of a variety ?

Ans. In India, a farmer or group of farmers or village community shall not be liable to pay any fee in any proceeding before the Authority or Registrar or the Tribunal or the High Court under Plant Variety Protection and Farmers' Rights Act or the rules made there under.

Q.6. What are the Rights granted to farmers ?

Ans. The Farmers' Rights Act provide following rights to the farmers in India even for the variety protected under PVP and FR Act.

(i) Rights to save, use, sow or replant seeds of his farm produce.

(ii) Rights to exchange or share his farm produce, and

(iii) Rights to sell seeds of his farm produce

However, the farmer cannot sell the seeds or planting material under a brand name.

Q.7. What are Rights of Communities in India ?

Ans. In India, any of the following can file application for registration on behalf of any village or community.

(i) Any person or group of persons (whether actively engaged in farming or not), and

(ii) Any governmental or non-governmental organization.

After completing all formalities, if the Authority is satisfied, the variety can be registered. It due credit has not been given to the farmers or communities who have helped in the development of the variety, the breeder has to pay compensation to the concerned party. Or part of compensation will be deposited in the gene fund.

Q. 8. Is Authorization of Farmers' Variety possible ?

Ans. The authorization for an essentially derived variety from farmers' variety cannot be given by the breeder of such farmers; variety except with the consent of the farmers or group of farmers or community of farmers who have made contribution in the preservation or development of such variety.

Q.9. What are the Materials Covered by Farmers' Rights ?

Ans. Farmers' Rights include new varieties of all self pollinated, cross pollinated and asexually propagated varieties.

Q.10. What is the Duration for Farmers' variety ?

Ans. The protection is provided for a definite period. The duration of protection for different plant species is as follows'

(i) For Trees and Vines: Initially for nine years and maximum for 18 years.

(ii) For extant varieties: Initially six years and maximum for 15 years fro the date registration of the variety .

(iii) In other cases: Initially six years and maximum for 15 years fro the date registration of the variety .

Q.11. What is the Validity of protection of Farmers' Rights ?

Ans. The protection is valid only in the country where the variety has been registered.

Q.12. What is Benefit Sharing ?

Ans. If the variety has been developed by a particular farmer, he can get benefits of his variety. In case of community rights, the sharing of benefits becomes difficult and the money is deposited in the gene fund.

Q.13. What is Gene Fund ?

Ans. The central government may constitute a Fund to be called National Gene Fund. The money earned from following will be deposited in this fund;

(i) The benefit sharing received fro the breeder of a variety

(ii) The annual payable to the Authority in the form of royalty.

(iii) The amount of compensation received by the Authority .

(iv) Contribution received from any National or International Organization and other sources.

The gene fund may be utilized for benefit sharing, rewards and other expenditure related to Plant Variety Protection and Farmers' Rights Act.

Q.14. Who will be entitled for Rewards ?

Ans. A farmer who is engaged in the conservation of genetic resources of land races and wild relatives of economic plants and their improvement through selection shall be entitled for recognition and reward. The expenses toward rewards will be met out from the National Gene Fund.

Q.15. What do you mean by Acknowledgement by Breeders ?

Ans. It is compulsory for the breeder of a variety to reveal the origin of the parental lines plant genetic resources used in the development of a variety .It will help in deciding the share of different persons, viz., farmers and other communities in benefits earned by breeder of such variety .

Q.16. Is Infringement is exempt for Indian farmers ?

Ans. If a farmer was not aware of the PVP and FR Act at the time of using any protected variety, he will be exempt from the infringement and will be asked not to do such act in future. This right has been provided to Indian farmers by PVP and FR Act 2001.

Q.17. What the purpose of farmers' rights ?

Ans. Farmers' Rights, as endorsed by *FAO* in 1989, recognize that farmers and rural communities have contributed and continue to contribute to creation, conservation, exchange and enhancement of genetic resources and they should be recognized and strengthened in their work.

Q.18. What the significance of farmers' rights ?

Ans. The important points about nature and scope of Farmers' Rights are briefly presented as follows:

(i) Ancient Rights. Farmers' Rights have a deep historic character. They existed since humans created agriculture to serve their needs, and played vital role in conservation of biodiversity. Farmers are the guardians of these genetic resources, which support the evolution of species. Farmers are the inheritors of the skills and knowledge of the generations that have created this biological wealth. For these contributions, it is necessary to recognize Farmers' Rights.

(ii) Rights Covered. Farmers' Rights include the right over resources and associated knowledge, united indivisibly, and mean the acceptance of traditional knowledge, respect for cultures and recognition that these are the basis of the creation of knowledge.

(iii) Property Rights. The farmers deserve right to control, the right to decide the future of genetic resources, the rights to define the legal framework of property rights of these resources.

(iv) Collective Nature. Farmers' Rights are of an eminently collective nature and therefore are protected by sui generic rights.

(v) Application. These rights should have a national application, respecting the sovereignty of each country, to establish local laws based on these principles.

(vi) Rights to Means. It includes territorial rights, right to land, right to water and air and rights to conserve biodiversity and achieve food security .

(vii) Active Participation. The farmers deserve right to participation in the definition, elaboration and execution of policies and programmes linked to genetic resources.

(viii) Rights to Technology. The farmers deserve right to appropriate technology as well as participation in the design and management of research programmes.

(ix) Rights to Benefit Sharing. The farmers deserve right to define the control and handling of benefits derived from the use, conservation and management of these resources.

(x) Rights on Genetic Resources. The farmers deserve right to use, choose, store and freely exchange of genetic resources.

(xi) Rights for Sustainable Agriculture. The farmers deserveright to develop models of sustainable agriculture that protects the biodiversity and to influence the policies that support it.

Q.19. What is the Role of Farmers Related to PGR ?

Ans. Farmers are both men and women who have domesticated, developed, conserved and made available plant genetic and other natural resources (land; water; vegetation; animal, bird and fish life) to which they have access in order to obtain a livelihood and ensure the well-being of their family through the provision of basic requirements such as food, fuel and water and income from the land.

Farmers have played key role in conservation of Plant Genetic Resource for thousands of years. The role of farmers in relation to PGR is briefly presented below:

(i) Farmers both men and women have conserved plant genetic resources over the past thousands of years.

(ii) Farmers are rightly called as on site managers of plant genetic resources.

(iii) Farmers have in depth knowledge of the land races and wild relatives of cultivated plant species.

(iv) All modern plant varieties contain only those genes that have originated from farmers' traditional varieties (landraces) or wild crop relatives.

(v) The women and men have historically provided the germplasm upon which scientific plant breeding is based.

(vi) Generally, germplasm is provided by small-scale subsistence farmers to plant breeders whose varieties are then adapted/ developed for medium to large scale commercial farmers (Spillanel, 1996).

Q.20. What is the role of women related to PGR ?

Ans.

(i) Women's special knowledge of the value and diverse uses of plants for nutritional, health and income has important implications for the conservation of plant genetic resources.

(ii) The women have played significant role not only in terms of labour and skills, SI but also their decision-making about how natural resources are used to satisfy the multiple needs of rural households.

(iii) In much of the developing world, the on-farm and in-situ conservation and use of plant genetic resources begins with women. As farmers, . they are responsible for growing crop plants and collecting seed for next year sowing.

(iv) It is estimated that 90 % of the planting materials used in developing countries is produced by the farmers themselves, i.e. from saved seed. The selection of wild genetic resources for home planting, of seeds for conserving for next year's planting and of asexually propagated material is usually carried out by women. This selection is a sophisticated process that takes into account many different criteria such as taste, color, palatability and texture; resistance to pests and diseases; adaptation to soil and agro-climatic conditions and so on. The selection criteria are built up over years of experience.

Q.21. What are the similarities between Plant Breeders' rights and farmers' Rights ?

Ans. Plant Breeders' Rights and Farmers' Rights have some similarities which are presented in Table 40.1

TABLE 41. Similarities between Plant Breeders' Rights and Farmers Rights

SN	Particulars	Plant Breeders' Rights	Farmers' Rights
1.	Registration	Essential	Essential
2.	Duration Trees and vines Extant varieties etc.	18 years 15 years	18 years 15 years
3.	Enforcement	After Registration	After Registration
4.	Requirements	Novelty, Distinctiveness, uniformity & stability	Same as in PBR
5.	Validity	In country of registration	In country of registration

Q.22. What are the differences between Plant Breeders' rights and farmers' Rights ?

Ans. Plant Breeders' Rights and Farmers' Rights have some differences which are presented in Table 40.2.

Table 40.2 Differences between Plant Breeders' Rights and Farmers' Rights

S.N.	Particulars	Plant Breeders' Rights	Farmers' Rights
1.	Registration Fee	Charged	Not charged
2.	Annual renewal fee	Charged	Not charged
3.	Penalty for infringement	Charged	Not charged
4.	Community Rights	Not covered	Covered
5.	Authorization	Easy	Not permitted

Q.23. Discuss the Future Outlook Related to PGR .

Ans. Plant Genetic Resources are the base materials for modern plant breeding programmes. Hence their proper conservation and sustainable use are essential. In future, the following points should be given due importance for proper conservation and utilization of plant genetic resources.

(i) Assistance to Farmers. Farmers in all regions of the world should be assisted in their conservation of natural resources and more specifically the selection, conservation, improvement and sustainable use of PGR. Non Governmental Organizations should assist in the entire process as they are close to the communities and have the capacities to provide long-term support. The establishment of strong village level institutions with adequate representation of both women and men farmers can be key in promoting successful programmes.

(ii) Incentives. Appropriate incentives should be developed, with the active participation of those concerned, to convince communities, industry and governments of the benefits of conservation and men and women at all levels should participate in the design and development efforts and to provide financial resources, legal support, and technical capability to develop agricultural systems that are suitable for local agro ecological conditions and, where possible, based on locally available resources. Incentives should be used to enable farmers to divert land, labour and capital towards conserving biological resources and to facilitate the participation of certain groups in activities that promote the sustainable use of those resources.

(iii) Participation of Women Farmers. The participation of women farmers and the recognition of their contributions in the design and elaboration of such policies and programmes at all levels is essential. Women farmers should be included in decision making processes at community and regional levels. Farmers, communities and countries, and men and women at levels should participate in the design and implementation of programmes policies and agreements related to PGR.

(iv) Self Decision. Those farmers who develop breeding materials should be empowered to decide, at their discretion, if the materials will be made available to other (i.e. external) users.

(v) Equal Rights to Women. There should be equal rights for conservation and participation in PGR related programmes and policies, which is often lacking in developing countries. There should not be discrimination of gender. In the past, women farmers could get little benefit from support in terms of information and technology, inputs and credit, extension and research.

(vi) Participatory Plant Breeding. Participatory research and plant breeding should be promoted with different socio-economic groups in rural communities in order to improve the productivity of local varieties and to promote marketing of special products derived from these materials.

(vii) Use of Land Races. Modern plant breeding has concentrated on developing a small number of varieties that are widely adapted to many environments. Land races should also be used for developing varieties suitable for diverse ecological conditions.

(viii) International Funds. There is need for the creation of an Intentional Fund to support in-situ conservation and to benefit rural men and women who produce and conserve PGR. In general farmers are interested in conserving resources but lack of financial support acts as a barrier in their way.

(ix) Integrated Programs. PGR conservation should be integrated into other community development programmes. It will promote and employment in activities related to PGR building up of local institutions, capacity building) and collecting and dissemination information on the status and trends of PGR.

(x) Benefit Sharing. There should be equitable sharing of benefits arising from the utilization of PGR, knowledge, innovations and practices.

Q.24. What are the advantages of Farmers' Rights ?

Ans. There are several advantages of farmers' rights which are listed below.

1. Provide recognition to farmers, communities or village for their role in conserving Plant Genetic Resources.
2. Provide respect to farmers.
3. Provide opportunities to farmers to participate in the policies, projects and programmes related to plant Genetic Resources (PGR).

4. Lead to benefit sharing when farmers material is used in developing new variety by breeders.
5: Farmers permission is required to use their plant material.
6. Provide rewards to those farmers who are actively engaged in conserving Plant Genetic Resources.

Q.25. What are the Limitations of Farmers' Rights ?

Ans. There are some limitations of farmers' rights which are listed below.

1. Mostly farmers' rights are of collective nature.
2. The benefit sharing is difficult.
3. Farmers do not get any financial support from government for the conservation of Plant Genetic Resources.
4. Farmers' Rights are of territorial nature.

41

GEOGRAPHICAL INDICATION

Q.1. What is geographical Indication ?

Ans. Geographical Indication may be defined as the name of a region, a specific place or, in exceptional cases a country, used to describe an agricultural, natural or manufactured goods, product or a food stuff. In other words, a geographical indication is a sign used on goods that have a specific geographical origin and often possess qualities or a reputation that are due to that place of origin. In other words, a geographical indication is a name or sign used on certain products or which corresponds to a specific geographical location or origin (e.g. a town, region, or country)

Q.2. What is Indian Legislation for geographical indication ?

Ans. The World Intellectual Property Organization (WIPO) helps in protection of Geographical Indications (GI). About 170 geographical indications have been registered by Lisbon Agreement members up to 1997. Now this number would be much more. In India, the Parliament has passed the Geographical Indications of Goods (Registration and Protection) Act, 1999 (in December 1999). This Act seeks to provide for the registration an better protection of geographical indications relating to goods in India. The Act would be administered by the Controller General of Patents, Design and Trade Marks who is the Registrar of Geographical Indications. The Geographical Indications Registry is located at Chennai.

Q.3. How many Geographical Indications have been registered so far ?

Ans. Till 1997, 170 geographical indications have been registered in some advanced countries because they have developed the system of protecting GI. In the past, wines, spirits, cheeses, tobacco, which account for 88%, have been registered as GI. Wines and spirits account for almost 71 % of all registrations. In addition to wines and spirits, there is need for protection of agricultural products as GI. For this there is need to establish a multilateral svstem of identification and registration of GI. Geographical Indication may be an agricultural, natural or manufactured goods or product.

Q.4. What is the duration for protection of geographical indication ?

Ans. The registration of a geographical indication is valid for a period of 10 years. It can be renewed from time to time for further period of 10 years each. If a registered geographical indication is not renewed it liable to be removed from the register.

Q. 5. Is Registration compulsory for protection of geographical indication ?

Ans. The registration is not compulsory. The registration provides better legal protection to take action against infringement. In other words, it confers legal protection to geographical indications. And prevents unauthorized use of a registered geographical indication by others.

Q.6. What is the Validity of geographical indication ?

Ans. The registration of the geographical indication is valid in the country of its origin.

Q.7. Is it possible to Transfer geographical indication ?

Ans. A geographical indication cannot be sold, purchased, transferred, licensed, pledged, mortgaged or put to such other agreement.

Q.8. What are the Matters Covered geographical indication ?

Ans. It is used to identify agricultural, natural and manufactured goods.

Q.9. Who has the Control over geographical indication ?

Ans. An authorized user has the exclusive rights to use the geographical indication in relation to goods for which it is registered.

Q.10. When does geographical indication become effective ?

Ans. The protection of geographical indication becomes effective after its registration.

Q.11. What is the Infringement of geographical indication ?

Ans. Infringement of a geographical indication is possible. The registered proprietor and authorized users can initiate action against infringement.

Q.12. What are the Rights Protected by geographical indication ?

Ans. The proprietor or authorized user has the rights to use the geographical indication and prevent other from using such material without permission.

Q.13. What are the Requirements for registration of a geographical indication ?

Ans. It originates from a definite geographical territory and should have a special quality or reputation or other characteristics. The main features are as given below.

(i) It is an indication.

(ii) It originates from a definite geographical territory.

(iii) It is used to identify an agricultural, natural or manufactured goods/ product

(iv) The manufactured goods should be processed or processed in that territory.

(v) It should have a special quality or reputation or other characteristics.

Q.14. What is the Procedure for Registration of geographical indication ?

Ans. The registration is done by the registrar of Geographical Indications. For registration any association of persons, producers, organization or authority established by or under the law can apply. The applicant must represent the interest of the producers. The registration consists of following steps:

1. Filing Application. The application in the prescribed form is filed to the registrar of Geographical Indications along with prescribed Fee and other details. .

2. Examination. The application is examined by the registrar office by conducting search. The application is accepted, if it fulfils the requirements. Then it is advertised in the GI journal for objection if any.

3. Acceptance. If there is no objection, the application is accepted for grant of registration.

4. Registration. Then the GI is entered in the Register. The register of geographical indication is divided into two parts, viz. part A and part B. Part A consists of particulars relating to registered geographical indication and part B consists of particulars of the registered authorized users.

5. Issue of Certificate. Finally a certificate is issued to the registered user. The registration of GI is for a period of 10 years which is renewable for further periods of 10 years each.

Q.15. What are the differences between geographical indication and patents

Ans. There are several differences between geographical indication and patents which have been presented in Table 41.1

Table 41.2 Differences between Geographical Indication and Patents

SN	Particulars	Geographical Indication	Patents
1.	Requirement	Reputed product of a specific place or territory.	It is valid in all countries.
2.	Period	10 years but renewable.	15-20 years
3.	Registration	Not compulsory	Compulsory
4.	Legal rights	After registration only	There are legal Rights
5.	Transfer	Not allowed	Admissible
6.	Items covered	Agricultural, natural and manufactured goods	Products/process/ invention/innovation
7.	Control	By authorized user or proprietor	By inventor

Q.16. What are the similarities between geographical indication and patents?

Ans. There are some similarities between geographical indication and patents which have been presented in Table 41.2

Table 42.2 Similarities between Geographical Indication and Patents

S.N.	Particulars	Geographical indication	Patents
1.	Validity	In the country of registration.	In original country.
2.	Fee	Fee is charged if registered	Fees is charged.
3.	Enforcement	After registration	After Registration
4.	Infringement	Possible	Possible

Q.17. What are the differences between Geographical Indication and Trademarks ?

Ans. There are several differences between geographical indication and trademarks as given below.

1. Number. The total number of registered trademarks is more than 16 millions. However, only about 6 millions trademarks are currently in use worldwide. Whereas about 170 GI have been registered so far. Thus only limited number of trademarks are useful.

2. Beneficiaries. In case of GI, only local people or people of a region are beneficiaries, whereas in case of trademarks large number of people are beneficiaries.

3. Role of State. The State Government plays important role in protection of. Trademarks are promoted by individual company.

4. Production. The production of GI is collective by producers of a place or region, whereas the production of trademarks is by individual company.

5. Protection. The protection of GI is difficult as compared to trademarks.

6. Creation. The creation of new trademarks is easier than creation of new GI.

Besides these, there are several other differences between geographical indications and trademarks which are presented in Table 41.3

Table 41.3 Differences between Geographical Indication and Trademarks

S.N.	Particulars	Geographical Indication	Trademarks
1.	Requirement	Reputed product of a specific place or territory.	New & distinctive.
2.	Period	10 years but renewable	Renewable indefinitely.

3.	Registration	Not compulsory	Is compulsory.
4.	Legal rights	After registration only	Legal rights are there
5.	Transfer	Not allowed	Possible
6.	Items covered	Agricultural, natural and manufactured goods.	Names, images & short phrases.
7.	Enforcement	After registration	With its commercial use.
8.	Control	By authorized user or proprietor.	By owner.

Q.18. What are the similarities between Geographical Indication and Trademarks ?

Ans. There are some similarities between geographical indication and trademarks as given in Table 41.4

Table. 41.4 Similarities between Geographical Indication and Trademarks

SN	Particulars	Geographical Indication	Trademarks
1.	Validity	In the country of registration.	In original country
2.	Fee	Fee is charged if registered.	Fee is charged.
3.	Infringement	Possible	Possible

Q.19. What are Advantages of Geographical Indication ?

Ans. There are several advantages of geographical indications which are presented below:

1. GI protection provides legal rights to the users.
2. It prevents unauthorized use of registered GI.
3. It boosts exports.
4. It promotes economic prosperity of producers of goods in a geographical territory.
5. Geographical indications bring name to place of their origin or territory .

Q.20. What are disadvantages of Geographical Indications ?

Ans. There are some disadvantages of geographical indications which are listed below.

1. Prevents others for use.
2. It leads to monopoly
3. Users are limited to place of origin

Q.21. What are Examples of Indian Geographical Indications ? .

Ans. In India some geographical indications have been identified which are listed below

(i) Basmati Rice
(ii) Darjeeling Tea
(iii) Kanchipuram Silk Saree
(iv) Alphanso Mango
(v) Nagpur Orange
(vi) Kolhapuri Chappal
(vii) Bikaneri Bhujia
(viii) Agra Petha
(ix) Banarasi Saree

The above geographical indications can be classified into three groups, viz. (i) agricultural, (ii) natural, and (iii) manufactured.

(i) Agricultural GI. This group includes Basmati Rice, Nagpur Orange and Alphanso mango.

(ii) Natural GI. This group includes Darjeeling Tea.

(iii) Manufactured GI. This group includes Kanchipuram Silk Saree, Banarasi Silk Saree, Kolhapuri Chappal, Bikaneri Bhujia and Agra Petha.

42

EVOLUTION OF CROP PLANTS

Q.1. What is evolution ?

Ans. A divergent process which increases genetic diversity and leads to change in allelic frequencies in a population is known as evolution. In other words, a process which leads to significant deviation in the characteristic features of existing individuals as compared to their pre-existing individuals is termed evolution. Thus, evolution leads to change in genetic composition of a population or individuals of population.

Q.2. What are types of evolution ?

Ans. The evolution is of two types, viz., (I) natural evolution, and (2) man-made evolution. In case of natural evolution, natural selection operates, while in man made evolution human selection operates. Thus, both natural and human selections play significant role in the process of evolution. Plant breeding is considered as current phase of crop evolution.

Changes which are brought out as a result of evolution are measured in terms of morphological, anatomical, embryological, physiological, biochemical and genetic modifications in the present forms of individuals as compared with their past forms.

Q.3. What are genetic bases of evolution ?

Ans. Selection plays a key role in the process of evolution. Selection either by nature or by human has been responsible for evolution of various crop plants. However, selection is effective in changing the features of species only when vast variability exists in the population of that species. Three genetic factors or forces, *viz.*, (I) polyploidy, (2) introgression, and (3) mutations have played significant role in the evolution of various crop plants. These three factors aid in the process of evolution by way of inducing additional genetic variability, which is a basic requirement to selection to operate.

Q. 4. What is the role of polyploidy in evolution ?

Ans. Polyploidy refers to numerical change in the genome (A basic set of chromosomes). Polyploidy is of two types, *viz.*, (I) autopolyploidy, and (2) allopolyploidy. Role of each type of polyploidy in evolution is discussed below:

Autopolyploidy

This is also known as simple polyploidy or single species polyploidy, because the increase in chromosome number relates to the same species. This type of polyploidy can occur in nature as well as can be induced by colchicine treatment. Increase in chromosome from diploid to tetraploid state leads to increase in vigour and size of flowers and fruits over diploid forms. However, polyploid plants exhibit slow growth rate and reduced fertility due to chromosomal imbalance. Autopolyploidy has been used in crops like banana, apples, sugarbeet, watermelon, potato, oranges, tulips, etc. The commercial banana is autotriploid (3n), which has seedless and larger fruits than diploid forms. Some varieties of apples in USA are triploids, which are propagated asexually by budding and grafting. Triploid varieties of sugarbeet have higher sugar content than diploids. Triploid watermelons, which are produced from a cross between tetraploid and diploid, are seedless, early maturing and resistant to diseases. In USA, triploid watermelons cover about 10% of the area under this crop. The commercial potato is regarded as an autotetraploid though interspecific hybridization may also be involved. Triploid oranges (which are seedless) are produced by pollinating diploid plants with pollen of tetraploid plants. Triploid tulips are propagated asexually. Autopolyploidy in Petunia, an ornamental plant, leads to brighter flower colour than diploid forms.

Allopolyploidy

This is also known as hybrid polyploidy or bispecies or multispecies polyploidy depending upon the species involved. Such polyploidy is obtained by doubling of chromosome number by colchicine treatment. Allopolyploidy has been more instrumental in evolution of crop plants, because 50% of the crop plants are allopolyploids. Hybrid polyploidy has played significant role in the evolution of crops like wheat, tobacco, cotton, *Brassica,* oat, etc. Examples of artificially produced allopolyploids include triticale, strawberry and loganberry. Triticale is a man made new cereal (between wheat and rye), which combines high yield of wheat and disease and drought resistance of rye. Strawberry is a polyploid between North and South American species, and loganberry is a polyploid between raspberry and blackberry.

Q.5. Define Introgression and describe its role in evolution ?

Ans. Incorporation of gene of one species into the genetic background of another species by means of interspecific hybridization and backcrossing is known as introgression (Anderson, 1949). The interspecific hybrid backcrosses in nature with one of the parental species. As a result of introgression, genes from two divergent species are combined. A true breeding recombinant form favoured by natural selection may give rise to a new species.

Introgressive hybridization between primitive maize and wild grass **Tripsacum** is considered to be responsible for the evolution of modern forms of maize.

Q.6. Explain role of mutations in evolution.

Ans. Mutations are important sources of creating variability in a genetic population. Mutations can occur in nature as well as can be induced by the use of physical or chemical mutagenic agents. In hexaploid wheat, a natural mutation is responsible for homologous pairing. Spontaneous mutations have played significant role in the evolution of crop plants. Spontaneous mutations can be used either as a cultivar or as a parent in the hybridization programme. Induced mutations have played key role in improving yield, quality, earliness, adaptation, and disease and insect resistance in various crop plants.

Q.7. How bread wheat has evolved ?

Ans. Wheat is a cereal crop of global importance. It belongs to the genus **Triticum** of the family **Poaceae** (old **Gramineae**). There are three types of species in the genus **Triticum**, *viz.*, diploid, tetraploid and hexaploid. The somatic chromosome number of these species is 14, 28 and 42, respectively (Table 10.1). Bread wheat **(Triticum aestivum)** is the predominantly cultivated species, which belongs to the hexaploid group. Other cultivated species are ***T. monococcum*** in diploid group and ***T. turgidum*** in tetraploid group.

Table 42.1 Different species of the genus Triticum (Sears, 1974)

Common name and ploidy	Species	2n number	Genome symbol	Remarks
Einkom-Diploid (2N = 2XJ)	**Triticum monococcum**	14	AA	Cultivated
	Unknown species	14	BB	Wild
	Triticum dichasians	14	CC	Wild
	Triticum tauschii	14	DD	Wild
Emmer/Durum	**Triticum turgidum**	28	AABB	Cultivated
Tetraploid (2n = 4x)	**Triticum timopheevii**	28	AAGG	Wild
Bread/Club Wheat Hexaploid (2n = 6x)	**Triticum aestivum**	42	AABBDD	Cultivated

Triticum dichasians (Aegilops caudata), Triticum tauschii (Aegilops squarrosa), Triticum turgidum (T. dicoccum, T. durum, T. polonicum, T. turgidum), and Triticum aestivum (T. vulgare, T. compactum, T. spelta). Parentheses denote old names.

The tetraploid species developed as an amphidiplod between two diploid species, and hexaploid species originated from a cross between tetraploid and diploid species. It is believed that tetraploid species **Triticum turgidum** evolved as an amphidiploid between **Triticum monococcum** (AA) and an unknown species (now probably extinct) with BB genome. The hexaploid bread wheat originated as an amphidiploid between **Triticum turgidum** (AABB) and T. tauschii (DD). The overall process can be represented as follows (Fig. 42.1).

Parents	**Triticum monococcum**		×	Unknown species
Genome	AA (2n = 14)		↓	BB (2n = 14
F_1		Sterile	AB (2n = 14)	
Chrpomosome doubling		Fertile	↓	
			AABB (24 = 28)	
			Triticum turgidum	
Parents	**Triticum turgidum**		×	**Triticum tauschii**
Genome	AABB (2n = 28)		↓	DD (2n = 14
F_1		Sterile	ABD (2n = 21)	
Chromosome doubling		Fertile	AABBDD (2n = 42)	
			Triticum aestivum	

Fig. 42.1 Probable origin of hexaploid bread wheat (Triticum aestivum)

Thus, A genome of bread wheat has derived from **T. monococcum,** B genome from an unknown species which is now probably extinct, and D genome from **T. tauschii.** The F_1 was sterile at both the stages, which became fertile through chromosome doubling in nature. Thus, interspecific hybridization and polyploidy have played key role in the evolution of bread wheat.

ROLE OF MUTATION

The bread wheat is hexaploid, combining diploid chromosome complements from three different species. In nature, hexaploid bread wheat behaves as diploid (n = 21 and 2n = 42). It has been found that hexaploid wheat has acquired this property of diploid pairing from a mutation on chromosome number 5B, which inhibits pairing between homologous chromosomes (chromosomes of different genomes). This mutant gene is present in the long arm of chromosome number 5 in B genome. The mutant gene permits only homeologous pairing (pairing between the chromosomes of the same species). Thus, this mutant gene enforces hexaploid wheat to have homologous pairing and behave like a diploid species. This system acts as a barrier to pairing between the genomes of different species. This phenomenon is called diploidization. The 5B system in wheat was discovered by Riley in 1954.

Q.8. Explain genetic origin of upland cotton.

Ans. Cotton is one of the major fibre crops of global importance. It is grown in more than sixty countries in the world. Cotton belongs to the genus **Gossypium** of the family **Malvaceae.** There are about 50 species in the genus **Gossypium.** Some of them are diploid (2n = 26) and some tetraploid (2n = 52). Out of 50 species, only four species are cultivated, viz., **G. arboreum, G. herbaceum, G. hirsutum** and **G. barbadense** (Table 42.2)

Table 42.2 Some important species of Gossypium

Common name and ploidy	Species	2n number	Genome symbol	Remarks
Asian Diploid	**Gossypium arboreum**	26	AA	Cultivated
	Gossypium herbaceum	26	AA	Cultivated
African Diploid	**Gossypium africanum**	26	AA	Wild
American Diploid	**Gossypium thurberi**	26	DD	Wild
	Gossypium raimondii	26	DD	Wild
Upland Tetraploid	**Gossypium hirsutum**	52	AADD	Cultivated
Egyptian Tetraploid	**Gossypium barbadense**	52	AADD	Cultivated

This is generally believed by the evolutionists that **Gossypium africanum** (native of South Africa) is the ancestor or progenitor of all cultivated species of cotton. It reached India possibly by sea and over a long time branced Into two species, viz., **G. arboreum** and **G. herbaceum**. These species have 13 large chromosomes in their haploid complements. The American diploids have 13 small chromosomes in their diploid complements. Skovsted (1937) proposed that tetraploid cottons have developed from natural crossing between diploid species with small and large chromosomes and natural chromosome doubling in the long past resulted in the evolution of present day tetraploid species. There are two opinions about the origin of upland cotton **(Gossypium hirsutum).** According to Beasley (1940) a possible origin of **G. hirsutum** is from the cross between Asiatic cultivated cotton, **G. arboreum** and American wild species, **G. thurberi** followed by chromosome doubling of F_1 in nature (Fig. 42.2a).

Parents	**Gossypium arborum**	×	**Gossypium thurberi**
Genome	AA (2n = 26, large)		DD (2n = 26, small)
F_1		Sterile AD (2n = 26, 13 large and 13 small) ↓	
F_1 Chromosome doubling fertile		AADD (4n = 52, 26 large and 26 small)	Like **G. hirsutum**

Fig. 42.2a Probable origin of G. hirsutum cotton as proposed by Beasley(1940)

According to more recent scheme **G. hirsutum** has originated from the cross between **G. herbaceum** Var **africanum** and **G. raimondii** followed by chromosome doubling of F_1 in nature (Fig. 42.2b). Now this theory is widely accepted.

Parents	**G. africanum**	×	**G. raimondii**
Genome	AA (n = 13, large)		DD (n = 13, small)
F_1		Sterile	AD (2n = 26)
Chromosome doubling		Fertile	AADD (2n = 52)
			Like **G. hirsutum**

Fig. 42. 2b. Probable origin of G. hirsutum as proposed by Phillips (1963).

Q.9 Explain the genetic origin of tobacco.

Ans. Tobacco is a narcotic plant which belongs to the genus **Nicotiana** in the family **Solanceae**. It is a native of America, but now it is grown in all the countries of South and South East Asia. There are two cultivated species of tobacco, *viz.,* **Nicotiana tabacum** and **N. rustica.** Both these species are tetraploid (2n = 48). The wild species are diploids (Table 42.3).

Table 42.3. Some important species of the genus Nicotiana

Name of species	2n number	Ploidy level	Remarks
Nicotiana sylvestris	24	Diploid	Wild
Nicotiana tomentosa	24	Diploid	Wild
Nicotiana paniculata	24	Diploid	Wild
Nicotiana undulata	24	Diploid	Wild
Nicotiana tabacum	48	Tetraploid	Cultivated
Nicotiana rustica	48	Tetraploid	Cultivated

It is believed that **Nicotiana tabacum** has originated as an amphi-

(Fig. 42.3a). Similarly, **N. rustica** is believed to be an amphidiploid between wild diploid species **N. paniculate** and **N. undulata** (Fig. 42.3b)

(a) Parents	**N. sylvestris**	×	**N. tomentosa**
Chromosome	(n= 12)		(n = 12)
F_1		↓	
		Sterile (2n = 24)	
		↓	
Chromosome doubling		Fertile (4n = 48)	**N. tabacum**
(b) Parents	**N. paniculata**	×	**N. undulata**
Chromosome	(n = 12)		(n=12)
F_1	Sterile	(2n = 24)	
		↓	
Chromosome doubling	Fertile	(4n = 48)	**N. rustica**

Fig. 42.3 Probable origin of cultivated tetraploid species of tobacco : (a) N. tabacum, and (b) N. rustica.

The crossing of diploid species, *viz.,* **N. sylvestris** and **N. tomentosa** with **N. tabacum** and **N. paniculata** and **N. undulata** with **N. rustica** leads to bivalent formation in F_1 during meiosis (metaphase I), which confirms involvement of respective diploid species in the evolution of respective tetraploid species.

Q.10. Explain genetic origin of tetraploid species of Brassica.

Ans. The genus **Brassica** belongs to the family **Cruciferèae** and has several oil bearing species. There are three basic species of Brassica from which three different tetraploid species have originated (Table 42.4). The tetraploid species have originated through interspecific hybridization between diploid species and chromosome doubling of the F_1 in nature (Fig. 42.4).

Table 42.4: Some *Important species of* Brassica

Species and ploidy	2n number	Genome	Common Name
Diploid Species :			
Brassica campestris	20	AA	Rapeseed
Brassica nigra	16	BB	Black Mustard
Brassica oleracea	18	CC	Cabbage
Tetraploid Species :			
Brassica juncea	36	AABB	Indian Mustard
Brassica napus	38	AACC	Rapeseed
Brassica carinata	34	BBCC	Abyssinian Cabbage

Brassica Juncea is natural amphidiploid between **B. campestris** and **B. nigra**, **B. napus** is amphidiploid between **B. oleracea** and **B. campestris**, and **B. carinata** is amphidiploid between **B. nigra** and **B. oleracea.** The origin of different tetraploid species from three diploid basic species can be represented by triangle of U (Fig. 42.4).

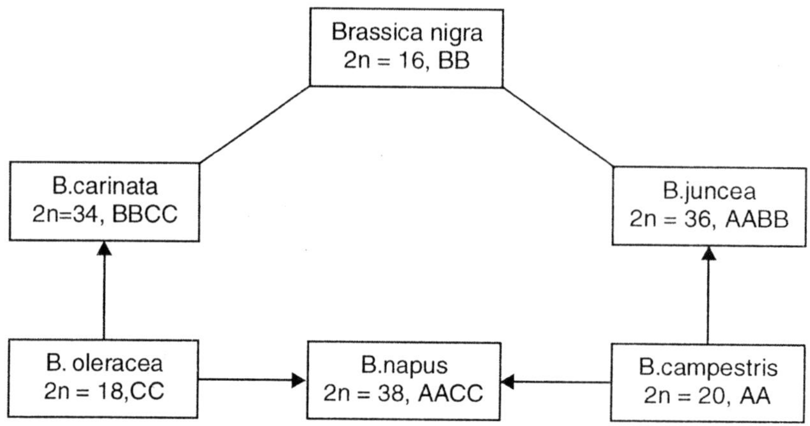

Fig. 42.4 A simple diagram demonstrating origin of three tetraploid species of Brassica, viz, B. juncea, B. carinata and B. napus.

Q.11. Explain genetic origin of potato.

Ans. Potato is an important vegetable crop of global importance. It belongs to the genus **Solanum** in the family **Solanaceae.** The commercially cultivated potato **(Solanum tuberosum)** is native of Central and South America from where it has spread to other parts of the world. Andes region of South Peru and Bolivia is believed to be the center of origin of cultivated potato. **S. tuberosum** is a tetraploid species (2n = 48), which is believed to have originated from the diploid species **S. stenotomum** which is also found growing in the Andes region of South Peru and Bolivia (Table 42.5).

First it was believed that **S. tuberosum** is an autotetraploid from the species **S. stenotomum.** But according to more recent views, it is probably a segmental allopolyploid derived from crosses between cultivated diploid species **S. stenotomum** and wild diploid species **S. sparsipilum** (Hawkes, 1956,58).

Table 42.5. Some tuber bearing species of the genus Solanum

Name of Species Number	2n level	Ploidy	Remarks
Solanum stenotomum	24	Diploid	Cultivated
Solanum sparsipilum	24	Diploid	Wild
Solanum tuberosum	48	Tetraploid	Cultivated

I. Earlier View

Solanum stenotomum —Chromosome doubling→ **Solanum tuberosum**

(2n = 24) (2n = 48)

2. Recent View

Parents	**S. stenotomum**	×	**S. sparsipilum**
Chromosome	(2n = 24)		(2n = 24)
F_1		Sterile (2n = 24)	
		↓	
Chromosome doubling		Fertile (4n = 48)	

Fig. 42.5 Probable origin of cultivated tetraploid potato.

Q.12. How cultivated maize has evolved ?

Ans. Maize is a cereal crop of global importance. It belongs to the genus **Zea** of the family Poaceae (old gramineae). It is grown both for food and fooder purposes. Maize is native of America from where it has spread to other parts of the world. Maize (**Zea** mays) is the only species in the genus Zea. It has two close relatives, *viz.,* Gamagrass **(Tripsacum)** and Teosinte **(Euchlaena)**. Teosinte is considered to be the closest relative of maize because maize crosses readily with Teosinte. Moreover, chromosome number is also similar in maize and annual forms of Teosinte (2n = 20). The present day commercially cultivated forms of maize are believed to have originated through introgression of genes from its wild relative Teosinte, and Teosinte is considered to have originated from a cross between primitive corn and *Tripsacum*. This is the most widely accepted theory about the genetic origin of maize crop.

Q.13. How cultivated rice has evolved ?

Ans. Rice is a staple food of global importance. It belongs to the genus **Oryza** in the family **Poaceae (old Gramineae).** It is grown in India and China from time immemorial. South and South East tropical Asia is considered to be the native place of predominantly cultivated species of rice, Oryza sativa, because vast diversity of this species is found in this region. The other cultivated species is **Oryza glaberrima.** which is native of West tropical Africa and also grown there. It is generally believed that the Asian rice **(O. sativa)** has originated in Asia from the perennial Asian wild species **Oryza rufipogon.** The African cultivated species **(O. glaberrima)** is believed to have originated in West Africa from wild diploid species **O. breviligulata.** Still there is controversy about the progenitor of Asian rice. Some researchers believe that Asian rice has originated from perennial wild species, while others consider annual form as the ancestor of cultivated rice (Fehr., 1987).

43

GREEN REVOLUTION

Q. 1. What is green revolution ?

Ans. Green revolution refers to the quantum jump in the productivity of cereals especially wheat and rice through the use of dwarfing genes.

Q. 2. Who is the Owner of green revolution ?

Ans. Green revolution is publicly owned technology. The ownership of the green revolution was vested in the farmer.

Q. 3. What are the funding agencies of green revolution ?

Ans. Green revolution was funded by Government. It was funded mainly by Rockefeller Foundation, Ford Foundation, Mexican Government, International Centre for Wheat and maize Improvement (CIMMYT) and International Rice Research Institute, Philippines. In the beginning, the green revolution program was funded by the Rockefeller Foundation. Later on other organizations such as the World Bank, Food and Agricultural Organization of the United Nations (FAO), United States Agency for International Development (USAID), and other national and international organizations joined the Rockefeller Foundation to make this program successful. It succeeded in terms of increasing food production in Asia, Latin America, and even parts of the industrialized world such as Great Britain.

Q. 4. What was the accessibility of green revolution ?

Ans. The green revolution was accessible to all the farmers and all Nations.

Q. 5. Whether IPR are applicable to green revolution ?

Ans. Intellectual Property Rights are not applicable i.e. patenting is not done in case of green revolution.

Q. 6. When did the green revolution take place ?

Ans. Though the term green revolution was coined in 1958 in USA, the actual green revolution took place in 1970s after the release of high yielding varieties of wheat and rice.

Q. 7. What was the source of gene transfer in green revolution?

Ans. In case of green revolution, the gene transfer was within the species. The dwarfing gene was transferred from the genotypes of the same species.

Q. 8. What are the methods involved in green revolution ?

Ans. Conventional methods such as selection, hybridization and mutation are involved in green revolution.

Q. 9. What's about the transparency of green revolution ?

Ans. There was 100 per cent transparency in case of green revolution.

Q. 10. Who are the beneficiaries of green revolution ? .

Ans. The green revolution has brought lot of benefits to the farmers, nation and consumers.

Q. 11. Whether bio-safety measures are required for green revolution ?

Ans. Bio-safety measures are not required for green revolution.

Q. 12. Whether bio-ethical measures were required for green revolution ?

Ans. No, bio-ethical measures were not required for green revolution.

Q. 13. Whether organic farming of end product is permitted ? '

Ans. Yes, organic farming of end product of green revolution is Permitted.

Q. 14. What the end product of green revolution ?

Ans. The end product of green revolution are simple varieties.

Q. 15. Whether evolution of super weeds is possible in case of green revolution ?

Ans. In case of green revolution evolution of super weeds is not possible.

Q. 16. How much technical skill is required for green revolution ?

Ans. In case of green revolution moderate technical skill is required.

Q. 17. How much initial cost of infra structure is required for starting conventional breeding ?

Ans. In case of green revolution the initial cost of infra structure is very low.

Q. 18. How much time is requirement for developing varieties through conventional methods ?

Ans. In case of green revolution the time required to develop new variety is very long as compared to gene technology. It takes 10-12 years for developing new variety .

Q. 19. What type of genes are involved in case of green revolution ?

Ans. In case of green revolution, dwarfing genes are involved in developing high yielding varieties.

Q. 20. What was the major emphasis in green revolution ?

Ans. In case of green revolution, the major emphasis was on improving the productivity and plant stature.

Q. 21. What were the crops covered by green revolution ?

Ans. The green revolution mainly covered cereals, viz. wheat and rice.

Q. 22. What are the advantages of green revolution ?

Ans. There are several advantages of green revolution such as (i) higher yield, (ii) food security , (iii) earliness, (iv) lodging resistance, and (v) uniform produce.

Q. 23. What are the disadvantages of green revolution ?

Ans. The main disadvantages of green revolution include (i) loss of biodiversity (ii) poor quality, (iii) high input requirements, (iv) danger, of uniformity, (v) depletion of resources exploitation of resources, and (vi) poor recovery of straw.

Q. 24. Who is known as father of green revolution ?

Ans. Dr, Norma E. Borlaug , a Mexican scientist was responsible for green revolution in wheat. Hence, he is rightly called as father of green revolution.

Q. 25. Who is responsible for green revolution in India ?

Ans. Dr. M. S. Swaminathan, Ex-Directer General of ICAR is responsible for green revolution in India. Hence, he is called as father of green revolution in India.

Q. 26. What was the source of dwarfing gene in wheat ?

Ans. The genotype Norin 10 was the source of dwarfing gene in wheat which was used? for developing high yielding varieties of wheat.

Q. 27. What was the source of dwarfing gene in rice ?

Ans. The genotype Dee Geo-Woo-gen was the source of dwarfing gene in rice which was used for developing high yielding varieties of rice.

Q. 28. Where the high yielding varieties of wheat were first developed ?

Ans. The high yielding varieties of wheat were first developed at CIMMYT, Mexico and later spread to other countries.

Q. 29. Where the high yielding varieties of rice were first developed ?

Ans. The high yielding varieties of rice were first developed by International Rice Research Institute, Philippines and later spread to other countries.

Q. 30. When did the green revolution reach its peak ?

Ans. The green revolution reached its peak in 1970s.

Q. 31. What are the features of high yielding varieties ?

Ans. The important features of high yielding varieties of wheat and rice include

(i) short stature, (ii) photo-insensitivity , (iii) wider adaptability (iv) responsive to fertilizers, (v) responsive to irrigation, (vi) high Co2 fixation efficiency, (vii) higher tillering capacity, and (viii) uniform produce.

Q. 32. What are the features of traditional varieties of wheat and rice ?

Ans. The important features of traditional varieties of wheat and rice include (i) tall stature, (ii) photo-sensitivity , (iii) narrow adaptation, (iv) poor response to fertilizers, (v) poor response to irrigation, (vi) poor Co2 fixation efficiency, (vii) and poor tillering capacity.

44

GENE REVOLUTION

Q. 1. What is gene revolution ?

Ans. Gene revolution refers to quantum jump in the productivity of various field crops through use of transgenic cultivars. It has been achieved through the application of Agricultural Biotechnology. Hence agricultural biotechnology is referred to as the 'Evergreen Revolution' or the 'Gene Revolution'. Modern biotechnology plays a vital role in the' "Gene Revolution".

Q. 2. Describe main points related to gene revolution.

Ans. Main points related to gene revolution are listed below.

1. Gene revolution is also known as ever green revolution.
2. Gene revolution has been achieved through the application of agricultural biotechnology.
3. It has been achieved by private seed companies.
4. It is a privately owned technology .
5. It has been funded by private seed companies.
6. Production and marketing of GM crops is controlled by private seed companies.
7. Private companies do not disclose the parentage of the hybrids.

Q. 3. What are the main features of gene revolution ?

Ans. Presently, the genetically modified crops (GM crops) are gaining increasing importance the world over. The GM crops are expected to spread around the world and lead to a global Gene Revolution. The main features of gene revolution or ever green revolution in terms of ownership, funding source, accessibility of the technology, application of IPR, gene transfer, method employed, transparency, benefits, bio-safety, bioethical measures, etc. in relation to green revolution are briefly discussed hereunder .

Q. 4. Who is the Owner of gene technology ?

Ans. It is privately owned technology. The ownership of the gene revolution is vested in the Private seed Companies or Corporations. Genetically modified crops are largely the product of private industry. This is partly because new technologies are far more costly than existing ones, and the biotechnology industry was able to gather the necessary funds to develop these technologies long before public awareness of GM crops could lead to publicly generated funding for GM crop development. Successful companies typically focus on their markets with the intent of generating profit. With regard to agricultural biotechnology, companies in the United States and elsewhere have thus far created primarily seeds that farmers in developed countries can and will purchase: corn and soybeans that can tolerate a particular herbicide, corn and cotton that are resistant to particular pests, and food crops that last longer on the supermarket shelf.

Q. 5. What are the funding agencies of gene technology ?

Ans. Gene technology is funded by private seed companies. Six seed companies viz., Monsanto, Syngenta, Bayer Crop Science, DuPont, Dow and BASF Plant Science control practically the entire research and output in the field of transgenic plants. In the beginning, the green revolution program was funded by the Rockefeller Foundation. Later on other organizations such as the World Bank, Food and Agricultural Organization of the United Nations (FAO), United States Agency for International Development (USAID), and other national and international organizations joined the Rockefeller Foundation to make this program successful. It succeeded in terms of increasing food production in Asia, Latin America, and even parts of the industrialized world such as Great Britain.

Q. 6. What is the accessibility of gene technology ?

And. The transgenic seed belongs to -various seed companies, which strictly control its production and marketing. Developed countries grow such costly seeds because of their high yield under their environmental conditions. Thus, in developed nations, GM crop technology has could revolutionize farming. However, the current GM crop seed varieties are neither affordable nor useful to most of the poorer farmers in the world; hence, their revolutionary impact in the developing world has been limited so far.

Q. 7. Whether IPR are applicable to gene technology ?

Ans. Intellectual Property Rights are applicable to the products of transgenic technology i.e. patenting is done. A fundamental challenge in this newest agricultural movement that did not arise during the Green Revolution is the definition and treatment of intellectual property (IP). IP issues are central to the Gene Revolution that can often limit the valuable diffusion of such technology .

Q. 8. What are the possibilities of gene transfer ?

Ans. The gene transfer is possible between two totally unrelated organisms or across any species. The gene transfer is possible from animals to plants and vice versa. The Gene Revolution combined with genetic engineering, allows unique trait combinations across species. For example, daffodil and bacterial genes can be introduced into the rice genome so that the rice produces betacarotene, the precursor of vitamin A.

Q.9. What types of methods are involved in gene technology ?

Ans. Highly sophisticated biotechnological approaches are used for development of transgenic plants. The science and technology required to create GM crop seeds are far more complicated than the science and technology used to create Green Revolution agricultural advancements.

Q. 10. What about the transparency of gene technology ?

Ans. The information is kept secret by the seed companies responsible for developing the technology. GM seeds are created largely through private enterprises rather than through public-sector efforts.

Q.11. What are benefits of gene technology ?

Ans. The gene revolution is expected to bring even greater benefits to the small and marginal farmers than green revolution. The benefits of the current varieties of GM crops include yield increase, reduced agricultural inputs such as pesticides and fertilizers, reduced vulnerability to the vagaries of nature, and improved nutritional content. For the most part, these benefits have been limited to parts of the developed countries to which current GM crop development and marketing have been targeted and also in those countries that have allowed cultivation of such crops. .

Q. 12. Whether bio-safety measures are required for gene technology ?

Ans. Yes, bio-safety measures are required in case of genetically modified crop plants.

Q. 13. Whether bio-ethical measures are required for gene technology ?

Ans. Yes, bio-ethical measures have to be adopted in case of genetically modified crop plants.

Q. 14. Whether organic farming of end product is permitted ?

Ans. No, organic farming of genetically modified crop plants is not permitted.

Q.15. What the end product of gene technology ?

Ans. The end product of agricultural biotechnology is a transgenic cultivar.

Q. 16. Whether evolution of super weeds is possible in case of transgenic Plant ?

Ans. Evolution of super weeds is possible after out crossing of transgenic plants with wild relatives of cultivated plants.

Q. 17. How much technical skill is required for gene technology ?

Ans. High technical skill is required in developing genetically engineered crop plants.

Q. 18. How much initial cost of infra structure is required for starting transgenic breeding ?

Ans. The initial cost of infra structure required for evolving cultivars through agricultural biotechnology is very high. Costly equipments, glassware and chemicals are required to carryout research in the biotech laboratory .

Q. 19. How much time is requirement for developing transgenic plants ?

Ans. It is a rapid method of crop improvement. The development of new transgenic variety takes 4-5 years as compared to 10-12 years taken by conventional methods of plant breeding.

Q. 20. What type of genes are involved in gene technology ?

Ans. Trans gene are involved in the development of transgenic cultivars. In case of green revolution, the high yielding cultivars were developed by using dwarfing genes.

Q. 21. What is the major emphasis in gene technology ?

Ans. In the gene revolution, major emphasis has been laid in the development of cultivars resistant to insects and herbicides.

Q. 22. What are advantages of GM technology ?

Ans. The cultivation of GM crops has played significant role in improving yield, in some cases food quality and reducing use of pesticides or particularly harmful herbicides. It is expected that GM crops can revolutionize world agriculture, particularly in developing countries. This technology will substantially improve food security, reduce malnutrition, and increase rural income, and in some cases even reduce environmental pollutants.

Q. 23. What criteria are considered important in evaluation of a technology for agricultural revolution ?

Ans. Five criteria are considered important in evaluation of a technology *for* agricultural revolution. These criteria include (i) benefit to farmers, (ii) increase in production, (iii) adoption by farmers and acceptance by the consumers, (iv) coordination among technology provider, policy makers and users, and (v) the technology should be affordable by users.

1. The new technology should provided a net benefit to farmers.
2. The technology should substantially improve agricultural production, food nutrition, or both; or substantially decrease necessary inputs such as fertilizer or water.

3. People should be willing to adopt the new technologies, and consumers should accept the products of the such technology.
4. There should b e cooperation among technology provider, policy makers and users.
5. The technology should be affordable by the users and sustainable without public subsidization.

The GM crops meet all five criteria for an agricultural revolution on regional scale. In the United States, Canada, China, and Argentina for example, genetically modified varieties of soybeans, corn, and cotton now constitute about 30 to 80 percent of total plantings of these crops and are beneficial to the growers. In other nations, such as India and South Africa, farmers have recently begun to plant GM crops and have made the beginnings of realizing, benefits of Gene Revolution. However, the revolution has yet to occur on a global scale.

Q. 24. What is the future of gene technology ?

Ans. In future, to achieve agricultural revolution through gene technology, the following points should be given due importance.

1. Affordable Technology: The new technology should be affordable by the farmers of developing-world. Moreover, the farmers must understand how to use them.

2. Research by Public Sector: There is a need for larger investments in research in the public sector. Partnerships between the public and private sectors can result in more efficient production of GM crops that are useful to the developing world and can expand the accessibility of those crops and their associated technologies to developing-world farmers.

3. Due Importance to Agriculture: Agricultural development must be given due importance from a policy perspective in both donor and recipient nations. In view of increasing global population, agricultural development is more necessary than ever to eliminate malnutrition and prevent famine, particularly in sub-Saharan Africa. GM crops seem to be effective means for addressing these problems. The policymakers should work jointly on such issues.

4. Proper Regulatory System: Policymakers in the developing world must set regulatory standards that take into consideration the risks as well as the benefits of foods derived from GM crops. This goal is crucial to the cooperation of the many stakeholders that are affected by GM crops and also for the sustainability of the GM crop movement in the near future. Without regulations that explicitly take into account potential benefits to both farmers and consumers, those nations that might stand to benefit most from

GM crops may be discouraged from allowing them to be plated. Revised regulations on genetically modified crops must accompany widespread collective policy efforts to. revitalize agricultural development.

If the GM crops are free of adverse effects on health and environment, they have the potential to provide benefits to farmers and consumers around the globe.

Such crops will reduce the use of potentially harmful chemicals or scarce water supplies for agriculture. It can then indeed become a blue "Gene Revolution."

The effects of GM crops′ on health and environment are uncertain. The cultivation of such crops can have unintended adverse effects on both animal health and the environment.

45

INTRODUCTORY STATISTICS

Q. 1. Define statistics ?

Ans. Statistics is a branch of applied mathematics which deals with collection, presentation, analysis and interpretation of numerical data. The statistics which is commonly used in genetics and plant breeding is divided into two parts, viz. (1) Statistical methods, and (2) Experimental designs. There are three types of statistics as given below :

(i) First degree statistics or First order statistics

(ii) Second degree statistics or Second order statistics

(iii) Third degree statistics or High order statistics

Q. 2. What do you mean by first degree statistics ?

Ans. It includes mean, which is used for the measurement of all type of parameters. The calculation of first degree statistics is simple. The estimates based on first order statistics are more robust and reliable due to high level of precision.

Q. 3. What are uses of first order statistics in plant breeding and genetics ?

Ans. Arithmetic mean is used in comparing varieties or breeding material. In biometrical genetics, first degree statistics in used for the study of generation means, heterosis and inbreeding depression, combining ability effects, metroglyph analysis, stability analysis etc.

Q. 4. What is second degree statistics ?

Ans. When variances and covariances are used for estimation of certain parameters, it is called second degree statistics. The parometers which are based on second degree statistics are not statistically every robust because of inaccuracies involved in the estimation of variances and covariances. Moreover, calculation of second order statistics is more difficult than first order statistics.

Q. 5. What are applications of second order statistics in genetics and plant breeding ?

Ans. In biometrical genetics, second order statistics is used for the estimation of correlations, path coefficients, discriminant function, D^2 statistics, heritability, coheritability, genetic advance, components of variance in diallel, partial diallel, line x tester cross, triple test cross, bioparental cross, triallel cross and quardriallel cross.

Q. 6. What is third degree statistics ?

Ans. It includes complex interactions like curtosis and skewness, which are used in fitting frequency curves, surfaces and flatness. In biometrical genetics, high order statistics is used to study skewness of F_2, F_3 and biparental progenies.

Q. 7. What is χ^2 test ?

Ans. χ^2 is a test of statistical significance which is used to test the significance of difference between observed and expected frequencies or ratios. The general formula of χ^2 test is as follows :

$\chi^2 = \Sigma(O - E^2)/ E$, where Σ = summation, O = observed frequencies and E = expected frequencies.

Q. 8. Who developed χ^2 test ?

Ans. The χ^2 test was developed by Karl Pearson who was an English mathematican. He applied statistics to biological problems of heredity and evolution.

Q. 9. What are applications of χ^2 test in genetics and plant breeding ?

Ans. χ^2 test is commonly used in Mendelian genetics and population genetics. In genetics, X^2 test is used for following three main purposes :

(i) To test the validity of various segregation ratios.

(ii) For detection of linkage.

(iii) For study of gene frequencies in population genetics.

Q. 10. What is degree of freedom ?

Ans. The degree of freedom refers to number of independent comparisons. Generally degree of freedom is one less than the total classes of comparison. In two class segregation (3 : 1), degree of freedom is 1 and in four class segregation (9 : 3 : 3 : 1), the degree of freedom is 3.

Q. 11. What is arithmetic mean ?

Ans. Arithmetic mean is defined as the sum of all observations in a simple divided by their number. It is denoted by $\overline{X}$ and is calculated as follows

Arithmetic mean $\overline{X} = (\Sigma X)/ N$ where Σ= summation, X = an observation and N = number of observations in a sample.

Q. 12. What are measures of dispersion ?

Ans. The degree to which numerical data tend to spread about the mean value is called dispersion. Thus dispersion is a measure of variation in a sample. The measure of dispersion include range, standard deviation, variance, standard error and coefficient of variation. These measures are also known as simple measures of variability.

Q. 13. What do you mean by range ?

Ans. Range is the difference between the highest and the lowest values present in the observations in a sample. Suppose in cotton 20 observations have been recorded on seed oil content. The highest value is 25% and the lowest value is 15%. The range will be 25–15 = 10. Thus it is a measure of the spread of variation in a sample.

Q. 14. Define standard deviation.

Ans. Standard deviation is the square root of the arithmetic mean of squares of all deviations measured from the mean. In other words, it is the square root of the variance. It is the best measure of variation in a population and is calculated as follows :

$$\text{Standard deviation} = \sqrt{\frac{\Sigma X^2 - (\Sigma X)^2 / N}{N-1}}$$

where, ΣX, X^2 and N = summation, an observation, square of an observation and number of observations, respectively.

Q. 15. What is variance ?

Ans. Variance is defined as the average of the squared deviation from the mean or it is the square of standard deviation. It is an effective measure of variability and is estimated by the following formula.

$$\text{Variance} = \frac{\Sigma X^2 - (\Sigma X)^2 / N}{N-1}$$

Q. 16. What is standard error ?

Ans. It is the measure of the mean difference between sample estimate mean ($\bar{X}$) and population parameter (μ). It is the measure of uncontrolled variation present in a sample. It is estimated as follows :

$\text{Standard Error} = \frac{SD}{\sqrt{N}}$ where SD = standard deviation and N = number of observations.

Q. 17. What is coefficient of variation ?

Ans. The ratio of standard deviation of a sample to its mean expressed in percentage is called coefficient of variation. Coefficient of variation does not have any unit and is estimated as follows :

$$\text{Coefficient of variation} = \frac{SD}{\bar{X}} \times 100$$

where, SD = Standard deviation, $\bar{X}$ = Mean.

Q. 18. Who evolved the coefficient of variation ? How it is useful in plant breeding ?

Ans. The coefficient of variation was evolved by Karl Pearson. A sample in which coefficient of variation is higher would have greater variation than the one in which it is lower. In plant breeding, phenotypic, genotypic and environmental coefficients of variation are estimated from corresponding variances.

Q. 19. What do you mean by tests of significance ?

Ans. Statistical procedures which are used to decide whether differences under study are significant or nonsignificant are known as tests of significance. The commonly used tests of significance include Z-test, t-test and F-test. These tests are used to decide whether the difference under study is significant or can be attributed to chance or the fluctuation of sampling.

Q. 20. What is null hypothesis ?

Ans. A hypothesis of no difference is known as null hypothesis. This hypothesis was given by Fisher which is devoted as H_0. This hypothesis states that no true difference exists in the sample result and population parameter. The observed difference is only due to sampling fluctuation.

Q. 21. Define Z-test.

Ans. It is a test of significance which is used to compare two means when the sample size is large (more than 30).

Q. 22. What is t-test ?

Ans. A test of significance which is used for comparing two means when sample size is small (upto 30). It is of two types : Student t—used with paired observations and Fisher's t—used when observations are unpaired.

Q. 23. What is analysis of variance ?

Ans. The statistical procedure which separates the total variation into different components is called analysis of variance. It has following four main advantages :

(i) It is useful in estimating components of variance.

(ii) It provides basis for test of significance.

(iii) It permits estimation of phenotypic, genotypic and environmental coefficients of variability.

(iv) It also permits estimation of broad sense heritibility and expected genetic advance.

Q. 24. Define analysis of covariance and explain its uses.

Ans. The statistical technique which splits simultaneously the variation of two variables (characters) into various components is called analysis of covariance. It has following main advantages along with analysis of variance.

(i) It permits estimation of coheritability

(ii) It also permits estimation of correlations, regression, path coefficients, selection indices etc.

Q. 25. What are joint applications of variances and covariances in plant breeding ?

Ans. The estimates of variances and covariances permit estimation of following statistics :

(i) Phenotypic, genotypic and environmental correlations

(ii) Above three types of path coefficients.

(iii) Regression coefficient

(iv) Coheritability

(v) D^2 statistics

(vi) Selection indices, etc.

Q. 26. What is regression coefficient ?

Ans. Regression coefficient is a statistical measure of the average functional relationship between two or more variables. In regression analysis, one variable is considered as dependent and other is independent. Thus it measures the degree of dependence of one variable on the other. Regression coefficient was first used by Sir Francis Galton for estimating the relationship between the heights of fathers and their sons. It is expressed in terms of original unit of data and is denoted by *b*.

Q. 27. What is correlation ?

Ans. The statistical measure that measures the degree and direction of association between two or more than two variables is called correlation. It is represented by r and is independent of the unit of measurement. Its value lies between—1 and 1. The general formula for calculation of correlation is given below :

Correlation r × y =Cov. (Xy)/ $\sqrt{vX.v.Y}$

where r × y = Correlation between X and *y*

Cov(Xy) = Covariance between X and *y*

v X = Variance of X, and

vy = Variance of *y*

Q. 28. What are experimental designs ?

Ans. Experimental designs are various types of plot arrangements which are used to test a set of treatments to draw valid conclusions about a particular problem. Experimental designs are used for following purposes :

(i) For testing the significance of difference among various objects of comparison.
(ii) For testing of treatments in scientific manner.
(iii) For partitioning of variation in different components.
(iv) For estimation of various statistical estimates.
(v) For proper interpretation of scientific results and drawing valid conclusions.

Q. 29. What is an experiment ?

Ans. Experiment is a scientifically planned method. The experiement is conducted to draw valid conclusion about a particular problem. The conclusion is drawn based on statistical observations.

Q. 30. Define treatment.

Ans. Treatments are nothing but various objects of comparison. In plant breeding, treatments include a set of varieties, a set of hybrids, a set of advanced breeding cultures. In agronomic trials, it may include various levels of fertilizers, sowing dates, spacing, seed rate, number of irrigations etc.

Q. 31. What is experimental unit ?

Ans. Experimental unit refers to the group of material to which a treatment is applied in the experiment. It may be a plot of land, a patient in a hospital, a group of cattle in a dairy etc.

Q. 32. What are basic principles of experimental designs ?

Ans. There are three basic principles of experimental designs, viz. replication, randomization and local control. All these principles help in reducing the experimental error and increasing accuracy of results.

Q. 33. What is replication ?

Ans. The repetition of treatments under investigation is known as replication. There are three main advantages of replication as follows :

(i) Increase in replication increases the precision by reducing the experimental error to a great extent.
(ii) Replication helps in detection of variation in a treatment and thus helps in comparison of various treatments.
(iii) It helps in estimation of average performance of various treatments in an experiment.

Q. 34. Define randomization and explain its advantages.

Ans. Randomization refers to allocation of treatments to different plots, in an experiment, by a random process. Randomization is done with the help of random statistical table. Randomization gives equal chance to all treatments for being allotted to a more fertile plot as well as less fertile plot.

Q. 35. What is local control ?

Ans. The principle of making use of greater homogeneity in group of experimental units for reducing the experimental error is called local control. It is also known as error control. The fertility variation may be either as gradient (in strips) or in patches (sporadic). The first type of fertility effects can be reduced by dividing the field into homogeneous blocks and second type of fertility is minimized by randomization of treatments within the block.

Q. 36. Define experimental error.

Ans. The variation due to environmental factors or uncontrolled factors is called experimental error. It can be reduced by adopting, replication and randomization.

Q. 37. What is correction factor ?

Ans. In the analysis of variance and covariance, the square of grand total divided by number of observations is called correction factor. It is calculated as follows :

$$\text{Correction factor} = \frac{(\text{Grand Total})^2}{N}$$

where N = number of observations.

Q. 38. What is F-test ?

Ans. It is a test of significance which is used for testing the significance of differences among several treatments. It differs from Z-test and t-test which are applied to test the significance of difference between two treatment means or between sample mean and population mean.

Q. 39. Define F-value ?

Ans. It is the ratio between the treatment variance and error variance. It is also called variance ratio and is estimated as follows :

F-Value = Treatment variance/Error variance

The observed F value is compared with table F value at appropriate degrees of freedom and at desired level of probability (5% or 1%). If the observed F value is more than Table F value, the differences among treatments are considered significant and vice-versa.

Q. 40. What is critical difference ?

Ans. The least significant difference, greater than which all the differences are significant is known as critical difference (CD) or least significant difference (LSD). It is estimated as follows :

Critical difference = SE difference × t

SE difference = $\sqrt{2VE/r}$ where Ve = error variance,

r= replications, t = table value at error degrees of freedom.

Critical difference is used to compare the observed differences among different treatments. If the difference is greater than critical difference, it is considered as significant and vice-versa.

Q. 41. What is the difference between χ^2 test and F-test ?

Ans. χ^2 test is used to test the significance of difference between observed and expected frequences, whereas F test is used to test the significance of difference among several means. Moreover, χ^2 test is based on single variance, whereas F test is based on ratio of two variances.

Q. 42. What are different types of experimental designs used in plant breeding ?

Ans. Various types of experimental designs which are used in genetics and plant breeding are listed below :

(i) Completely randomized design
(ii) Randomized block design
(iii) Latin square design
(iv) Lattice design
(v) Split plot design, and
(vi) Augmented design.

Q. 43. What is completely randomized design ?

Ans. The design which is used when the experimental material is limited and experimental unit is homogeneous is called completely randomized design. This design is specially used for pot culture experiments. The principle of local control is not adopted in this design because homogeneous blocks are not formed in this design.

Q. 44. What is randomized block design ?

Ans. The experimental design which controls fertility variation in one direction only is called randomized block design. Adoption of this design is useful where the fertility variation between the blocks is significant. The principle of local control is adopted in this design because the experimental field is divided into homogeneous blocks. This design can test upto 20 treatments Beyond 20, the efficiency of error control is decreased due to increase in heterogeneity within the block.

Q. 45. Explain Latin Square design.

Ans. This experimental design simultaneously controls fertility variation in two directions. In this design the number of treatments, replications, rows and columns is equal. This design adopts principl of local control of forming rows and columns. In this design, maximum 12 treatments can be tested.

Q. 46. What is lattice design ?

Ans. Lattice designs are incomplete block designs in which number of treatments or varieties forms square.

Q. 47. What is augmented design ?

Ans. This an experimental design which is used to test a large number of germplasm lines in a limited experimental area. In this design, standard or check varieties are replicated among the germplasm lines after a definite number of germplasm lines. The concept of augmented design was developed by Federer (1956).

Q. 48. What is split plot design ?

Ans. The experimental design in which experimental plots are divided into main plots, sub-plots and ultimate plots is called split plot design. In this design several factors are studied simultaneously with different levels of precision. It includes factors like irrigation, sowing date, seed rate etc.

46

INTRODUCTION TO COMPUTERS

Q. 1. What do you mean by computer ?

Ans. Computer refers to an electronic device that can transmit, store and process information or data in a desired form.

Q. 2. Define Bioinformatics.

Ans. The computer aided study of biology particularly genetics and molecular biology is called bioinformatics. Now the science of bioinformatics is gaining increasing importance in life science especially in the field of molecular biology and plant genetic resources. It is the interface between computer and biology. In other words, it is the application of information technology in the study of biology. It has wide applications in handling data related to plant genetic resources.

Q. 3. What are types of computers ?

Ans. Based on size and capacity, computers are of four types, viz., (1) super computers. (2) mainframe computers, (3) minicomputers, and (4) micro-computers.

Q. 4. What is super computer ?

Ans. Super computers have extremely large storage capacity. Its speed is at least 10 times faster than other computers. Such computers are generally used for large scale numerical problems in scientific and engineering disciplines. These computers can process billions of instructions per second.

Q. 5. What are mainframe computers ?

Ans. Mainframe computers can also process data at very high speeds i.e., hundreds of million instructions per second and they are quite expensive. Normally, they are used in banking, airlines and railways, etc. for their applications.

Q. 6. What do you mean by mini computers ?

Ans. Mini computers are lower to mainframe computers in terms of speed and storage capacity. They are also less expensive than mainframe computers. Some of the features of mainframes will not be available in mini computers. Hence, their performance also will be less than that of mainframes.

Q. 7. What are micro computers ?

Ans. Micro-computers are smallest general purpose computers. Such computers have been designed for use by one person at a time and, therefore, are known as personal computers. A computer designed for use by an individual or single person is called personal computer. In price, personal computers are relatively cheaper. They are used by businessman for word processing, accounting, desktop publishing, and for running spreadsheet and database management applications. Personal Computers include desktop computers and portable computers. Portable computers include desktop computers, notebook computers, laptop computers, palmtop computers and tablet computers.

Q. 8. What is desktop computer ?

Ans. A computer designed to keep on top of a desk and work is referred to as desktop computer. Desktop models designed to be very small are sometimes referred to as slim line models.

Q. 9. Define notebook computer.

Ans. An extremely light weight (less than 6 pounds) small personal computer which can fit easily in a briefcase is called notebook computer. In terms of computing power, modern notebook computers are nearly equivalent to personal computers. They have the same CPUs, memory capacity, and disk drives. Notebook computers cost about twice as much as equivalent regular-sized computers. Notebook computers come with battery packs that enable you to run them without plugging them in. However, the batteries need to be recharged every few hours.

Q. 10. What is laptop computer ?

Ans. A small, portable computer designed to keep on your lap and work is called laptop computer. Nowadays, laptop computers are more frequently called notebook computers. Laptop computers are portable computers. They are lightweight computers with a thin screen. They are also called as notebook computers because of their small size. They can operate on batteries and hence are very popular with travelers. The screen folds down onto the keyboard when not in use.

Q. 11. Define palmtop computer.

Ans. A small sized computer designed to keep on palm and work is known as palmtop computer. Palmtops are in limited use, but they are used as phone books and calendars. Palmtops that use a pen rather than a keyboard for input are often called hand-held computers or personal digital assistant (PDA). Because of their small size, most palmtop computers do not include disk drives.

Q. 12. What is a tablet computer ?

Ans. A tablet computer, or a tablet, is a mobile computer, larger than a mobile phone or personal digital assistant. A tablet PC is a wireless, portable personal computer with a touch screen interface. It is smaller than a notebook computer but larger than a smart phone. It is a type of notebook computer that has an LCD screen on which the user can write using a special-purpose pen, or stylus. The handwriting is digitized and can be converted to standard text through handwriting recognition, or it can remain as handwritten text. Tablet PCs also typically have a keyboard and/or a mouse for input. A tablet computer may be of three types, viz., (a) convertible, (b) hybrid, and (c) rugged. These are defined as follows.

Q. 13. What is convertible tablet computer ?

Ans. It has typically a display that rotates 180 degrees and can be folded to close, screen up, over the integrated keyboard. Convertible models may allow user Input through a variety of methods in addition to the hardware keyboard, including natural handwriting with a stylus or digital pen and typing through a screen-based software keyboard.

Q. 14. What is hybrid tablet computer ?

Ans. It is like a regular notebook but with a removable display that functions independently as a slate.

Q. 15. Define rugged tablet computer.

Ans. It is a slate-like model that is designed to withstand rough handling and extreme conditions. Rugged tablets are usually encased in a protective shell and have shock-protected hard drives.

Q. 16. Describe architecture of a microprocessor.

Ans. A microcomputer consists of four components, viz., microprocessor, input, output and memory. These components are organized around a common communication path called bus. The entire group of components is called a system or a microcomputer system. A brief description of various computer components ia presented as follows.

Q. 17. Define microprocessor.

Ans. It is a semiconductor device consisting of electronic logic circuits. It is manufactured by using either large or very large scale integration technique. It performs computing functions and consists of three units, viz., (i) arithmetic logic unit, (ii) register unit and (iii) control unit. These are defined as follows.

Q. 18. What is Arithmetic Logic Unit ?

Ans. In this area of microprocessor, computing functions are performed on data. The ALU performs arithmetic operations such as addition and subtraction and logic operations. Results are stored either in registers or in memory or sent to output devices.

Q. 19. Define Register Unit.

Ans. This area of microprocessor consists of various registers which are used primarily to store data temporarily during execution of program.

Q. 20. What is control unit ?

Ans. The control unit provides necessary timing and control signals to all the operations in microprocessor. It controls the flow of data between the microprocessor and peripherals (including memory).

Q. 21. What do you mean by memory ?

Ans. This is a medium or device that stores binary information such as instructions and data, and provides that information to the microprocessor as and when necessary. The memory is of two types, viz., read only memory (ROM) and read and write memory. These are defined as follows.

(i) ROM(Read Only Memory): A memory that stores binary information during the operation of computer. This memory is used as writing pad to write user programs and data. The information stored in this memory can be read only.

(ii) Read/ Write Memory): It is also known as random access memory (RAM) and user memory. The information stored in this memory can be read and altered easily.

Q. 22. What is input device ?

Ans. A device that transfers information from outside world to the computer. It includes key board.

Q. 23. What is output device ?

Ans. Computer devices that transfers information from computer to outside world. It includes monitor and printer. The input and output devices of computer are jointly known as peripherals.

Q. 24. Define bit.

Ans. A binary digit, 0 and 1.

Q. 25. What do you mean byByte ?

Ans. A group of eight bits.

Q. 26. What is Bus.

Ans. A group of lines used to transfer bits between the microprocessor and the components of computer system.

Q. 27. What is hardware ?

Ans. The physical components of computer such as key board, processing unit, monitor, mouse and printer.

Q. 28. What do you mean by software ?

Ans. A group of program written for the computer to perform a task.

Q. 29. What is nibble ?

Ans. A group of four bits.

Q. 30. What is a program ?

Ans. A set of instructions written in a specific sequence for the computer to accomplish a given task.

Q. 31. What is word ?

Ans. A group of bits the computer can recognize and process at a time.

Q. 32. What are external storage devices ?

Ans. Various media that are used to store information temporarily and transfer the same from one computer to another are referred to as external storage devices. Such devices are not permanently fixed inside a computer.

Q. 33. What are advantages of external storage devices ?

Ans. There are several advantages of external storage devices as given below:

(i) External storage devices provides additional storage other than that available in computer.

(ii) Data can be transported easily from one place to another.

(iii) It is useful to store software and data that is not needed frequently.

(iv) External storage also works as data back up.

(v) This back up may prove useful at times such as fire or theft because important data is not lost.

Q. 34. What are commonly used external storage devices ?

Ans. The commonly used external storage devices include floppy disk, compact disk, digital versatile disk, USB storage devices, SD cards, etc.

Q. 35. What is Floppy Disk ?

Ans. It is a thin plastic disk coated with magnetic oxide and looks like a phonograph. It is available in various sizes. Floppy disk has been replaced by other better storage devices such as CD, DVD, USB, SD card and external storage hard disk. Now floppy disks are no more in common use.

Q. 36. What do you mean by compact disk ?

Ans. Compact disk is commonly known as CD. The speed of compact disk is much less than a hard disk. The storage capacity is 700 MB. There are three types of CDs as discussed below.

(i) CD-ROM: It is compact disk read only memory. It can be read only.

(ii) CD-Recordables: It was invented in 1990s. Using CD-R it is possible to write data once on a disk.

(iii) CD-ReWritables: There is a limit on how many times a CD-RW can be written. Presently this limit is 1000 times. CD-RW drives are compatible with CD-ROM and CD-R.

Q. 37. What do you mean by Digital Versatile Disk ?

Ans. It is commonly known as DVD. Its speed is much faster than CD but not as fast as hard disk. The standard DVD-5 technology has a storage capacity of 4.7 GB. The storage capacity changes with the standard used. Its storage capacity (4.7 GB) is much higher than a CD (700 MB). It is achieved by a number of design changes.

Q. 38. What is USB Storage Device ?

Ans. This is a very handy storage device which has replaced floppy disk and is eásy to carry. It is known as pen drive, handy drive and thumb drive. Now pen drives are widely used by all type of users such as students, teachers, scientists and business man. Pen drive is available in various capacities such as 1 GB, 2GB,4GB, 8GB, 16Gb and 32GB.

Q. 39. What do you mean by Secured Digital Card ?

Ans. This is commonly called as SD card. It is an ultra small flash memory card designed to provide high-capacity memory in a small size. SD cards are used in many small portable devices such as digital video camcorders, digital cameras, handheld computers, audio players and mobile phones. SD cards are in use since 1999. They are available in capacities between 16 Megabytes and 1 Gigabyte. An SD card typically measures 32 x 24 x 2.1 mm and weighs approximately 2grams. SD cards are of different capacities and some of them combine input /output functions with data storage. USB pen drives and other storage like High Capacity SD cards (SDHC) use flash memory to store information.

Q. 40. What do you mean by external hard disk ?

Ans. A hard disk drive (HDD; also hard drive, hard disk, or disk drive) is a device for storing and retrieving digital information, primarily computer data. Hard disks have very high speed, storage capacity and reliability. Hard disk is generally permanently fixed in the computer. Now external hard disks are available in various capacities. Such data storage disks are now being used by students, teachers and researchers.

Q. 41. What are various computers programs used in genetics and plant breeding ?

Ans. Various micro-soft programs which are commonly used in crop improvement (genetics and plant breeding) include Microsoft Word (M.S.

Word), Microsoft Excel (M.S. Excel), Microsoft Power Point (M.S. Power Point) and Internet.

Q. 42. What do you mean by M. S. Word ?

Ans. It is a very useful program for preparation of annual reports, project reports, writing research papers, varietal information system, plant genetic resources data base, etc.

Q. 43. What do you mean by M.S. Excel ?

Ans. It is a software program which is useful for various types of statistical and biometrical analyses. It can also be used for graphical and diagrammatic display of experimental results.

Q. 44. What do you mean by M.S. Power Point ?

Ans. It is a software program which is widely used for preparation of slides and presentation of results in various scientific meetings.

Q. 45. What is Internet ?

Ans. It is very much useful in searching literature, chatting and online transactions.

Q. 46. What are various applications of computers in genetics and plant breeding ?

Ans. Bioinformatics or computer aided programs have wide practical applications in genetics and plant breeding. Some important applications of bioinformatics in plant breeding and genetics include (i) varietal information system, (ii) plant genetic resources data base, (iii) biometrical analysis, (iv) storage and retrieval of data, (v) studies on plant modeling, (vi) pedigree analysis, (vii) preparation of reports, (viii) updating of information, (ix) diagrammatic representation, and (x) planning of breeding programs.

Q. 47. What are advantages of computers ?

Ans. There are several advantages of bioinformatics in crop improvement programs. Some important advantages are listed as follows.

(i) High Speed: The computers have very high speed. The calculations that take months together can be performed in few minutes with the help of computers.

(ii) High Accuracy: The computer aided programs have very high level of accuracy. A computer practically never commits mistake on its own. The computer error results due to error in the programming and handling.

(iii) Better Planning: Computer aided programs help in better planning of experiments and systematizing the information of PGR and breeding experiments.

(iv) Large Storage capacity: Computers have very large storage capacity. The huge data of PGR and breeding experiments can be easily stored in computer memory and other storage devices. Results of long term experiments can be stored in systematic manner.

(v) Pictorial Presentation: Computer aided programs permit pictorial presentation of various reports and research papers which make them more attractive.

Q. 48. Define key board.

Ans. It is an input device that is used for entering alpha-numeric characters from outside world into computer memory.

Q. 49. What is mouse ?

Ans. It is an input device that is used for pointing and selecting options. It controls the movement of cursor and execution of various programs.

Q. 50. What is printer ?

Ans. It is an output device that is used for transferring data from computer memory to the outside world that is paper.

Q. 51. What is monitor ?

Ans. It is an output device or video screen that displays output data to the user.

INTRODUCTION TO GENOMICS AND BIOINFORMATICS

(A) GENOMICS

Q. 1. What is genomics ?

Ans. The term genomics was first used by Thomas Roderick in 1986. It refers to the study of structure and function of entire genome of a living organism.

Q. 2. Define Genome.

Ans. Genome refers to the basic set of chromosomes. In a genome, each type of chromosome is represented only once. Now genomics is being developed as a sub discipline of genetics which is devoted to the mapping, sequencing and functional analysis of genomes.

Q. 3. Explain main points related to genomics.

Ans. Main points related to genomics are listed below.

(i) It is a computer aided study of structure and function of entire genome of an organism.

(ii) It deals with mapping and sequencing of genes on the chromosomes.

(iii) It is a rapid and accurate method of gene mapping. It is more accurate than recombination mapping and deletion mapping techniques.

(iv) The genomic techniques are highly powerful, efficient and effective in solving complex genetic problems.

(v) Now use of genomic techniques has become indispensible in plant breeding and genetics.

Q. 4. What do you mean by structural genomics ?

Ans. The discipline of genomics consists of two parts, viz. structural genomics and functional genomics. Structural Genomics deals with the study of the structure of entire genome of a living organism. In other words, it deals with the study of the genetic structure of each chromosome of the genome. It determines the size of the genome of a species in megabases (Mb) and also the genes present in the entire genome of a species.

Q. 5. Define Functional Genomics.

Ans. The study of function of all genes present in the entire genome is known as functional genomics. It deals with transcriptome and proteome. The transcriptome refers to complete set of RNAs transcribed from a genome and proteome refers to complete set of proteins encoded by a genome.

Q. 6. How would you classify genomics ?

Ans. The genomics can be classified as plant genomics, animal genomics, eukaryotic genomics and prokaryotic genomics. These are defined as follows.

(i) Plant Genomics: It deals with the study of structure and function of entire genome of plant species.

(ii) Animal Genomics: It deals with the study of structure and function of entire genome of animal species.

(iii) Eukaryotic Genomics: It deals with the study of ructure and function of entire genome of higher (multi-cellular) organisms.

(iv) Prokaryotic Genomics: It deals with the study of structure and function of entire genome of unicellular organisms.

Q. 7. Give a brief description of genome mapping.

Ans. The discipline of genomics is of recent origin. The genome mapping was first completed in a free living bacteria *Haemophillus influenza* in 1995. Later on genome sequencing work was intensified both in prokaryotes and eukaryotes. In plants, genome sequencing was first completed in *Arabidopsis thaliana* (a weedy relative of mustard followed by rice *(Oriza sativa)*. Now genome sequencing work has been completed in more than 40 crop plants.

Q. 8. Give genome size in ten crop plants

Ans. The list of some field crops, fruit crops and other plant species in which genome sequencing work has been completed is presented in Table 47-1.

Table 47.1. Estimated genome size and gene number in some crop plants.

S. No.	Species	Botanical Name	Genome Size (Mb)	Gene No.	Year/Reference
A. Field Crops					
1.	Arabidopsis	*Arabidopsis thaliana*	120	27416	2000
2.	Rice	*Oryza sativa*	370	40577	Goff, et al., 2002
3.	Corn	*Zea mays*	2500	>32000	Schnable, P et al. 2009

4.	Sorghum	*Sorghum bicolor*	700	34496	Paterson, A et al. 2009.
5.	Cucumber	*Cucumis sativa*	243.5	26682	Haung et al. 2009
6.	Soybean	*Glycine max*	950	46430	Schmutz, et al. 2010
7.	Caster bean	*Ricinus communis*	320	31237	AP Chan, et al, 2010
8.	Pigeon pea	*Cajanus cajan*	833	48680	Varshney, el al, 2011
9.	Potato	*Solanum tuberosum*	844	39031	PGS Consortium, 2011
10.	Cannabis	*Cannabis sativa*	534	30,000	Harm van Bakel, et al. 2011
11.	Sugar beet	*Beeta vulgaris*	760	28000	Bernd Weisshaar, et al, 2011
12.	Arabidopsis	*Arabidopsis lyrata*	125	32,670	Tina T. Hu et al., 2011
13.	Brassica	*Brassica rapa*	468-516	41174	Wang, et al. 2011
14.	Foxtail Millet	*Setaria italica*	406	35471	Zheng G et al. 2012
15.	Tomato	*Solanum lycopersicum*	781	31760	Tomato Genome Consortium, 2012
16.	Melon	*Cucumis melo*	375-450	33372	Garci- Mas J et al., 2012
17.	Flax	*Linum usitatissimum*	686	43384	Wang, et al. 2012
18.	Cotton	*Gossypium raimondii*	750	40996	Kunbo Wang et al. 2012
19.	Lotus	*Lotus japonicus*	315	19848	Sato, et al. 2008
20.	Brachypodium	*Brachypodium distachyon*	272	25532	Vogel, J et al.,2010.
21.	Thellungiella	*Thellungiella parvula*	140	25,784	2011 (Unreleased)

B. Fruit Crops					
22.	Grapes	*Vitis vinifera*	500	26346	Jaillon, et al. 2007
23.	Papaya	*Carica papaya*	372	28629	Ming, et al., 2008
24.	Apple	*Malus domesticus*	742.3	57386	Reccardo Velasco, et al, 2010
25.	Peaches	*Prunus persica*	227	27852	2010
26.	Date palm	*Phoenix dactylifera*	380	>25,000	Eman K Al-Dous et al., 2011
27.	Banana	*Musa acuminata*	523	36542	Angelique D'Hont et al., 2012
28.	Sweet Orange	*Citrus sinensis*	380	25376	Wang & Jhou, 2012
29.	Clementine Orange	*Citrus clementina*	370	25385	Wang & Jhou, 2012
C. Trees					
30.	Poplur	*Populus tricocrpa*	550	45555	Tuskan, et al. 2006
31.	Rose gum tree	*Eucalyptus grandis*	691	41204	Unpublished

Q. 9. Name the species with smallest and largest genome size.

Ans. In plants, the smallest genome size has been reported in *Arabidopsis thaliana* (120Mb) and the largest in corn (2500). Thus, in plants the genome size investigated so far varies from 120Mb to 2500Mb. After gene sequencing, function is assigned to individual gene through various molecular techniques.

Q. 10. What are the genes to be mapped ?

Ans. In genome research, both types of genes, viz., major genes (oligogenes) and minor genes (polygenes) can be easily mapped. The mapping of polygenic traits is possible by genome mapping techniques which is not possible by conventional gene mapping techniques such as recombination mapping and deletion mapping. The genome mapping is done for morphological, productivity, resistance, quality, agronomic and some special traits as discussed below.

(i) Morphological Characters: It includes highly heritable traits such as shape, size and color of leaf, flower, calyx, corolla, etc. It also includes surface of leaf and stem (hairiness and smoothness).

(ii) Productivity traits: Such characters differ from species to species.

(iii) Resistance Traits: Such characters include resistance to diseases, insects, drought, soil salinity, soil alkalinity, soil acidity, heat, frost, water logging, cold, etc.

(iv) Quality Traits: Such traits include nutritional quality, market quality and keeping quality.

(v) Agronomic Traits: Such traits include earliness, plant height, plant type,etc.

(vi) Special Characters: Such characters include genes controlling male sterility, self incompatibility, photo and thermo insensitivity, toxic substances, apomixis, adaptation, etc.

Q. 10. Give a brief description of genome mapping laboratories

Ans. The genome mapping work is carried out by collaborative efforts of various International and National Research Laboratories of public domain. The list of some organizations where genome sequencing and mapping work on different crop species is carried out is presented in Table 47-2.

TABLE 47-2. List of some Genome Sequencing Organization

S. No.	Name of Crop	Scientific name	Name of Genome Project	In Collaboration with
1.	Arabidopsis	*Arabiddopsis thaliana*	Arabidopsis Genome Initiative.	USA, China and Japan.
2.	Brassica	*Brassica Spp.*	Brassica Genome Gate way	UK, USA, France, Germany, Australia, South Korea, Poland and China.
3.	Wheat	*Triticum aestivum*	International Wheat genome Sequencing Consortium	UK Scientists from Uni. Of Liverpool, Bristol and John Innes.
4.	Rice	*Oriza Spp.*	International Rice Genome Sequencing Project	Mexico, USA, China and japan
5.	Cotton	*Gossypium Spp*	International Cotton Genome Initiative Project	USA, France, Australia

6.	Soybean	*Glycine max*	Uni. Purdue, North Carolina Uni, and 18 other centres	United States of America
7.	Potato	*Solanum tuberosum*	Potato Genome Sequencing Consortium	USA, UK, Russia, China, India, Poland, Chile,New Zealand, Nether land, Peru, etc.
8.	Tomato	*Lycopersicum esculentum*	Tomato Genome Consortium	China, japan, Germany,Korea, Spain, UK, USA, Israel, Netherland & Others
9.	Pigeon pea	*Cajanus cajan*	International Initiative for Pigeon pea Genomics	India (ICRISAT), China, USA, Mexico, etc.
10.	Apple	*Malus domesticus*	Apple Genome sequencing Initiative	Washington State Uni. USA and NRI COMPETITIVE GRANT

Q. 11. Give a brief account of genome mapping in India.

Ans. In India, the functional genome research projects are looked after by the Depart of Biotechnology (DBT) and ICAR. The DBT has initiated such work on several crops such as rice, wheat, maize, chick pea, banana, tomato, Brassica, etc. The ICAR has created genome mapping facilities for rice at NRCPB, IARI, New Delhi. In India, the genome mapping work is carried out at the following centres.

(i) National Research Centre for Plant Biotechnology, IARI, New Delhi.

(ii) International Centre for Genetic Engineering and Biotechnology, New Delhi.

(iii) Jawahar Lal Nehru University, New Delhi.

(iv) National Botanical Research Institute, Lucknow.

Q. 12. What are applications of genomics in crop improvement ?

Ans. Genomics has several practical applications in crop improvement. Genome mapping is useful in several ways. It is useful or provides information about genome size, gene number, gene mapping, gene sequencing, evolution of crop plants, gene cloning, identification of DNA markers, marker assisted selection, transgenic breeding, construction of linkage maps and QTL mapping. All these aspects are briefly discussed as follows.

(i) Genome Size: Genome mapping is a very useful technique for determining the genome size in various plant species. In the plant species studied so far, the largest genome size has been reported in maize (2500Mb) and the smallest in *Arabidopsis thaliana* (120 Mb).

(ii) Gene Number: Genome mapping provide information about gene number in a species. In crop plants studied so far, the maximum number of genes have been reported in rice (56,000).

(iii) Gene mapping: Genome research is very much useful in mapping/ tagging of genes on the different chromosomes of a genome. In other words, it helps in large scale discovery of new gene in a genome.

(iv) Gene Sequencing: Genome mapping helps in determining the order of genes on the chromosomes. The order of genes is determined on each chromosome of a genome.

(v) Evolution: Genome mapping provide information about the evolution of different species. It measures the association between different genomes and thus provides information about the relatedness or evolutionary biology of crop plants.

(vi) Gene Cloning: Genome research is very much useful in making multiple copies of a gene and transfer of the same from one genotype to another. Thus, it aids in specific gene transfer.

(vii) Identification of DNA Markers: The genome mapping techniques are useful in identification of DNA markers which can be used in molecular breeding i.e. marker assisted selection. The mapping populations developed from inter-specific crosses have higher polymorphism for DNA markers than those mapping populations derived from intra-specific crosses.

(viii) Marker Assisted Selection: Marker assisted selection refers to indirect selection for a desired phenotype based on banding pattern of linked DNA markers. The improvement of crop plants using such selection is called molecular breeding. Various DNA markers used for such purpose include RFLP, AFLP, RAPD, SSR, etc. The effect of DNA marker is correlated with morphological markers and then selection is made for particular trait. The selection based on DNA markers is more reliable because DNA markers are not influenced by environmental factors.

(ix) Transgenic Breeding: Genome mapping is useful in gene cloning. The gene of interest can be cloned and used in developing transgenic plants (genetically engineered plants). Transgenic breeding permits direct gene transfer bypassing sexual process.

(x) Construction of Linkage maps: Genome mapping helps in construction of linkage groups. The linkage groups can be constructed from the information of gene mapping and gene sequencing.

(xi) QTL Mapping: The genome mapping techniques is widely used for mapping of quantitative trait loci (QTL). The mapping of QTL or polygenic traits is not possible by conventional methods, viz., recombination mapping and deletion mapping techniques.

Q. 13. What are practical achievements of genome mapping ?

Ans. In crop plants, genome mapping work has been completed in more than 40 field and fruit crops. Important crop plants in which genome sequencing has been completed include *Arabidopsis, Brassica rapa,* Rice, Corn, Cucumber, Soybean, Pigeon pea, Tomato, Potato, Melon, Flax, Cotton from field crops: and oranges, apple, peaches, papaya, Grapes, Date palm, Banana, etc. from fruits crops. The genome sequencing of many more crop plants is in progress.

Q. 14. What are limitations of genome mapping ?

Ans. These days, the genome mapping of crop plants is gaining increasing importance. It has several useful applications as discussed above. However, there are some limitations of genome mapping such high cost, high technical skill, laborious work, availability of limited genes and lake of proper markers. These are briefly presented below.

(i) Expensive Technique: The genome research requires well equipped sophisticated laboratory with costly chemicals and glassware. Thus lot of funds are required for carrying genomic research. Lack of adequate funds sometimes becomes limiting factor in the progress of such project.

(ii) High Technical Skill: The genome mapping work requires high technical skill. It requires training of scientists in the specialized field of genomics. It also requires International collaboration with other leading genome research laboratories which sometimes becomes limiting factor. The international collaboration is possible if the crop on which genome research work is to be carried out is of global significance.

(iii) Laborious Work: The genome mapping requires detection of various DNA markers (RFLP, AFLP, RAPD, SSR, etc.) which is a laborious and time consuming work. Huge populations related to F2, RILs, NILs and doubled haploids need to be screened for such purpose. This limits the progress of the work.

(iv) Limited Genes Available: Firstly, limited number of genes and promoters are available for development of transgenics. Secondly, such genes are protected under Intellectual Property Rights and, therefore, cannot be used for developing transgenic plants.

(v) Lack of Proper Markers: Most of the useful agronomic traits are governed by polygenes and are complex in nature. Tightly linked DNA markers are yet to be identified for such characters.

Q. 15. What points should be taken in to account for rapid progress of genome mapping work ?

Ans. Considerable research work on genome sequencing and mapping has been done on various crop plants so far. In future, for rapid progress of genome sequencing work several points need consideration. Important points include funding, training, material sharing, research priorities, important traits, selection of species, etc. These are briefly discussed as follows.

(i) Funding: Since genome mapping projects are very expensive, there is need of International collaboration for supporting such prestigious projects.

(ii) Training: Some laboratories are well equipped for genomic research. Such Laboratories should impart training to scientists from various countries for human resource development in genomic research.

(iii) Material Sharing: The leading laboratories should develop and distribute frame work DNA markers to other research laboratories for their use and further research.

(iv) Research Priorities: In order to get International collaboration, areas of common interest or global significance should be identified for genomic research work.

(v) Important Traits: In genome mapping, the major emphasis needs to be given to characters of economic importance such as productivity, quality and resistance to biotic and abiotic stresses. The experience gained in genome mapping of one species may be helpful in the study of related species.

(B) BIOINFORMATICS

Q. 16. Define Bioinformatics.

Ans. The computer aided study of biology particularly genetics and molecular biology is called bioinformatics. Now the science of bioinformatics is gaining increasing importance in life science specially in the field of molecular biology and plant genetic resources.

Q. 17. Explain main points related to bioinformatics.

Ans. Main points related to bioinformatics are listed as under.

(i) It is the interface between computer and biology. In other words, it is the application of information technology in the study of biology.

(ii) It utilizes information science for the study of biology.

(iii) It is used for computer based analysis of large scale data derived from genome sequencing.

(iv) It is used for analysis of data related to genomics, proteomics, metabolomics and other biological aspects.

(v) It has wide applications in handling data related to plant genetic resources.

Q. 18. What are branches of bioinformatics ?

Ans. Bioinformatics is broadly divided in to two groups, viz. animal bioinformatics and plant bioinformatics. These are defined as follows.

(i) Animal Bioinformatics: It deals with computer aided study of various animals. It includes gene mapping, gene sequencing, animal breed information data base, animal genetic resources data base. It can be further divided as bioinformatics of mammals, reptiles, insects, birds, fishes, etc. It also deals with the study of genomics, proteomics and metabolomics in various animal species.

(ii) Plant Bioinformatics: It deals with computer aided study of plant species. It includes gene mapping, gene sequencing, varietal information data base, plant genetic genetic resources data base. It also deals with the study of genomics, proteomics and metabolomics in various plant species. It can be further divided as agricultural bioinformatics, horticultural bioinformatics, medicinal plants bioinformatics, forest plants bioinformatics, etc.

Q. 19. What are computer programs used in biology ?

Ans. Various micro-soft programs which are commonly used in crop improvement (genetics and plant breeding) include Microsoft Word (M.S. Word), Microsoft Excel (M.S. Excel), Microsoft Power Point (M.S. Power Point) and Internet. These are briefly defined as follows.

(i) M. S. Word: It is a very useful program for preparation of annual reports, project reports, writing research papers, varietal information system, plant genetic resources data base, etc.

(ii) M.S. Excel: It is a software program which is useful for various types of statistical and biometrical analyses. It can also be used for graphical and diagrammatic display of experimental results.

(iii) M.S. Power Point: It is a software program which is widely used for preparation of slides and presentation of results in various scientific meetings.

(iv) Internet: It is very much useful in searching literature, chatting and online transactions.

Q. 20. What are applications of bioinformatics in genetics ?

Ans. Bioinformatics or computer aided programs have wide practical applications in genetics and plant breeding. Some important applications of bioinformatics in plant breeding and genetics include (i) varietal information system, (ii) plant genetic resources data base, (iii) biometrical analysis, (iv) storage and retrieval of data, (v) studies on plant modeling, (vi) pedigree analysis, (vii) preparation of reports, (viii) updating of information, (ix) diagrammatic representation, and (x) planning of breeding programs. These are briefly discussed as follows.

Q. 21. Discuss use of bioinformatics in varietal information system .

Ans. Bioinformatics has useful application in developing varietal information system. In connection with plant variety protection (PVP) Act, various terms such as extant variety, candidate variety, reference variety, example variety and farmer's variety are frequently used. Hence, knowledge of these terms is essential. These are defined below:

Extant Variety: All released, notified and unprotected varieties.

Candidate variety: A variety to be registered under Plant Variety Protection Act is referred to as candidate variety.

Reference Variety: All released and notified extant varieties of common knowledge which are in seed production chain.

Example Variety: A variety that is used for comparison for a particular character is called example variety.

Farmers' Variety: A variety that has been developed by a farmer and used for commercial cultivation for several years is called farmers variety.

The detailed information about various types of varieties can be developed using highly heritable characters. Such information can be used in various ways as given below.

(i) For varietal identification in DUS testing.
(ii) In grouping of varieties on the basis of various highly heritable characters.
(iii) In sorting out of cultivars for use in pre-breeding and traditional breeding.

The information can be stored in the computer memory and be retrieved as and when required.

Q. 21. Explain application of bioinformatics in PGR data base ?

Ans. The genetic material of plants which is of value for present and future generations of people is referred to as plant genetic resources. It is also known as gene pool, genetic stock and germplasm. The germplasm is evaluated for several characters such as highly heritable morphological, yield contributing characters , quality characters, resistance to biotic and abiotic stresses and characters of agronomic value.

International Plant Genetic Resources Institute (IPGRI), Rome Italy has developed descriptors and descriptor states for various crop plants. Such descriptors help in uniform recording of observations on germplasm of crop plants all over the world. Thus huge data is collected on crop genetic resources for several years. Bioinformatics plays an important role in systematic management of this huge data. Bioinformatics is useful in handling such data in several ways as follows.

(i) It maintains the data of several locations and several years in systematic way.

(ii) It permits addition, deletion and updating of information.

(iii) It helps in storage and retrieval of huge data.

(iv) It also helps in classification of PGR data based on various criteria.

(v) It helps in retrieval of data belonging specific group such as early maturity, late maturity, dwarf types, tall types, resistant to biotic stresses, resistance to abiotic stresses, superior quality, marker genes, etc.

All such data can be easily managed by computer aided programs and can be manipulated to get meaningful results.

Q. 22. Define proteomics.

Ans. Proteomics refers to the study of structure and function of all proteins found in an individual. In fact, proteomics is a new sub-discipline of functional genomics. It is the study of proteomes which refer to complete set of proteins encoded by a genome. A variety of techniques are used for the study of proteomics. Now computer aided programs are available for the study of proteomics.

Q. 23. What are types of proteomics ?

Ans. The discipline of proteomics consists of two parts, viz. structural proteomics and functional proteomics. These are defined as under.

(i) Structural Proteomics: It refers to the study of the structure of all proteins found in a living organism.

(ii) Functional Genomics: It deals with the function of all proteins found in a living organism.

Q. 24. Define transcriptone and proteome ?

Ans. It deals with transcriptome and proteome. The transcriptome refers to complete set of RNAs transcribed from a genome and proteome refers to complete set of proteins encoded by a genome.

Q. 25. Give a brief description of metabolomics.

Ans. Metabolomics refers to the study of all metabolic pathways in a living organism. In other words, it is the computer aided information of all metabolic pathways of a living organism.

Q. 26. Explain main points related to metabolomics.

Ans. Main points related to metabolomics are listed below.

(i) It deals with study of all metabolic pathways in a living organism.

(ii) It is the computer aided information of all metabolic pathways of a living organism.

(iii) It helps in detection and correction of metabolic disorders in an organism.

(iv) It helps in selection of individuals with normal metabolic pathways.

(v) It helps in early detection of genetic disorders associated with metabolic pathways.

Q. 27. Describe applications of bioinformatics in biometrical analysis.

Ans. In crop improvement, various biometrical analyses are performed. Important biometrical analyses that are performed in plant breeding and genetics are given below.

(1) Simple measures of variability such as mean, standard deviation, standard error, coefficient of variation, etc.

(2) Correlations: It includes genotypic, phenotypic and environmental correlations. It also includes simple, partial and multiple correlations.

(3) Path Coefficients: It includes analysis of genotypic, phenotypic and environmental paths.

(4) Discriminant function analysis.

(5) Metroglyph analysis and D2 statistics

(6) Stability analysis

(7) Diallel, partial diallel, line x tester triallel, quadriallel, biparental and triple test cross analysis.

(8) Generation mean analysis, etc.

All these analyses can be easily performed through computer aided programs.

Q. 28. Explain applications of bioinformatics related to storage and retrieval of data.

Ans. In crop improvement, huge data is collected on the following aspects:

(i) Segregating populations: Single plant selections are made in segregating populations and data are recorded on various characters such as yield components, quality characters, resistance to biotic and abiotic stresses, etc.

(ii) Multilocation Experiments: Such experiments are conducted mainly for identification and release of new varieties and hybrids and also for assessment of varietal stability.

(iii) Multi-seasonal Experiments: Such experiments are conducted for several years (3-5 years) for identification of new varieties and hybrids.

The above data remain in active use generally for two decades. Handling of such a huge data is a difficult task. However, such data can be easily stored in various storage devices such as hard disks, compact disks, pen drive, data cards, etc. Storage of data in computers require less space and is very safe as compared to storage of data in paper registers and files.

Q. 29. Discuss use of computers in plant modeling.

Ans. Computers are useful tools for undertaking studies on modeling of plants. First the theoretical model can be prepared with the help of computer keeping in view various plant characters. Then such model plants can be developed through hybridization and directional selection. This type of studies are useful in developing crop ideotypes or ideal plant types in different field crops. First the conceptual model is prepared and then efforts are made to achieve such model by combining desirable genes from different sources into a single genotype through appropriate breeding procedures. Such studies have been made in field pea.

Q. 30. Explain use of computers in pedigree analysis.

Ans. Computer aided studies are useful in pedigree analysis of various cultivars and hybrids. Information about the parentage of cultivars and hybrids is entered into the computer memory which can be retrieved any time. The list of parents which are common in the pedigree of various cultivars and hybrids can be sorted out easily. It helps in the pedigree analysis which in turn can be used in planning plant breeding programs especially in the selection of parents for use in hybridization programs. The study of proteomics also helps in pedigree analysis.

Q. 31. Explain use of computers in preparation of reports.

Ans. After biometrical analysis of data, results are interpreted and various types of reports or documents are prepared. In crop improvement following types of reports are prepared.

(i) Research Project Report: The annual progress repot of each project is prepared and salient findings are documented.

(ii) Monthly, quarterly, half yearly and annual progress reports of all the research projects are also prepared.

(iii) Sometimes, bulletin and booklets are prepared to document specific information for adoption and benefit of farmers.

(iv) Research papers and popular articles are prepared based on research findings.

(v) Germplasm catalogues are prepared for various characters.

Such reports can be easily be prepared with the help of computers using MS

Word program. This information can be stored in computer memory and reused as and when required. The editing and updating of reports can be done any time without much extra efforts.

Q. 32. Discuss use of computers in updating of information.

Ans. In plant breeding and genetics, results of multi-seasonal and long term experiments require continuous updating. Computers have made this task very simple. The information related to any experiment which is already stored in the computer memory, can be updated any time by editing the concerned file. Any portion of information can be deleted or revised easily.

Q. 33. Explain use of computers in diagrammatic representation.

Ans. Inclusion of diagrams makes the reports, research papers, articles, bulletins, etc. more attractive, informative and easily understandable. The following types of diagrams are made in plant breeding.

(i) Line diagrams, bar diagrams, histograms and pie diagrams.

(ii) Cluster diagram: It is prepared when data is subjected to D2 analysis.

(iii) Path diagram: It is prepared when data is subjected to path coefficient analysis.

(iv) Vr-Wr Graph: It is prepared when data is subjected to Hayman's graphical approach of diallel cross analysis.

(v) Metroglyph Chart: It is prepared when data is subjected to Metroglyph analysis.

All these diagrams can be easily prepared with the help of computer using specific program.

Q. 34. Describe use of computers in planning of breeding programs.

Ans. Plant breeders have to plan various breeding programs every year. Computers are useful tools in such planning. The following activities can be easily planned with the help of computers.

(i) Sowing plans of various breeding experiments.

(ii) Selfing and crossing plans.

(iii) Breeder seed production plan.

(iv) Hybrid seed production plan.

(v) Germplasm collection, conservation, evaluation, distribution, utilization and documentation plan.

(vi) Screening plan of breeding material against biotic and abiotic stresses.

(vii) Selection , quality evaluation and multi-location testing plans.

All the above plans can be easily prepared with the help of computer well in advance. This is very important for proper implementation of various breeding programs. Computers are also useful in printing out labels and list of observations to be recorded in various breeding experiments.

48

NANOTECHNOLOGY IN CROP IMPROVEMENT

Q. 1. What do you mean by nanotechnology ?

Ans. The term nanotechnology was defined in 1974 by a professor from Tokyo Science University, Norio Taniguchi. By definition nanotechnology deals with units that are smaller than 100 nanometers. Nanotechnology has been defined in various ways. Some definitions of nanotechnology are as follows:

(i) Nanotechnology, in short called nanotech, is the branch of engineering that deals with things smaller than 100 nanometers (especially with the manipulation of individual molecules).

(ii) Nanotechnology, sometimes shortened to nanotech, refers to a field of applied science whose theme is the control of matter on an atomic and molecular scale. Generally nanotechnology deals with structures 100 nanometers or smaller, and involves developing materials or devices within that size.

(iii) It is the science and technology of creating nanoparticles and of manufacturing machines which have sizes within the range of .1 to 100 nanometre.

Q. 2. What are the main features of nanotechnology ?

Ans. Nanotechnology deals with the creation and use of objects at the nanoscale, up to 100 nanometers in size. One nanometre is one-millionth of a millimeter or one billionth of a metre or one thousand-millionth of a meter. The main points related to nanotechnology are listed below:

(i) Nanotechnology creates and uses structures, devices, and systems that have novel properties.

(ii) It is the manipulation or manufacture of material at the molecular level, or one billionth of a metre.

(iii) It includes the ability to devise self-replicating machines, robots, computers that are molecular sized,

(iv) It also includes nano-delivery systems for drugs, and quantum and molecular computing.

(v) It is the science and technology of building electronic circuits and devices from single atoms and molecules.

(vi) Nanotechnology involves working with particles at nanometer scale (at the atomic or molecular level).

(vii) It makes if possible to arrange atoms and molecules in certain defined structures.

(viii) It deals with manipulating and exploiting the properties of matter at a molecular level.

(ix) It is the science of particles that are measured by a nanometer, which is one billionth of a metre.

Q. 3. Describe briefly history of nanotechnology.

Ans. The idea of the use of nanotechnology was first given in 1959 by a well-known physicist Richard Feynman in his speech at an American Physical Society meeting at California Institute of Technology (Caltech) on December 29, 1959. His speech was completely theoretical and seemingly fantastic because at that it was not possible to manipulate single atoms or molecules as they were far too small for manipulation.

He described that the laws of physics do not limit our ability to manipulate single atoms and molecules. However, it was our lack of the appropriate methods for doing so. He correctly predicted that the time would come in which atomically precise manipulation of matter would inevitably arrive.

Q. 4. What are various applications of nanotechnology ?

Ans. Nanotechnology has a number of potential applications in diversity of fields such as (i) medicine, (ii) filtration, (iii) energy , (iv) displays, (v) consumer goods, (vi) agriculture, and (vii) crop improvement. First applications of nanotechnology have been discussed in crop improvement and agriculture and then applications in other fields are presented in Table 48.1

Table 48.1 Applications of Nanotechnology

SN	Main Field	Applications
1.	Agricultural Biotechnology (Crop Improvement)	(i) Gene delivery (ii) Fine control (iii) Multiple gee delivery (iv) Study of gene function
2.	Agriculture	(i) Detection of disease (ii) Treatment of disease (iii) Better nutrient absorption (iv) Better use of insecticides and herbicides

		(v) Protection of environment (vi) Precision farming
3.	Medicine	(i) Drug delivery (ii) Tissue engineering.
4.	Filtration	Waste-water treatment, air purification.
5.	Energy	(i) Increasing the efficiency of energy production (ii) Recycling of batteries
6.	Displays	Carbon nanotubes
7.	Consumer goods	Foods, house hold, optics, textiles and cosmetics.

Q. 5. What are the application of nanotechnology in crop improvement ?

Ans. Nanotechnology has practical applications in crop improvement through its use in agricultural biotechnology. A team of Iowa State University plant scientists and chemists (Brian Trewyn, Francois Torney, Kan Wang and Victor Lin) first used nanotechnology to penetrate rigid plant cell walls and deliver DNA and chemicals with precise control. They successfully used nanotechnology to penetrate plant cell walls and deliver a gene and a chemical that triggers its expression. This application of nanotechnology to plant biology and agricultural biotechnology has resulted in a significant breakthrough in delivering specific gene into plant cells. The main applications of nanotechnology in agricultural biotechnology include (i) gene delivery, (ii) controlled gene delivery , (iii) multiple gene delivery , and (iv) study of gene function.

Q. 6. How nanotechnology is useful in gene delivery ?

Ans. Nanotechnology permits successful delivery of a gene into a plant cell. Till recently, we were at nature's mercy when we delivered a gene into a cell, "Lin said, "There has been no precise control as to whether the cells will actually incorporate the gene and express the consequent protein. With this technology, I we may be able to control the whole sequence in the future. And once the gene is inside the plant cell wall, it opens up whole new possibilities. In the *future,* scientists could use the new technology to deliver imaging agents or chemicals inside cell walls. This would provide plant biologists with a window into intracellular events.

Q. 7. How nanotechnology is useful in controlled gene delivery ?

Ans. The gene delivery is possible in a controlled manner. We can bring in a gene and induce it in a controlled manner at the same time and at the same location. That has never been done before" (Wang).

Q. 8. Describe multiple gene delivery using nanotechnology.

Ans. The most tremendous advantage is that several genes can be delivered into a plant cell at the same time and released them whenever required

(Torney said). With the mesoporous nanoparticles, it is possible to deliver two biogenic species at the same time.

Q. 9. How nanotechnology is used in the study of gene function ?

Ans. The controlled release of gene in to plant cell will improve the ability to study gene function in plants. In a separate process, chemicals are used to activate the gene's function. The process is imprecise and the chemicals could be toxic to the plant.

In plant, transformation mostly occurs with the use of a gene gun. In order to use the gene gun to introduce the nanoparticles to walled plant cells, the chemists made another useful modification on the particle surface. They synthesized even smaller gold particles to cap the nanoparticles. These "golden gates" not only prevented chemical leakage, but also added weight to the nanoparticles, enabling their delivery into the plant cell with the standard gene gun.

Q. 10. What are the major achievements of nanotechnology in crop improvement ?

Ans. The biologists have successfully used the nanotechnology to introduce DNA and chemicals to Arabidopsis, tobacco and corn plants.

"The team also found a chemical that made the nanoparticle look yummy to the plant cells so they would swallow the particles," The nanoparticles were swallowed by the plant protoplasts (naked cells).

Q. 11. What are applications of nanotechnology in Agriculture ?

Ans. There are new challenges in agricultural sector. Such challenges include a growing demand for healthy and safe food, an increasing risk of disease, and threats to agricultural and fishery production from changing weather patterns. Nanotechnology has the potential to revolutionize the agricultural and food industry with new tools. Important applications of nanotechnology in the field of agriculture include (i) treatment of diseases, (ii) detection of diseases, (iii) better nutrient absorption, (iv) better use of insecticides and herbicides, and (v) protection of environment. All these have been discussed below:

(i) Treatment of Diseases: It provides molecular treatment of diseases.

(ii) Detection of Diseases: It helps in rapid detection of diseases. Smart sensors and smart delivery systems will help the agricultural industry combat viruses and other crop pathogens.

(iii) Better Nutrient Absorption: It enhances the ability of plants to absorb nutrients etc.

(iv) Better Use of Insecticides and Herbicides: In the near future nano structured catalysts will be available which will increase the efficiency of pesticides and herbicides, allowing lower doses to be used.

(v) Protection of Environment. Nanotechnology will also protect the environment indirectly through the use of alternative (renewable) energy supplies, and filters or catalysts to reduce pollution and clean-up existing pollutants.

Q. 12. What is Controlled Environment Agriculture (CEA) ?

Ans. An agricultural methodology widely used in the USA, Europe and Japan, which efficiently utilises modern technology for crop management, is called Controlled Environment Agriculture (CEA). CEA is an advanced and intensive form of hydroponically-based agriculture. Plants are grown within a controlled environment so that horticultural practices can be optimized. The computerized system monitors and regulates localized environments such as fields of crops. CEA technology, as it exists today, provides an excellent platform for the introduction of nanotechnology to agriculture. In CEA, nanotechnological devices can tremendously improve the grower's ability to determine the best time of harvest for the crop, the vitality of the crop, and food security issues, such as microbial or chemical contamination.

Q. 13. What are applications of nanotechnology in precision farming ?

Ans. There are two main applications of nanotechnology in the precision farming, viz. (i) reduction in the cost of production, and (ii) reduction in pollution. These are discussed below:

(i) Reduction in Cost of Production: The major goal of precision farming is to maximize output (i.e. crop yields) and minimize input Le. fertilizers, pesticides, herbicides, etc) by monitoring environmental variables and applying targeted action. Precision farming makes use of computers, global satellite positioning systems, and remote sensing devices to measure highly localized environmental conditions. Thus it determines whether crops are growing at maximum efficiency or precisely identifies the nature and location of problems. By using centralized data to determine soil conditions and plant development, seeding, fertilizer, chemical and water use can be fine-tuned to lower production costs and potentially increase production- all benefiting the farmer .

(ii) Reduction in Pollution: Precision farming can also help to reduce agricultural waste and thus keep environmental pollution to a minimum. Although not fully implemented yet, tiny sensors and monitoring systems enabled by nanotechnology will have a large impact on future precision farming methodologies. One of the major roles for nanotechnology-enabled devices will be the increased use of autonomous sensors linked into a GPS system for real-time monitoring. These nanosensors could be distributed throughout the field where they can monitor soil conditions and crop growth. Wireless sensors are already being used in certain parts of the USA and Australia. For example, one of the Californian vineyards, Pickberry , in Sonoma Country has installed wifi systems with the help of the IT company, Accenture. The initial cost of setting up such a system is justified by the fact that it enables the best grapes to be grown which in turn produce finer wines, which command a premium price. The use of such wireless networks is of course not restricted to vineyards only. This technique can also be used in other crops.

Q. 14. What are applications of nanotechnology in Medicine ?

Ans. Nanotechnology has two major applications in the field of medicines, viz.(i) Drug delivery and (ii) tissue engineering. The overall drug consumption and side-effects can be lowered significantly by depositing the active agent in the morbid region only and in no higher dose than needed. This highly selective approach reduces costs and human suffering. An example can be found in dendrimers and nano-porous materials. They could hold small drug molecules and transporting them to the desired location.

Nanotechnology can help to reproduce or to repair damaged tissue. Tissue engineering might replace todays conventional treatments, e.g. transplantation of organs or artificial implants.

□□□□

CROP BREEDING INSTITUTES/CENTRES

Q. 1. What are crop breeding institutes ?

Ans. Various organisations where plant breeding research is carried out are known as crop breeding institutes. Crop breeding institutes are of two types, viz. National crop breeding institutes and International crop breeding institutes.

Q. 2. What are National crop breeding organisations ?

Ans. In India, crop improvement in carried by following organisations.

(i) State Agricultural Universities
(ii) Crop Research Institutes of ICAR
(iii) Private seed Companies

Q. 3. What are International crop breeding institutes ?

Ans. There are nine International Crop Research Institutes as given below.

(i) International Rice Research Institute, Philippines
(ii) International Wheat and Maize Improvement Centre, Mexico.
(iii) International Centre for Tropical Agriculture, Columbia.
(iv) International Institute for Tropical Agriculture, Nigeria.
(v) International Potato Centre, Peru.
(vi) International Crop Research Institute for Semi Arid Tropics, Hyderabad, India.
(vii) International Centre for Agricultural Research in Dry Areas, Syria.
(viii) International Plant Genetic Resources Institute, Rome, Italy.
(ix) Asian Vegetable Research and Development Centre, Tiawan.

Q. 4. What are functions of International Crop Research Institutes ?

Ans. There are four main functions of International Crop Research Institutes as given below :

(i) To carry out basic research on various aspects of concerned crops.
(ii) To develop improved varieties.

(iii) To improve cropping systems for increased production.
(iv) To collect and conserve global germplasm of concerned crops.

Q. 5. What is CGIAR ?

Ans. CGIAR refers to Consultative Group on International Agricultural Research.

Q. 6. When was CGIAR constituted ?

Ans. CGIAR was constituted in 1971.

Q. 7. What are functions of CGIAR ?

Ans. Main functions of CGIAR are given below :

(i) Supervision of the progress of International Crop Research Institutes.
(ii) Establishment of International Research Institutes.
(iii) Providing financial support to International Research Institutes.

Q. 8. What are funding agencies of CGIAR ?

Ans. CGIAR is supported by following funding agencies

(i) Food and Agricultural Organisation of United Nations.
(ii) The World Bank, and
(iii) United Nations Development Programme (UNDP)

Q. 9. What is FAO ?

Ans. FAO refers to Food and Agriculture Organisation of United Nations.

Q. 10. What is UNDP ?

Ans. UNDP refers to United Nations Development Programme. It is supported by member countries.

Q. 11. What are International Research Institues which were established before set up of CGIAR ?

Ans. The following four International crop research institutes were established before constitution of CGIAR

(i) International Rice Research Institute, Philippines (1960)
(ii) International Wheat and Maize Improvement Centre, Mexico (1966)
(iii) International Centre for Tropical Agriculture, Colombia, (1968)
(iv) International Institute for Tropical Agriculture, Nigeria (1968)

Q. 12. What are International Agriculture Research Institutes established by CGIAR ?

Ans. The following International Crop Research Institutes have been established by CGIAR.

(i) International Potato Centre, Peru (1972)

(ii) Internation Crop Research Institute for Semi Arid Tropics, Hyderabad, India (1972)

(iii) International Centre for Agriculture Research in Dry Areas, Syria (1976)

(iv) International Plant Genetic Resources Institute Rome, Italy (1994)

(v) Asian Vegetable Research and Development Centre, Tiawan.

Q. 13. What is IRRI ?

Ans. IRRI refers to International Rice Research Institute.

Q. 14. When was IRRI established ?

Ans. IRRI was established in 1960.

Q. 15. Where is IRRI located ?

Ans. IRRI is located at Loss Banos, Philippines.

Q. 16. Who established IRRI ?

Ans. IRRI was established with joint efforts of following agencies

(i) Ford Foundation (ii) Rockfeller foundation.

Q. 17. What are main activities of IRRI ?

Ans. IRRI deals with genetic improvement of tropical rice with special emphasis on Asian region.

Q. 18. What is CIMMYT ?

Ans. CIMMYT refers to International Wheat and Maize Improvement Centre.

Q. 19. Who established CIMMYT ?

Ans. CIMMYT was established by following organisations :

(i) Ford foundation (ii) Rockfeller foundation

Q. 20. When was CIMMYT established ?

Ans. CIMMYT was established in 1966.

Q. 21. Where is CIMMYT located ?

Ans. CIMMYT is located at Elbaton in Mexico.

Q. 22. What are main activities of CIMMYT ?

Ans. CIMMYT deals with genetic improvement of following crops on global basis :

(i) Wheat
(ii) Maize
(iii) Barley
(iv) Triticale

Q. 23. What is CIAT ?

Ans. CIAT refers to International Centre for Tropical Agriculture.

Q. 24. Who established CIAT ?

Ans. CIAT was established jointly by following organisations

(i) Ford foundation
(ii) Rockfeller foundation.

Q. 25. When was CIAT established ?

Ans. CIAT was established in 1968.

Q. 26. Where is CIAT located ?

Ans. CIAT is located at Palmira in Colombia.

Q. 27. What are main activities of CIAT ?

Ans. CIAT deals with genetic improvement of following crops with special emphasis on Latin America.

(i) Cassava
(ii) Beans
(iii) Rice
(iv) Maize

Q. 28. What is IITA ?

Ans. IITA refers to International Institute for Tropical Agriculture.

Q. 29. Who established IITA ?

Ans. IITA was established jointly by following organisations

(i) Ford foundation (ii) Rockfeller foundation.

Q. 30. When was IITA established ?

Ans. IITA was established in 1968.

Q. 31. Where is IITA located ?

Ans. IITA is located at Ibodan in Nigeria.

Q. 32. What are main activities of IITA ?

Ans. IITA deals with genetic improvement of following crops with emphasis on low land tropics of Africa.

(i) Cowpea (ii) Soybean (iii) Limabean (iv) Pigeonpea (v) Cassava (vi) Sweet potato (vii) Yam (viii) Rice, and (ix) Maize

Q. 33. What is CIP ?

Ans. CIP refers to International Potato Centre

Q. 34. Who established CIP ?

Ans. CIP was established by CGIAR.

Q. 35. When was CIP established ?

Ans. CIP was established in 1972.

Q. 36. Where is CIP located ?

Ans. CIP is located at Lima in Peru.

Q. 37. What are main activities of CIP ?

Ans. CIP deals with genetic improvement of potato for tropical and temperate climates.

Q. 38. What is ICRISAT ?

Ans. ICRISAT refers to International Crop Research Institute for Semi-Arid Tropics.

Q. 39. Who established ICRISAT ?

Ans. ICRISAT was established by CGIAR.

Q. 40. When was ICRISAT established ?

Ans. ICRISAT was established in 1972.

Q. 41. Where is ICRISAT located ?

Ans. ICRISAT is located at Hyderabad in India.

Q. 42. What are main activities of ICRISAT ?

Ans. ICRISAT deals with genetic improvement of following crops :

(i) Sorghum (ii) Pearl millet (iii) Pigeon pea (iv) Groundnut (v) Chickpea.

Q. 43. What is ICARDA ?

Ans. ICARD refers to International Centre for Agriculture Research in Dry Areas.

Q. 44. Who established ICARDA ?

Ans. ICARD was established by CGIAR.

Q. 45. When was ICARDA established ?

Ans. ICARDA was established in 1976.

Q. 46. Where is ICARDA located ?

Ans. ICARDA is located in Aleppo Syria.

Q. 47. What are main activities of ICARDA ?

Ans. ICARDA deals with genetic improvement of *Durum wheat,* barley and beans with main focus on Medeterranian area and North Africa.

Q. 48. What is IPGRI ?

Ans. IPGRI refers to International Plant Genetic Resources Institute.

Q. 49. Who established IPGRI ?

Ans. IPGRI was established by CGIAR.

Q. 50. When was IPGRI established ?

Ans. IPGRI was established in 1994.

Q. 51. Where is IPGRI located ?

Ans. IPGRI is located ir. Rome, Italy.

Q. 52. What are main activities of IPGRI ?

Ans. Main activities of IPGRI are collection, conservation and exchange of plant genetic resources on global bases. It coordinates various germplasm activities on global basis.

Q. 53. What was old name of IPGRI ?

Ans. The old name of IPGRI was IBPGR—International Board for Plant Genetic Resources which was established by CGIAR in 1974 in Rome, Italy. In 1994 IBPGR was renamed as IPGRI.

Q. 54. Which crop breeding institutes are located in Uttar Pradesh ?

Ans. The following ICAR crop breeding institutes are located in Uttar Pradesh.

(i) Indian Institute of Sugarcane Research, Lucknow.
(ii) Indian Institute of Pulses Research, Kanpur.
(iii) Indian Institute of Horticulture for Northern Plains, Lucknow.
(iv) Indian Grassland and Fodder Research Institute, Jhansi.
(v) Indian Institute of Vegetable Research, Varanasi.

Q. 55. What are ICAR institutes located in Uttranchal ?

Ans. The following ICAR Research Institutes are located in Uttranchal.

(i) Vivekanand Parvatiya Krishi Anusandhan Shala, Almora.
(ii) Central Soil and Water Conservation Research and Training Institute, Dehradun.

Q. 56. What are ICAR Crop research institutes located in Maharashtra. ?

Ans. The following ICAR crop research institutes are located in Maharashtra :

(i) Central Institute for Cotton Research, Nagpur
(ii) Central Institue for Research on Cotton Technology, Mumbai.
(iii) National Research Centre for Citrus, Nagpur.
(iv) National Research Centre for Onion and Garlic, Nashik.
(v) National Research Centre for Grapes, Pune.

Q. 57. What are ICAR crop research institutes located in Andhra Pradesh ?

Ans. The following ICAR crop research institutes are located in Andhra Pradesh :

(i) Central Tobacco Research Institute, Rajahmundry.
(ii) Central Research Institute for Dryland Agriculture, Hyderabad.
(iii) Directorate of Oilseed Research, Hyderabad.
(iv) Directorate of Rice Research, Hyderabad.
(v) National Research Centre for Sorghum, Hydrabad.
(vi) National Research Centre for Oil Palm, Ashoknagar.

Q. 58. What are ICAR crop research institutes located in Kerala ?

Ans. The following ICAR crop research institutes are located in Kerala.

(i) Central Planoatation Crop Research Institute, Kasargod.

(ii) Central Tuber Crops Research Institute, Trivandrum.

(iii) Indian Institute of Spices Research, Calicut.

Q. 59. What are ICAR crop research institutes located in Delhi ?

Ans. The following ICAR crop research institutes are located in Delhi.

(i) Indian Agricultural Research Institute, New Delhi.

(ii) National Bureau of Plant Genetic Resources, New Delhi.

(iii) Project Directorate on Maize, New Delhi.

(iv) National Research Centre on Plant Biotechnology, New Delhi.

Q. 60. What are ICAR crop research institutes located in Haryana ?

Ans. The following ICAR crop research institutes are located in Haryana.

(i) Central Soil Salinity Research Institute, Karnal.

(ii) Directorate of Wheat Research, Karnal.

Q. 61. What are ICAR Crop research institutes located in Karnataka ?

Ans. The following ICAR crop research institutes are located in Karnataka.

(i) Indian Institute of Horticultural Research, Bangalore.

(ii) National Research Centre for Cashew, Puttur.

Q. 62. What are ICAR Crop research institutes located in Rajasthan ?

Ans. The following ICAR crop research institutes are located in Rajasthan.

(i) Central Arid Zone Research Institute, Jodhpur.

(ii) National Research Centre for Rapeseed and Mustard, Bharatpur.

Q. 63. What are ICAR crop research institutes located in Tamil Nadu ?

Ans. The following ICAR crop research institutes are located in Tamil Nadu.

(i) Sugarcane Breeding Institute, Coimbatore.

(ii) National Research Centre for Banana, Tiruchirapalli.

Q. 64. What are ICAR crop research institutes located in West Bengal ?

Ans. The following ICAR crop research institutes are located in West Bengal.

(i) Central Research Institute for Jute and Allied Fibres, Barrack pore.

(ii) Jute Technological Research Institue, Kolkata.

Q. 65. Where is National Research Centre for groundnut located ?

Ans. NRC for Groundnut is located at Junagarh, Gujarat.

Q. 66. Where is NRC for Soybean located ?

Ans. NRC for Soybean is located at Indore, Madhya Pradesh.

Q. 67. Where is Central Rice Research Institute located ?

Ans. Central Rice Research Institute is located at Cuttack in Orissa.

Q. 68. Where is Central Potato Research Institute (CPRI) located ?

Ans. CPRI is located at Shimla in Himachal Pradesh.

Q. 69. Where is Central Agriculture Research Institute (CARI) located ?

Ans. CARI is located at Port Blair in Andaman and Nicobar.

Q. 70. What are ICAR institutes known as deemed universities ?

Ans. The following ICAR institutes are considered as deemed universities.

(i) Indian Agricultural Research Institute, New Delhi.

(ii) Indian Veterinary Research Institute (IVRI), Izatnagar, Bareilly, U.P.

(iii) National Dairy Research Institute (NDRI), Karnal, Haryana,

(iv) Central Institute of Fisheries Education (CIFE), Mumbai, Maharashtra.

Q. 71. Where is Central Agricultural University located ?

Ans. Central Agricultural University is located at Imphal in Manipur.

Q. 72. Where is Sardar Vallabi Bhai Patel University of Agriculture and Technology located ?

Ans. Sardar Vallabh Bhai Patel University of Agriculture and Technology is located at Meerut in Uttar Pradesh.

Q. 73. Where is Maharana Pratap University of Agriculture located ?

Ans. Maharana Pratap University of Agriculture and Technology is located at Udaipur in Rajasthan.

Q. 74. Which Agriculture University is located in Uttranchal ?

Ans. G.B. Pant University of Agriculture and Technology is located at Pantnagar in Uttranchal.

Q. 75. Name Agricultural Universities of Uttar Pradesh.

Ans. The following Agricultural Universities are located in Uttar Pradesh.

(i) C.S. Azad Univeristy of Agriculture and Technology, Kanpur.

(ii) N.D. University of Agriculture and Technology, Kumarganj, Faizabad.

(iii) Sardar Vallabh bhai Patel University of Agriculture and Technology, Meerut.

Q. 76. What are Agricultural Universities of Maharashtra ?

Ans. The Agricultural Universities of Maharashtra are given below :

(i) Panjabrao Deshmukh Krishi Vidyapeeth, Akola.

(ii) Marathwada Agricultural University, Parbahni.

(iii) Mahatma Phule Krishi Vidyapath, Rahuri.

(iv) Konkan Krishi Vidyapeeth, Dapoli.
(v) Maharashtra University of Animal Sciences, Nagpur.

Q. 77. Which Agriculture University is located in Chhatisgarh ?

Ans. Indira Gandhi Krishi Vishwa Vidyalaya is located at Raipur in Chhatishgarh.

Q. 78. Which Agriculture University is located in Jharkhand ?

Ans. Birsa Agricultural University is located at Ranchi in Jharkhand.

Q. 79. What are Agricultural Universities located in West Bengal ?

Ans. The following Agricultural Universities are located in West Bengal.

(i) Bidhan Chandra Krishi Vishwa Vidyalaya, Haringhatta.
(ii) West Bengal University of Animal and Fishery Sciences, Belgachia Kolkata.

Q. 80. Where is Rajindra Agricultural University Located ?

Ans. Rajindra Agricultural University is located at Pusa, Bihar.

Q. 81. Where is Jawaharlal Nehru Krishi Vishwa Vidyalaya (JNK VV) located ?

Ans. JNK VV is located at Jabalpur in Madhya Pradesh.

WORLD TRADE ORGANISATION

Q.1. What is World Trade Organization ?

Ans. The World Trade Organization (WTO) can be defined as an international organization that regulates trade and tariffs between nations in order to ensure that trade flows smoothly, predictable and as freely as possible.

The WTO was established in 1995 as the successor to the General Agreement on Tariffs and Trade (GATT), which was itself established after the Second World War. The overall system was developed through a series of trade negotiations, or rounds, held under GATT. The main points are listed below.

1. It was created by the Uruguay Round of GATT to enforce free trade of manufactured goods, services, and intellectual property.
2. It is an international organization that monitors and enforces rules governing global trade.
3. It is designed to supervise and liberalize international trade.
4. It provides the ground rules for international trade and commerce.
5. It was created through a multilateral agreement on trade and investment.
6. It has been formed by the agreement of member nations.

Q. 2. How many countries have become member of 'WTO ?

Ans. The membership is open to all countries. Interested countries become member of WTO. India has joined membership of WTO. Till July 27, 2007, 151 countries have become of WTO.

Q. 3. What is the Organizational Structure of WTO ?

Ans. The World Trade Organization came into force on January 1, 1995, fully replacing the previous GATT Secretariat as the organization responsible for administering the international trade regime. The basic structure of the WTO includes the (i) the Ministerial Conference, (ii) the General Council, (iii) the Trade Policy Review Body, (iv) the Dispute Settlement Body, (v) the Councils on Trade in Goods and Trade in Services, (vi) the Secretariat and Director General, and (vii) the Committees.

Q. 4. What id the Ministerial Conference ?

Ans. This is composed of international trade ministers from all member countries. This is the governing body of the WTO, responsible for setting the strategic direction of the organization and making all final decisions on agreements under its wings. The Ministerial Conference meets at least once every two years. Although voting can take place, decisions are generally taken by consensus all member countries.

Q. 5. What the role of the General Council ?

Ans. This is composed of senior representatives (usually ambassador level)' of all members. It is responsible for overseeing the day-to-day business and

Management of the WTO, and is based at the WTO headquarters in Geneva. In practice, this is the key decision-making arm of the WTO for most issues. Several of the bodies described below report directly to the General Council.

Q.6. What are the functions of the Trade Policy Review Body ?

Ans. This is also composed of all the WTO members, and oversees the Trade Policy Review Mechanism, a product of the Uruguay Round. It periodically reviews the trade policies and practices of all member states. These reviews are intended to provide a general indication of how states are implementing their obligations, and to contribute to improved adherence by the WTO parties to their obligations.

Q.7. What is the role of the Dispute Settlement Body ?

Ans. This is also composed of all the WTO members. It oversees the implementation and effectiveness of the dispute resolution process for all WTO agreements, and the implementation of the decisions on WTO disputes. Disputes are heard and ruled on by dispute resolution panels chosen individually for each case, and the permanent Appellate Body that was established in 1994. Dispute resolution is mandatory and binding on all members. A final decision of the Appellate Body can only be reversed by a full consensus of the Dispute Settlement Body.

Q.8. What is the function of the Councils on Trade in Goods and Trade in Services ?

Ans. These councils operate under the mandate of the General Council and are composed of all members. They provide a mechanism to oversees the details of the general and specific agreement on trade in goods (such as those on textiles and agriculture) and trade in services. There is also a Council for the Agreement on Trade-Related Aspects of Intellectual Property Rights, dealing with just that agreement and subject area.

Q.9. What is the role of the Secretariat ?

Ans. The Director General of the WTO reside in Geneva, in the old home of GATT. The Secretariat now numbers just under 550 people, undertakes the administrative functions of running all aspects of the organization. The

Secretariat has no legal decision-making powers but provides vital services, and often advice, to those who do. The Secretariat is headed by the Director General, who is elected by the members.

Q.10. What is the role of the Committees ?

Ans. There are several committees, but two committees, viz., (a) committee on trade and development (CTD), and (b) Committee on Trade and Environment (CTE) are very important. They have specific mandates to focus on these relationships, which are especially relevant to how the WTO deals with sustainable development issues. The Committee on Trade and Development was established in 1965. The forerunner to the Committee on Trade and Environment (the Group on Environmental Measures and International Trade) was established in 1971, but did not meet until 1992. Both Committees are now active at discussion grounds but do not actually negotiate trade rules.

Q.11. Where is the Headquarters of WTO ?

Ans. The World Trade organization is based at Geneva in Switzerland In other words, Geneva is the headquarters of WTO. The WTO is headed by the Director General who lives at Geneva in Switzerland in the old home of GATT.

Q.12. What are the Objectives of WTO ?

Ans. The main objective of World Trade Organization is to ensure smooth tree, fair and predictable global trade flow.

Q.13. What are the Functions of WTO ?

Ans. There are four main functions of World Trade Organization as given below :

(i) Administering trade agreements (enrolment of members)

(ii) Acting as a forum for trade negotiation

(iii) Settling trade disputes and

(iv) Reviewing national trade policies to improve transparency and to assess their impact.

Q.14. What are the Principles of WTO ?

Ans. The WTO agreements are based on certain principles. Some important principles of WTO are briefly presented below :

(i) WTO rules have been framed by consensus among all member countries. The modification in any rule is also done by consensus among all member countries.

(ii) WTO permits non discriminating trading system for all member countries.

(iii) The membership is open for all countries of the world. Any country can become member of WTO.

(iv) In WTO, all countries have equal rights irrespective of their status.

(v) WTO functions purely on democratic pattern. All decisions are taken by consensus among member countries.

(vi) Each country gets guarantee for its export in the market other member countries.

(vii) WTO is the government organization. The private sector can participate only in specific activities such as seminar and symposiums.

(viii) WTO does not force any country to become its member. Interested countries join WTO.

(ix) WTO agreements keep into account various issues such as environmental safety, job opportunities and development. It does not ignore economic development.

(x) WTO improves standard by enhancing per capita income of those engaged with WTO activities.

Q.15. What are the Materials Covered for Trading under WTO ?

Ans. WTO permits a variety of items for trading across the nation. The list of such items is presented in Table 50.1.

Table 50.1 List of items permitted for trading by WTO

SN	Group of items	Actual Items
1.	Agriculture	Cereals, pulses, oilseeds, oils, sugar etc.
2.	Horticulture	Fruits, vegetables, cut flowers etc.
3.	Textiles	Fibre, yarn, fabrics and garments
4.	Automobiles	Cars, buses, tractors, trucks, road making machines and their spare parts
5.	Electronic Goods	Television, computers, and instruments and equipments.
6.	Electrical goods	Refrigerators, electric iron, heaters, air, coolers, deep fridges etc.
7.	Consumer Goods	Soaps, cosmetics, creams, tooth pastes, tooth brush etc.
8.	Household Goods	Utensils, furniture, sofa, bed, dinning table etc.
9.	Literary Goods	Books, papers, pens etc.
10.	Services	Banks, Insurance firms, phone companies, transport companies, hotels etc.
11.	Defense Items	Guns, rifles, fighter aircrafts, missiles, ships, helicopters etc.
12.	Intellectual Property	Copyrights, trademarks, industrial designs, Integrated circuit layout designs, trade secret geographical indications, etc.

Q.16. What are the Advantages of WTO ?

Ans. There are several advantages of WTO which are listed as follows.

(i) Improvement in the quality of product or service due to competition among companies.

(ii) There will be reduction in the price again due to competition among companies.

(iii) WTO provide job opportunities by creating new jobs.

(iv) WTO provides a wide range of variety in various items for choice.

(v) WTO improves relationship among trading countries due to frequent interactions.

(vi) WTO leads to faster development and growth of member countries by floating their goods in international market.

(vii) Since each item is marketed by several countries, there will be no monopoly of any country.

(viii) WTO operates on democratic pattern, each country has equal rights and all decisions are taken with consensus of member countries.

(ix) WTO provides atmosphere for free and fair deal. It curbs corruption.

(x) WTO leads to increase in the income by permitting trade across the nations.

Q.17. What are the Disadvantages of WTO ?

Ans. There are two main disadvantages of WTO. Firstly it leads to several trade disputes and secondly, it has adverse effects on the growth of small industries.

GLOBALIZATION

Q.1. What is Globalization ?

Ans. Globalization is the process of organizing the whole world into a single integrated marketing unit. It is also defined as the process of trans border free flow of products, services, people, culture, technology, and finance. It leads to the integration of economic, cultural, political, and social systems across national borders Globalization is also referred to as internationalization by some persons. Both these terms are used as synonyms. However, some people use these terms separately. This chapter deals with various globalization such as phases of globalization, types (kinds) of globalization, global connectivity, measurement of globalization, factors affecting globalization, advantages of globalization and disadvantages of globalization.

Q.2. What are Phases of Globalization ?

Ans. Globalization is not a new phenomenon. It started with human civilization. In the past 130 years, modern historians have identified three stages or phases of globalization, viz., (i) first phase, (ii) second phase, and (iii) third phase.

Q.3. Describe the First Phase of globalization.

Ans. This phase of globalization started from 1870 and ended in 1913 with outbreak of the First World war. The main features of first phase are given below.

(i) There was marked mobility of capital.
(ii) The labour mobility was high.
(iii) The magnitude of free trade was limited.
(iv) The global institutions were non existent.
(v) The National Institutions were heterogeneous and were not well organized.

The First World War had adverse effects on the process of globalization i.e. flow of products, services, labour, and technology across the countries. It started resuming since 1930 gain and got setback from 1940 to 1944 due to second World War.

Q.4. Explain briefly the Second Phase of globalization.

Ans. The second phase of globalization started from 1945 and culminated in 1973. The main features of this phase are given below.

(i) The mobility of capital was low than first phase.
(ii) The labour mobility was low.
(iii) The magnitude of free trade was low.
(iv) The global Institutions were getting created during this period.
(v) The National Institutions were heterogeneous and were not standardizes.

Q.5. What are the main features of the Third Phase of globalization ?

Ans. The third phase of globalization started from 1974 and is still in progress. It is also known as the current phase of globalization. Main features of this phase are given below.

(i) The mobility of the capital is high.
(ii) The labour mobility is low.
(iii) The magnitude of free trade is extreme.
(iv) The International Institutes started functioning.
(v) The National Institutions have been standardized.
(vi) The on line global transactions are possible.
(vii) Information technology is being used extensively.

Q.6. What are different Types of Globalization ?

Ans. There are five main types of globalization, viz., (i) economic globalization, (ii) production globalization, (iii) cultural globalization, (iv) information globalization, and (v) ecological globalization.

Q.7. What is Economic Globalization ? Give its main features.

Ans. It refers to trans country flow of capital/ finance or money. Main effects of economic globalization are:

(a) Enhancement in world wide economic relationships.
(b) Increase in international trade at a faster rate than the growth in the world economy.
(c) Increase in international flow of capital including foreign direct investment.
(d) Creation of international agreements leading to organizations like the WTO and OPEC.
(e) Development of global financial systems.
(f) Increased role of international organizations such as WTO, WIPO, IMF that deal with international transactions.
(g) Increase of economic practices like outsourcing, by multinational corporations.

Q. 8. Define Production Globalization.

Ans. It refers to trans country flow of goods or products. It leads to trans national production of various goods or products. In such system, a product can be manufactured in several countries of the world with same quality .

Q.9. Give a brief account of Cultural Globalization.

Ans. It refers to trans country flow of culture

The main effects of cultural globalization are:

(a) Greater international cultural exchange,

(b) Spreading of multiculturalism, and better individual access to cultural diversity,

(c) Greater international travel and tourism ,

(d) Greater immigration, including illegal immigration

(e) Spread of local foods such as pizza and Indian food to other countries

(f) Development of a global telecommunications infrastructure and greater trans border data flow, using such technologies as the Internet, communication satellites and telephones.

(g) Increase in the number of standards applied globally; e.g. copyright laws and patents.

(h) Formation or development of a set of universal values.

(i) Spread of local goods, dresses and language to other countries.

Q.10. What is Information Globalization ?

Ans. It refers to trans border flow of knowledge, ideas and information. It is also known as communication globalization or technological globalization. It makes use of information technology and permits on line global transactions.

Q.11. Explain briefly Ecological Globalization.

Ans. It refers to global protection of ecosystem from degradation and pollution. Main features of ecological globalization are given below.

(a) It prevents ecosystem from various types of risks.

(b) It requires global collective action.

(c) It is also known as ecosystem globalization or environmental globalization.

(d) It leads to protection of environment globalization.

Q.12. What are different types of global connectivity ?

Ans. Five types of global connectivity have been identified (Arjun Appadurai), viz., (i) ethnoscapes, (ii) financescapes, (iii) ideoscapes, (iv) media scapes, and (v) technoscapes. These are defined as follows.

Table 51.1 Types of global connectivity

S.N.	Type of Global Connectivity	Refers to or deals with
1.	Ethnoscapes	Movements of people including tourists, immigrants, refugees and business travelers.
2.	Financescapes	Global flow of capital or finance or money.
3.	Ideo scapes	Global spread of ideas and political ideologies.
4.	Media scapes	Global distribution of media images that appear on our computer screen, in news papers, television and radios.
5.	Technoscapes	Global flow of technologies such green revolution in wheat and rice and information technology.

Q.13. How the Globalization is measured ?

Ans. The rate or extent of globalization is measured on yearly basis. Different types of globalizations are measured separately as follows.

Table 51.2. Measurement of the rate of globalization.

S.N.	Type of Globalization	Measured from
1.	Economic Globalization	The data of trans border flow of capital or finance or money and direct foreign investment.
2.	Production Globalization	The data of trans country mobility of goods and products.
3.	Cultural Globalization	The mobility of tourists, travelers, and traders across the countries.
4.	Information Globalization	The data of information flow across the borders.
5.	Ecological Globalization	The work done for the protection of global ecosystem.

Q.14. What are the Factors affecting Globalization ?

Ans. The rate of globalization is affected by several factors such as (i) global atmosphere, (ii) natural calamities, (iii) international relationships, (iv) means of transportation, (v) means of communications, (vi) tourists places, (vii) demand of goods, services, etc. These are briefly discussed below.

1. **Global Atmosphere.** The peaceful global atmosphere promotes globalization, whereas the war situation restricts globalization. The globalization was adversely affected during First and Second World war period.

2. **Natural Calamities.** Natural calamities such as earth quake, tornadoes, floods and disease epidemic have adverse effects on the rate of globalization.
3. **International Relationships.** Harmonious relationships among countries enhance the rate of globalization, whereas disharmonious relationships restrict the process of globalization.
4. **Means of Transportation.** Better means of transportation among countries promotes globalization, whereas poor transportation system restricts the globalization.
5. **Means of Communication.** Better means of communication promotes globalization, whereas poor means of communication restricts the globalization.
6. **Tourist Places.** Good and large number of tourist places in a country will attract tourists and travelers and vice versa.
7. **Demand.** The demand of goods, services and information in other countries will enhance the international trade and the globalization.

Q.15. What are the Advantages of Globalization ?

Ans. Globalization has several advantages. Some important advantages of globalization are listed below.

1. **Connectivity.** People around the world are more connected to each other than ever before. Global mass media connects all the people in the world.
2. **Relationships.** The relationships between counties improve and the possibility of war between the developed countries decreases. The interdependence among Nations increases.
3. **Freedom of Trade.** It increases free trade between countries and reduces the international barriers.
4. **Investment opportunity**. As the liquidity of capital increases, developed countries can invest in developing ones. The flexibility of corporations to operate across borders increases.
5. **Integration.** It leads to integration or consolidation of global markets. In other words the markets are interlinked. It is much easier for people to travel, communicate and do business internationally.
6. **Quality and Price.** There is improvement in quality and reduction in price due to competition among different companies.
7. **Flow of Goods and Services.** Information, money, technology and products flow across the border quicker than ever before. Products produced in one part of a country are available to the rest of the world.

There is increase flow of communication between the individuals and corporations in the world. The movement of goods and people across the border is faster than ever before.

8. **Standard of Living.** Globalization offers a higher standard of living for people in rich countries and is the only realistic route out of poverty for the world's poor. It is claimed that globalization increases the economic prosperity and opportunity in the developing world. All the countries involved in the free trade are at a profit. As a result, there are lower prices, more employment and a better standard of life in these developing nations.
9. **Efficient use of Resources.** The civil liberties are enhanced and there is a more efficient use of resources. The environmental protection in developed countries increases.
10. **Spread of Culture.** Globalization leads to better cultural understanding and tolerance. Due to improved transport facilities, more and more people are traveling to different countries, thereby spreading their culture to other parts of the world. Reduction of cultural barriers increases the global village effect. There is spread of democratic ideals.

Q.16. What are the Disadvantages of Globalization ?

Ans. There are some disadvantages of globalization which are listed below.

1. **Increase in population.** Trans country flow of people will lead to increase in the population of certain countries especially in developed countries due to better facilities.
2. **Small Industries.** It will have adverse effects on small scale industries which can not compete in global market in terms of quality and price. Thus there will be hold of big industries.
3. **Employment.** There will be adverse effect on employment due to close down of small industries.
4. **Monetary Gain.** This will lead to tough competition among companies leading to loss in monetary gain.
5. **Terrorism.** Trans border flow of people may lead to increase in criminal activities and terrorism.
6. **Spread of Diseases.** There is greater risk of unintentional transmission of diseases between nations.
7. This may lead to widening of gap between rich and poor countries.
8. This may lead to exploitation of workers specially labours.

Q.17. What are the effects of globalization ?

Ans. The globalization has effects on movement of goods, services, information, finance, people, spread of cultures and ideas, markets, export and intellectual properties etc. These are briefly discussed below.

1. **Flow of Goods, Services Information etc.** There is enhancement in the information flow between geographically remote locations and more trans border data flow using communication satellites, the Internet, wireless telephones etc.
2. **Markets.** The global common market has a freedom of exchange of goods and capital. The worldwide production and financial markets emerge. The free trade zones are formed having less or no tariffs.
3. **Access to Goods and Finance.** There is a broad access to a range of goods for consumers and companies. Corporate, national and sub-national borrowers have a better access to external finance.
4. **Solution of Global Problems.** Global environmental problems like cross-boundary pollution, over fishing on oceans, climate changes are solved by discussions International criminal courts and international justice movements are launched.
5. **Uniform Standards.** The standards applied globally like patents, copyright laws and world trade agreements increase.
6. **Spread of Culture.** Globalization leads to spread of cultures as there is individual access to cultural diversity. This diversity decreases due to hybridization or assimilation. There is enhancement in worldwide fads and pop culture. The cross-cultural contacts grow and cultural diffusion takes place.
7. **Movement of People.** The International travel and tourism increases and immigration between countries increases. The worldwide sporting events like the Olympic Games and the FIFA World Cup are held. The free circulation of people of different nations leads to social benefits.
8. **Export.** The local consumer products are exported to other countries. There is an increase in the desire to use foreign ideas and products, adopt new practices and technologies and be a part of world culture.
9. **Reduction in Transport Cost and Subsidies.** Due to development of containerization for ocean shipping, the transportation costs are reduced . The subsidies for local businesses decrease and there is reduction in capital controls.
10. **Recognition of Intellectual Property.** There is supranational recognition of intellectual property restrictions i.e. patents authorized by one country are recognized in another.

INTELLECTUAL PROPERTY RIGHTS

Q.1. What is a property ?

Ans. The property refers to wealth or valuable things earned by a person. It is estimated in terms of land, house, garden, industries, animals, gold, silver, diamond, money etc. accumulated by a person.

Q.2. What is the need to earn property ?

Ans. Each person wants to earn property for four main purposes as given below:

1. To get a better or higher status in the society, because of a person is measured in terms of property accumulated by him or her. On the basis of property, the person is called as rich, middle class or poor .
2. To lead a better quality of life in the society.
3. To get a name and fame in the society. In the society, rich people get more respect than others.
4. For better financial security of life for himself and for his children.

Q.3. What are different types of property ?

Ans. The various valuable belongings of a person are broadly classified into two groups, viz.,

1. immovable property, and 2. movable property.

Q.4. What is Immovable Property ?

Ans. It refers to fixed types of properties which can not move from one place to other. Such properties include land, buildings (house, flat, bungalow, farm house) and gardens. Such properties are protected by laws of land in each country .

Q.5. Define Movable Property.

Ans. It refers to the property which can be easily shifted from one place to other. Such property includes animals, farm machines, furniture, fixtures, gold, silver, diamond, and money. Such properties are also protected by public laws.

Q.6. What is Intellectual Property ?

Ans. There is third type of property what we call intellectual property. The product/ process/ idea which is outcome of the brain of a person and can be used on commercial scale for benefit of human kind is called intellectual property. In other *words,* Intellectual property refers to creations of the mind: inventions, literary and artistic works, and symbols, names, images, and designs used on commercial scale.

Q.7. Describe main features of Intellectual Property.

Ans. The main features of intellectual property are given below:

1. It is measured in terms of new ideas, processes, products, inventions and innovations developed by a person.
2. It requires lot of intellectual inputs in terms of thinking, planning and fine tuning of new ideas/products/processes etc.
3. It requires considerable amount of funds and other resources to develop new products/processes.
4. The main problem with intellectual property is that it can be copied, reproduced and used by others resulting is loss of inventor. Hence protection of intellectual property is essential so that the inventor can derive maximum benefits from his invention.

Q.8. What do you mean by Intellectual Property Rights ?

Ans. The rights to intangible property that is the product of the human intellect are referred to as intellectual property rights. Intellectual property may be protected by copyright, trademark or patent. The holder of intellectual property rights is usually the person or persons who developed the product or the organization that funded it.

Q.9. What are different forms of IPR ?

Ans. The Intellectual Property Rights are broadly divided into two groups, viz., Primary Rights and Sui-generis Rights. These are briefly discussed below.

Q.10. What do you mean by Primary Rights ?

Ans. Primary rights include ,Copyrights, Patents, Trademarks, Trade Secret, Trade names, Domain names, Industrial Designs ,Geographical Indications etc.

Q.11. Define Sui-generis Rights.

Ans. Sui-generis refers to things of their own kind or things with unique characteristics. Such rights include Database Rights, Mask work, Plant Breeders Rights, Traditional Knowledge, Moral Rights and Supplementary Protection Certificate.

Q.12. Give a list of different types of Intellectual Property Rights.

Ans. A list of different types of Intellectual Property Rights is presented in Table 52.1

Table 52.1 Forms of Intellectual Property Rights/Laws

S.N.	Primary Rights	Sui Generis Rights
1.	Copyrights	Database Rights
2.	Patents	Mask work
3.	Trademarks	Plant Breeders' Rights
4.	Industrial Design Rights	Farmers' Rights
5.	Utility Models	Moral Rights
6.	Geographical Indication	Supplementary Protection Certificate
7.	Trade Secret	Indigenous Intellectual Property
8.	Related Rights	
9.	Trade Names	
10.	Domain Names	

A. Primary Rights

Q.13. Define Copyrights.

Ans. Copyright has been defined in various ways by different authors. Some commonly used definitions of copyright are given below:

1. Copyright refers to a document which grants exclusive right to the author / creater to publish and sell literary or musical or artistic work.
2. The exclusive rights to reproduce, sell and distribute a work, prepare derivative works and display the work publicly is referred to as copyright.
3. The right of an author, artist, publisher etc. to retain ownership of works and to produce or contract others to produce copies is called copyright.
4. The legal right granted to an author, composer, playwright, publisher, or distributor to exclusive publication, production, sale, or distribution of a literary, musical, dramatic, or artistic work is called copyright.

Q.14. What are the main features of copyrights ?

Ans. Main features of copyrights are briefly presented below;

(i) It provides protection for a specific duration.

(ii) It is applicable in all countries.

(iii) The copyright holder has the rights to authorize others to use the protected work.

(iv) It is applicable to books, movies, music, paintings, photographs and software.

(v) The information cannot be reproduced as such without written permission. However, the information or idea can be used by anyone.

Q.15. What is a Patent ?

Ans. Patent has been defined in various ways by different authors. Patent refers to a document granting an inventor sole rights to an invention. It is an official document which grants sole rights to the inventor for manufacturing and marketing his product/process/invention to derive benefits. In other words, patent is a set of exclusive rights granted by a state to a patentee (the inventor or assignee) for a fixed period of time.

Q.16. Explain briefly main points of a patent.

Ans. The main points related to a patent are briefly presented as follows.

(i) It is a legal document issued by a federal government that grants exclusive rights for the production, sale and profit from the invention of a product or process for a specific period of time. Patents also grant the right to prevent others from copying the invention.

(ii) An official license is granted by the Patent Office to issue exclusive right to an individual or business for production or sale of a specific invention.

(iii) It is a grant made by a government to an inventor, assuring the sole right to make, use, and sell the invention for a certain period of time. Many audio and video technologies are covered by patents. .

(iv) It is a document that allows the patent owner to prevent others from making, using or selling the invention protected by the patent.

Q.17. What is a Trademark ?

Ans. A trademark or trade mark is a distinctive sign of some kind which is used by an individual, business organization or other legal entity to uniquely identify the source of its products and/or services to consumers, and to distinguish its products or services from those of other entities.

Q.18. What are main features of a trademark ?

Ans. The main features of trademarks are briefly presented below:

(i) A trademark can be a word, name, symbol, device or mark which is used to identify and distinguish the goods or services of one company from goods or services of another .

(ii) Trademark is used to identify its product and to distinguish them from others. It is the name of a product made by a particular person or company.

(iii) The period of protection for a trademark varies, but can generally be renewed indefinitely.

Q.19. What do you mean by Industrial design ?

Ans. An industrial design -or simply a design -is the ornamental or aesthetic aspect of an article produced by industry or handicraft; registration and renewals provide protection for, in most cases, up to 15 years.

Q.20. Describe main points related to an industrial design.

Ans. The main points about industrial designs are given below :

(i) Industrial design is an applied art whereby the aesthetics and usability of products may be improved for marketability and production. Industrial Designers often utilize 3D Industrial Design is the field of developing physical solutions to meet a particular need. These physical solutions might include products.

(ii) Industrial design refers to any original shape, picture, or some combination applied to a useful article of manufacture. In other words, it refers to the design of the mass-produced products of our everyday environment, from sinks and furniture to computers.

(iii) Industrial design refers to the professional service of creating and developing concepts and specifications that optimize the function, value, and appearance of products and systems for the mutual benefit of both user and manufacturer (Industrial Design Society of America).

Q.21. What are Utility Models ?

Ans. A utility model is an exclusive right granted for an invention, which allows the right holder to prevent others from commercially using the protected invention, without his authorization, for a limited period of time. In its basic definition, which may vary from one country (where such protection is available) to another, a utility model is similar to a patent. In fact, utility models are also referred to as "petty patents" or "innovation patents, minor patent and small patent."

Q.22. Describe different points related to utility models.

Ans. The important points about utility model are listed below.

(i) Utility model is an intellectual property rights to protect inventions.

(ii) Utility models are also known as petty patents, petty inventions, petty innovations, utility innovations, minor patents and small patent.

(iii) The rights conferred by utility models are similar to those granted by patent laws but has a shorter term.

(iv) They are registered with National Patent Office.

(v) They are more suited to incremental inventions.

Q.23. What do you mean by Geographical indication ?

Ans. Geographical Indications (GI) is the name of a region, a specific place or, in exceptional cases a country, used to describe an agricultural, natural or manufactured goods product or a food stuff. In other words, a geographical indication is a sign used on goods that have a specific geographical origin and often possess qualities or a reputation that are due to that place of origin. In other words, a geographical indication is a name or sign used on certain products or which corresponds to a specific geographical location or origin (eg. a town, region, or country)

Q.24. Write short notes on geographical indication.

Ans. Geographical Indication may be an agricultural, natural or manufactured goods or product. So far geographical indications have been registered in some advanced countries because they have developed the system of protecting GI. In the past, wines, spirits, cheeses, tobacco, which account for 88%, have been registered as GI. Wines and spirits account for almost 71% of all registrations. In addition to wines and spirits, there is need for protection of agricultural products as GI. For this there is need to establish a multilateral system of identification and registration of GI.

Q.25. Define Trade Secret.

Ans. Trade secret has been defined in various ways by different authors. Some commonly used definitions of trade secret are given below:

1. A "trade secret" is anything (a formula, process , method, mechanism, tool, pattern or device) which the disclosing party desires to keep secret. Trade secrets usually include such things as the manufacturing details for a product, variations or alternative uses.
2. It may refer to a formula or process or device used in business that is not published or divulged which gives an advantage over competitors.
3. A form of industrial property (See. I). Refers to a non-patented process, mechanism, or formula, known only to its owner, that is used in producing something of commercial value.

Q.26. Describe main points related to trade secret.

Ans. The main points about trade secret are briefly presented as follows:

(i) There is no specific period for trade secret. It may continue life long or for generations together.

(ii) There is no need of registration for trade secret.

(iii) It does not provide opportunity to others for improvement of innovation.

(iv) It is not applicable to books, equipments, plant varieties, designs which are openly used.

Q.27. What are Related Rights ?

Ans. Related rights is a term in copyright law, used in opposition to the term "authors' rights". The term neighbouring rights is exactly equivalent, and a more literal translation of the original French *droits voisins* Related rights in civil law are rights which are similar to authors' right but which are not connected with the actual author to the work. Both authors' rights and related rights are copyrights in the sense of English or U. S. law.

Q.28. Explain main points about related rights.

Ans. The main points about related rights are given below.

1. Related rights are similar to copyrights.
2. Related rights come under primary property rights
3. Related rights are also known as neighouring rights.
4. Related rights are closer to copyrights but are not covered by the Berne Convention.
5. Related Rights are covered by Rome Convention, 1961

Q. 29. Define Trade Names.

Ans. A trade name, also known as a trading name or a business name, is the name which a business trades under for commercial purposes, although its registered, legal name, used for contracts and other formal situations, may be another. Pharmaceuticals also have trade names (e.g. "Aspirin"), often dissimilar to their chemical names ("acetylsalicylic acid"). Trading names are sometimes registered as trademarks or are regarded as brands. A trade name is the name under which a product is commercially known The commercial name by which a chemical is known is called trade name. On chemical may have a variety of trade names depending on the manufacturers or distributors involved.

Q.30. Describe main points about Trade Names.

Ans. The main points related to trade name are given below.

1. It is also known as business name, trading name, assumed name, brand name and corporate name.
2. It may or may not be registered.
3. It distinguishes a particular business from others.
4. It can be exclusive or non-exclusive.
5. It is generally used on letter heads and bank accounts.
6. It is the name given to a particular substance by each company that manufactures it.
7. It is the name `under which a company conducts its business.
8. It is the business name of the person or organization making and/or selling a particular product. "General Motors, Inc" is the "trade name" of the company.

It is used by a company to describe and distinguish its brand of a generic product. Kleenex is a trade name for a brand of tissue; Xerox:, a single brand of copier. A "trade name" is the business name of the person or organization making and/or selling a particular product. "General Motors, Inc" is the "trade name" of the company. Trade names are registered by the state in which the person or company is based.

Q.31. What do you mean by Domain names ?

Ans. Domain refers to the unique name that identifies an internet site. Domain names have two or more parts separated by dots. For example www.ask-edi.com or **www.bcbsks.com.**

The term domain name has multiple related meanings:

1. It identifies a web site on the internet (eg, company.com); also referred to as a URL. (Uniform Resource Locator)
2. It allows reference to Internet sites without knowing the true numerical address.
3. It refers to the official name of a computer connected to the Internet. (eg, "www.microsoft.com").
4. It is the unique name of a computer on the Internet that distinguishes it from other systems around the world.
5. They are sometimes colloquially (and incorrectly) referred to by marketers as "web addresses".

Q.32. What are the Advantages of Intellectual Property Rights ?

Ans. Intellectual Property Rights refer to the legal rights provided to an inventor to derive economic benefits from his invention/innovation. The main advantages of IPR are as follows.

1. It promotes healthy competition for invention/innovation among the intellectuals.
2. It helps in improving the quality of the product.
3. It makes available new ideas/ technologies to different countries.
4. It leads to faster development of industries/ organizations engaged in research and development work.

Q.33. What the disadvantages of Intellectual Property Rights ?

Ans. There are some disadvantages of Intellectual Property Rights which are briefly presented below.

(i) The procedure of registration, particularly of patents, is very lengthy.
(ii) It involves lot of money transaction in registration, renewal and licensing.
(iii) It invites lots of court cases due to infringements.
(iv) It may lead to monopoly of right holders, etc.

B. Sui-generis Rights

Q.34. Define sui generis rights.

Ans. Sui-generis is a Latin expression, literally meaning of its own kind/genus or unique in its characteristics. In law, it is term of art used to identify a legal classification that exists independently of other categorizations because of its uniqueness or due to the specific creation of an entitlement or obligation. In

Intellectual Property there are rights which are known as being sui generic to owners of a small class of works. Such Intellectual Property Rights include (i) database Rights, (ii) mask work, (iii) plant breeders' rights, (iv) farmers' rights, (v) moral rights, (vi) supplementary protection certificate, and (vii) indigenous intellectual property .

Q.35. What do you mean by Database rights ?

Ans. Database rights are a form of exclusive right introduced by European Union Law in 1996 in those countries which follow EU Law. A 'database right' is an intellectual property right given to a computer database. The main points about database rights are as follows:

(i) In most countries database are covered by copyright law to some degree.

(ii) The database rights were introduced by European Union Law to provide greater protection to collection of information.

(iii) Database right is covered by sui generic rights.

(iv) **Duration.** Database rights last for 15 years, but can be extended if the database is updated.

(v) **Protection.** Database right prevents copying of substantial parts of database. The protection is over the information, not over the form of expression.

Q.36. What is Mask Work Right ?

Ans. A mask work is a two or three-dimensional layout of an integrated circuit (IC or "chip"), i.e. the arrangement on a chip of semiconductor devices such as transistors and passive electronic components such as resistors and interconnections. Main points related to mask work are listed as under.

(i) Mask work can not be protected under copyright law or patent law.

(ii) The mask work exclusive rights were first granted in USA by Semiconductor Chip Protection Act of 1984. Such legislations also exist in Canada and Australia.

(iii) Duration Rights for semiconductor mask work lasts only for two years (if unregistered) or 10 years (if registered).

(iv) The exclusive rights in mask work are somewhat like those of copyright: the right to reproduce the mask work or distribute an IC made using mask work.

(v) Reproduction for reverse engineering of a mask work is permitted by law.

(vi) Mask work rights exist when they are created, even without registration.

(vii) Protection under mask work cannot be provided to a mask that is not original.

Q.37. What do you mean by Plant breeders' rights ?

Ans. Plant breeders' rights, also known as plant variety rights (PVR), are intellectual property rights granted to the breeder of a new variety of plant. Plant Breeders Rights are granted to novel plant varieties that are distinctive, uniform, and stable (e.g, cultivars bred true-to-type for desired traits). The legal protection of a new plant variety is granted to the breeder or his successor. The effect of PBR is that prior authorization is required before the material can be used for commercial purposes.

Q.38. What are Advantages of PBR ?

Ans. There are several advantages of plant breeders' rights which are listed below

(i) Breeders get benefit of their variety.

(ii) PBR help in faster development of seed industry.

(iii) PBR lead to improvement in quality because of competition.

(iv) PBR are useful in procurement of good material on payment basis.

(v) PBR help in enrichment of genetic resources.

Q.39. What are Disadvantages of PBR ?

Ans. There are several advantages of plant breeders' rights which are listed below.

(i) It will promote monopoly.

(ii) It will encourage unhealthy practices.

(iii) It may lead to increase in prices.

(iv) There will be reduction in genetic variability.

(v) There will be compulsion to purchase fresh seed every year,

Q.40. What are Farmers' Rights ?

Ans. Farmers' rights refer to the rights arising from the past, present and future contributions of farmers in conserving, improving and making available plant or animal genetic resources, particularly those in the centres of origin/ diversity. In other words the legal rights provided to farmers to save, use, sow, replant, exchange, share or sell his farm produce including seed of a variety protected under Plant Variety Protection Act refer to Farmers. Rights. The purpose to these rights is to "ensure full benefits to farmers and support the continuation of their contributions.

The F AO, Conference held in Rome from 11-29 November, 1989 endorsed the concept of Farmers Rights with a view to:

(i) ensuring global recognition of the need for conservation and the availability of sufficient funds for these purposes;

(ii) assisting farmers and farming communities throughout the world, especially those in areas of original diversity of plant genetic resources, in the protection and conservation of their PGR and of the natural biosphere; and,

(iii) allowing the full participation of farmers, their communities, and countries in the benefits derived, at present and in the future, from the improved use of PGR.

Q.41. What are Moral rights ?

Ans. Moral rights are rights of creators of copyrighted works generally recognized in civil law jurisdictions and first recognized in France and Germany, before they were included in the Berne Convention for the Protection of Literary and Artistic Works in 1928. Moral rights are a special extension to copyright that gives the originator of a copyrighted work rights over its use. The originator can specify moral conditions on the use or exploitation of their creation, even when the rights are licensed or sold to others'.

Q.42. Define Supplementary protection certificate.

Ans. This term is used in European Countries. This is an extension of the term given to pharmaceutical or plant protection patent. A supplementary protection certificate (SPC) is a sui generis, patent-like, intellectual property right. This type of right is available for medicinal products, such as drugs, and plant protection products, such as insecticides, and herbicides.

Q.43. What do you mean by Traditional Knowledge ?

Ans. The Indigenous Knowledge also known as Traditional Knowledge (TK) refers to the knowledge, innovations and practices of indigenous people and local communities. In other words, it is the: knowledge gained through tradition from ancestors about plants and animals, medicines, agricultural practices and other matters.

Q.44. Define ecological knowledge.

Ans. The ecological knowledge is composed of both traditional knowledge and experimential knowledge(gained through personal experience. Traditional knowledge is very important almost in every sphere of life. Hence, its protection is very necessary otherwise this valuable knowledge will be gradually lost.

Q.45. What are differences between Database rights and copyrights ?

Ans. There are some differences between database rights and". copyrights which are presented in Table 52.2

Table 52.2 Differences between database rights and copyrights

S.N.	Particulars	Database Rights	Copyrights
1.	Governed by	Sui generic rights	Primary intellectual rights

2.	Period of protection	15 years	60-110 years
3.	Protection over	Information	Expression

Q.46. What are the similarities between database rights and copyrights ?

Ans. There are some similarities between database rights and copyrights which are presented in Table 52.3

Table 52.3 Similarities between database rights and copyrights

S.N.	Particulars	Database Rights	Copyrights
1.	Registration	No compulsory	Not compulsory
2.	Validity	Country of registration	Country of registration
3.	Enforcement	After creation	After creation

Q.47. What are differences between Database rights and Mask Work rights ?

Ans. There are some differences between database and mask work which are presented in Table 52.4

Table 52.4 Differences between database and Mask work

S.N.	Particulars	Database Rights	Mask Work
1.	Period of protection	15 years	Unregistered 2 Registered 10 years
2.	Protection over	Information	Integrated Circuits

Q.48. What are similarities between Database rights and Mask Work rights ?

Ans. There are some similarities between database and mask work which are presented in Table 52.5

Table 52.5 Similarities between database and Mask work

S.N.	Particulars	Database Rights	Mask Work
1.	Governed by	Sui generic rights	Sui generic rights
2.	Registration	Not compulsory	Not compulsory
3.	Validity	Country of registration	Country of registration
4.	Enforcement	After creation	After creation

MATCHING QUESTIONS

Q. 1. Match the information of group A with group B

Group A	Group B
(*a*) Homozygous homogeneous population	(*i*) Multilines
(*b*) Homozygous heterogeneous population	(*ii*) Purelines
(*c*) Heterozygous homogeneous population	(*iii*) Synthetics
(*d*) Heterozygous heterogeneous population	(*iv*) F_1 hybrids

Q. 2. Match the information of group A with group B

Group A	Group B
(*a*) Semidwarf wheat	(*i*) Dee Geo Woogen
(*b*) Semidwarf rice	(*ii*) Pearl millet
(*c*) Tiff 23 A	(*iii*) Norin 10
(*d*) Kafir 60	(*iv*) Sorghum

Q. 3. Match the information of group A with group B

Group A	Group B
(a) Kafir 60	(i) Wheat
(b) Tift 23A	(ii) Cotton
(c) Gregg 399	(iii) Sorghum
(d) Norin 10	(iv) Pearl millet

Q. 4. Match the information of group A with group B

Group A	Group B
(*a*) First Cotton Hybrid	(*i*) China
(*b*) First Rice Hybrid	(*ii*) India
(*c*) First Pigeon pea hybrid	(*iii*) USA
(*d*) First transgenic plant	(*iv*) ICRISAT

Q. 5. Match the information of group A with group B

Group A	Group B
(*a*) Father of hybrid cotton	(*i*) Donald
(*b*) Father of green revolution	(*ii*) CT Patel
(*c*) Father of hybrid rice	(*iii*) Norman E Borlaug
(*d*) Ideotype	(*iv*) Y.L. Ping

Q. 6. Match the books with authors

Books	**Authors**
(*a*) Principles of Plant Breeding	(*i*) J.M. Poehlman
(*b*) Methods of Plant Breeding	(*ii*) RW Allard
(*c*) Breeding Field Crops	(*iii*) NW Symmonds
(*d*) Principles of Crop Improvement	(*iv*) Hays, Immer, Smith

Q. 7. Match the breeders with their crops

Breeders	**Crops**
(*a*) T.S. Venketraman	(*i*) Wheat
(*b*) Puskarnath	(*ii*) Sugarcane
(*c*) BP Paul	(*iii*) Rice
(*d*) K Ramiah	(*iv*) Potato

Q. 8. Match the breeders with their crops

Breeders	**Crops**
(*a*) Bosisen	(*i*) Pearlmillet
(*b*) DS Attwal	(*ii*) Sorghum
(*c*) NGP Rao	(*iii*) Cotton
(*d*) CT Patel	(*iv*) Maize

Q. 9. Match the information of group A with group B

Group A	**Group B**
(*a*) Traitor gene	(*i*) Borlaug
(*b*) Sex in plants	(*ii*) Jahannsen
(*c*) Pure line	(*iii*) Camerarius
(*d*) Green revolution	(*iv*) Monsanto

Q. 10. Match the books with their authors

Books	**Authors**
(*a*) Breeding Asian Field Crops	(*i*) Walter Fehr
(*b*) Principles of Cultivar Development	(*ii*) D.S. Falconer
(*c*) Introduction to Quantitative Genetics	(*iii*) Phundan Singh
(*d*) Essentials of Plant Breeding	(*iv*) Poehlman and Borthakur.

Q. 11. Match the terms with scientists

Terms	**Scientists**
(*a*) Genepool	(*i*) Weismann
(*b*) Germplasm	(*ii*) Frankel
(*c*) Core Collections	(*iii*) Harland
(*d*) Micro-centre	(*iv*) Dobzhansky

Q. 12. Match the information of group A with group B

Group A	**Group B**
(*a*) Primary genepool	(*i*) GP_3

(*b*) Secondary genepool	(*ii*) GP_1
(*c*) Tertiary genepool	(*iii*) GP_2
(*d*) Land races	(*iv*) Primitive Cultivars.

Q. 13. Match the information with their headquarters

Organizations	**Headquarters**
(*a*) NBDGR	(*i*) Kolkata
(*b*) IPGRI	(*ii*) Dehradun
(*c*) BSI	(*iii*) New Delhi
(*d*) FRI	(*iv*) Rome, Italy

Q. 14. Match the information of group A with group B

Group A	**Group B**
(*a*) IBPGR	(*i*) 1976
(*b*) IPGRI	(*ii*) 1974
(*c*) NBPGR	(*iii*) Vavilov
(*d*) Centres of Origin	(*iv*) 1994

Q. 15. Match the terms with scientists

Terms	**Scientists**
(*a*) Apomixis	(*i*) Druery and Bower
(*b*) Amphimixis	(*ii*) Winkler
(*c*) Parthenogenesis	(*iii*) Weismann
(*d*) Apospory	(*iv*) Owen

Q. 16. Match the terms with scientists

Terms	**Scientist**
(*a*) Apogamy	(*i*) Strasburger
(*b*) Adventive embryony	(*ii*) Battaglia
(*c*) Semigamy	(*iii*) Darwin
(*d*) Heterostyly	(*iv*) Winkler

Q. 17. Match the terms with scientists

Terms	**Scientist**
(*a*) Monoecious	(*i*) Wright
(*b*) Gamogamy	(*ii*) Darwin
(*c*) Double fertilization	(*iii*) Navashin and Guignard
(*d*) Mating systems	(*iv*) Hartmann

Q. 18. Match the information of group A with group B

Group A	**Group B**
(*a*) Autogamy	(*i*) Sexual reproduction
(*b*) Allogamy	(*ii*) Asexual reproduction
(*c*) Apomixis	(*iii*) Inbreeding
(*d*) Amphimixis	(*iv*) Outbreeding

Q. 19. Match the information of group A with group B

Group A	Group B
(*a*) Homozygosity	(*i*) Apomixis
(*b*) Heterozygosity	(*ii*) Amphimixis
(*c*) Lack of Sexual fusion	(*iii*) Self pollination
(*d*) Sexual fusion	(*iv*) Cross pollination

Q. 20. Match the terms with scientists

Terms	Scientists
(*a*) Self Incompatibility	(*i*) Hughes & Babcock
(*b*) Gametophytic SI	(*ii*) Stout
(*c*) Sporophytic SI	(*iii*) Koelreuter
(*d*) Male sterility	(*iv*) Mangelsdorf

Q. 21. Match the information of group A with group B

Group A	Group B
(*a*) Male gametocides	(*i*) Male sterile line
(*b*) A-line	(*ii*) Maintainer line
(*c*) B-line	(*iii*) Restorer line
(*d*) R-line	(*iv*) Chemical hybridization agents

Q. 22. Match the information of group A with group B

Group A	Group B
(*a*) Functional pollen	(*i*) Heritable
(*b*) Nonfunctional pollen	(*ii*) Non-heritable
(*c*) Mutagen induced male sterility	(*iii*) Male sterility
(*d*) Gametocide induced male sterility	(*iv*) Self Incompatibility

Q. 23. Match the information of group A with group B

Group A	Group B
(*a*) Phenotypic variance	(*i*) Heritable fixable
(*b*) Additive variance	(*ii*) Total variance
(*c*) Dominance variance	(*iii*) Nonheritable
(*d*) Environmental variance	(*iv*) Heritable nonfixable

Q. 24. Match the terms with scientists

Terms	Scientists
(*a*) Selection	(*i*) Johannsen
(*b*) Pureline	(*ii*) Gardner
(*c*) Grid method of mass selection	(*iii*) Webber
(*d*) Clone	(*iv*) Darwin

Q. 25. Match the concepts with scientists

Concepts	Scientists
(*a*) Pure line theory	(*i*) Jensen

(*b*) Bulk breeding method	(*ii*) Johannsen
(*c*) Single seed discent	(*iii*) Nilsson Ehle
(*d*) Multiline breeding	(*iv*) Goulden

Q. 26. Match the information of group A with group B

Group A	**Group B**
(*a*) Recurrent parent	(*i*) Vilmorin
(*b*) Non-recurrent parent	(*ii*) Recipient parent
(*c*) Progeny test	(*iii*) Harrington
(*d*) Mass Pedegree method	(*iv*) Donor parent

Q. 27. Match the information of group A with group B

Group A	**Group B**
(*a*) Pureline	(*i*) one gene difference
(*b*) Multilines	(*ii*) Manifold effects of a gene
(*c*) Isogenic lines	(*iii*) Narrow genetic base
(*d*) Pleiotropy	(*iv*) Broad genetic base.

Q. 28. Match the information of group A with group B

Group A	**Group B**
(*a*) Recurrent selection	(*i*) Comstock & Robinson
(*b*) Disruptive selection	(*ii*) Hull
(*c*) Biparental Mating	(*iii*) Jensen
(*d*) Diallel selective mating	(*iv*) Mather

Q. 29. Match the terms with scientists

Terms	**Scientists**
(*a*) Centripetal selection	(*i*) Schmalhausen
(*b*) Directional selection	(*ii*) Grant
(*c*) Progressive selection	(*iii*) Mather
(*d*) Interpopulation selection	(*iv*) Simpson

Q. 30. Match the information of group A with group B

Group A	**Group B**
(*a*) Progressive selection	(*i*) Centripetal selection
(*b*) Stabilizing selection	(*ii*) Directional selection
(*c*) Disruptive selection	(*iii*) Stadler
(*d*) Gamete selection	(*iv*) Centrifugal selection

Q. 31. Match the information of group A with group B

Group A	**Group B**
(*a*) Recurrent selection for gca	(*i*) No tester
(*b*) Recurrent selection for sca	(*ii*) One homozygous tester
(*c*) Reciprocal recurrent selection	(*iii*) One heterozygous tester
(*d*) Simple recurrent selection	(*iv*) Two heterozygous testers

Q. 32. Match the terms with scientists

Terms	Scientists
(*a*) Heterosis	(*i*) Renner
(*b*) Luxuriance	(*ii*) Shull
(*c*) Useful heterosis	(*iii*) Hull
(*d*) Overdominance	(*iv*) Meredith and Bridge

Q. 33. Match the information of group A with group B

Group A	Group B
(*a*) Overdominance	(*i*) Economic heterosis
(*b*) Heterosis	(*ii*) Super dominance
(*c*) Luxuriance	(*iii*) Hybrid vigour
(*d*) Useful heterosis	(*iv*) Pseudo heterosis

Q. 34. Match the information of group A with group B

Group A	Group B
(*a*) Dominance hypothesis	(*i*) Davis
(*b*) Overdominance hypothesis	(*ii*) Jones
(*c*) Top cross	(*iii*) Devanport
(*d*) Dominance of linked genes hypothesis	(*iv*) Shull and East.

Q. 35. Match the information of group A with group B

Group A	Group B
(*a*) Inbred variety cross	(*i*) Intraspecific hybrid
(*b*) Intervarietal cross	(*ii*) Interspecific hybrid
(*c*) Inter specific cross	(*iii*) Triallel cross
(*d*) Three way crosses	(*iv*) Top cross.

Q. 36. Match the hybrid with their crops

Hybrids	Crops
(*a*) Hybrid 4	(*i*) Sorghum
(*b*) CSH 9	(*ii*) Pigeon pea
(*c*) ICPH 8	(*iii*) Pearl millet
(*d*) HB 3	(*iv*) Cotton

Q. 37. Match the hybrids with their crops

Hybrids	Crops
(*a*) Ganga 5	(*i*) Cotton
(*b*) Hybrid 6	(*ii*) Sunflower
(*c*) BSH1	(*iii*) Castor
(*d*) GCH2	(*iv*) Maizi

Q. 38. Match the hybrids with their crops

Hybrids	Crops
(*a*) Varalaxmi	(*i*) Maize

(*b*) Deccan 103	(*ii*) Cotton
(*c*) KBSH1	(*iii*) Castor
(*d*) GCH_3	(*iv*) Sunflower

Q. 39. Match the hybrids with their crops

Hybrids	**Crops**
(*a*) CSH 10	(*i*) Cotton
(*b*) JKHy1	(*ii*) Sorghum
(*c*) Ganga 2	(*iii*) Pearl millet
(*d*) BJ 104	(*iv*) Maize

Q. 40. Match the terms with scientists

Terms	**Scientists**
(*a*) Synthetics	(*i*) Rickey
(*b*) Convergent Improvement	(*ii*) Stadler
(*c*) Heterosis	(*iii*) Jenkins
(*d*) Gamete selection	(*iv*) Shull

Q. 41. Match the terms with scientists

Terms	**Scientists**
(*a*) Mutation	(*i*) Roentgen
(*b*) Non sense mutation	(*ii*) de Vries
(*c*) Transversion	(*iii*) Brenner et al.
(*d*) X-rays	(*iv*) Freese

Q. 42. Match the terms with scientists

Terms	**Scientists**
(*a*) Point mutation	(*i*) Goldschmidt
(*b*) Macro-mutation	(*ii*) Bridges
(*c*) Transition	(*iii*) Brenner et al
(*d*) Missense mutation	(*iv*) Freese

Q. 43. Match the terms with scientists

Terms	**Scientists**
(*a*) Muton	(*i*) Muller
(*b*) Mutational load	(*ii*) Wright
(*c*) Frame shift mutation	(*iii*) Benzer
(*d*) Mutation pressure	(*iv*) Brenner

Q. 44. Match the terms with scientists

Terms	**Scientists**
(*a*) Mutator gene	(*i*) Drake & Allen
(*b*) Antimutator gene	(*ii*) Helsings et al.
(*c*) Mutation repair	(*iii*) Muller
(*d*) Mutation rate	(*iv*) Demerec

Q. 45. Match the information of group A with group B

Group A	Group B
(*a*) Gamma rays	(*i*) Base analogue
(*b*) EMS	(*ii*) Acridine dye
(*c*) 5 Bromouracil	(*iii*) Alkylating agent
(*d*) Ethidium bromide	(*iv*) Irradition

Q. 46. Match the mutations with their survival rate

Mutations	Survival rate
(*a*) Lethal mutation	(*i*) 100%
(*b*) Sub-lethal mutation	(*ii*) < 50%
(*c*) Subvital mutation	(*iii*) > 50%
(*d*) Vital mutation	(*iv*) zero

Q. 47. Match the mutagens with their category

Mutagens	Category
(*a*) X-rays	(*i*) Alkylating agent
(*b*) EES	(*ii*) Acridine dye
(*c*) 2 Amino Purine	(*iii*) Irradiation
(*d*) Proflavin	(*iv*) Base analogue

Q. 48. Match the information of group A with group B

Group A	Group B
(*a*) Beta particles	(*i*) Neutral
(*b*) Alpha particles	(*ii*) Negatively charged
(*c*) Thermal neutrons	(*iii*) Crowther
(*d*) Target theory	(*iv*) Positively charged

Q. 49. Match the mutant varieties with their crops

Mutant varieties	Crops
(*a*) Biraj	(*i*) Sugarcane
(*b*) MCU 7	(*ii*) Barley
(*c*) DL 253	(*iii*) Cotton
(*d*) CO 6608	(*iv*) Rice

Q. 50. Match the mutant varieties with their crops

Mutant Varieties	Crops
(*a*) IIT 60	(*i*) Cotton
(*b*) MCU 10	(*ii*) Rice
(*c*) PL 56	(*iii*) Peanut
(*d*) TG3	(*iv*) Barley

Q. 51. Match the mutant varieties with their crops

Mutant varieties	Crops
(*a*) CO 8153	(*i*) Groundnut

(*b*) TG4	(*ii*) Sugarcane
(*c*) V16	(*iii*) Chickpea
(*d*) Pusa 408	(*iv*) Cow pea.

Q. 52. Match the mutant varieties with their crops

Mutant Varieties	**Crops**
(*a*) Pusa 413	(*i*) Groundnut
(*b*) CO 997	(*ii*) Chick pea
(*c*) Vikram	(*iii*) Mungbean
(*d*) CO4	(*iv*) Sugarcane

Q. 53. Match the mutant varieties with their crops

Mutant Varieties	**Crops**
(*a*) BP1	(*i*) Pigeonopea
(*b*) V240	(*ii*) Groundnut
(*c*) TAP7	(*iii*) Cow pea
(*d*) CO3	(*iv*) Mungbean

Q. 54. Match the mutant varieties with their crops

Mutant Varieties	**Crops**
(*a*) Jagannath	(*i*) Cotton
(*b*) Pusa Ageti	(*ii*) Rice
(*c*) V37	(*iii*) Peanut
(*d*) TG17	(*iv*) Cow pea

Q. 55. Match the mutant varieties with mutagen

Mutant Varieties	**Mutagens**
(*a*) TG3	(*i*) Gamma rays
(*b*) CO4	(*ii*) X-rays
(*c*) TAT5	(*iii*) EMS
(*d*) CO2	(*iv*) Fast Neutrons

Q. 56. Match the organizations with their locations

Organizations	**Locations**
(*a*) IAEA	(*i*) Mumbai
(*b*) BARC	(*ii*) Vienna
(*c*) IARI	(*iii*) KARNAL
(*d*) NDRI	(*iv*) New Delhi

Q. 57. Match the terms with scientists

Terms	**Scientists**
(*a*) Genome	(*i*) Winkler
(*b*) Heteroploid	(*ii*) Blakeslee
(*c*) Haploid	(*iii*) Kihara
(*d*) Monosomic	(*iv*) Strasburger

Q. 58. Match the terms with scientists

Terms	Scientists
(*a*) Trisomic	(*i*) Navashin
(*b*) Hypoploid	(*ii*) Strasburger
(*c*) Polyploidy	(*iii*) Belar
(*d*) Amphidiploid	(*iv*) Blakeslee

Q. 59. Match the species with their genomes

Species	Genomes
(*a*) Brassica nigra	(*i*) CC
(*b*) Brassica oleracea	(*ii*) AA
(*c*) Brassica compestris	(*iii*) AABB
(*d*) Brassica jumeea	(*iv*) BB

Q. 60. Match the species with their genomes

Species	Genomes
(*a*) Brassica carinata	(*i*) AABB
(*b*) Brassica napus	(*ii*) CC
(*c*) Brassica juncea	(*iii*) AACC
(*d*) Brassica nigra	(*iv*) BBCC

Q. 61. Match the information of group A with group B

Group A	Group B
(*a*) Raphanobrassica	(*i*) Rimpu
(*b*) Triticate	(*ii*) Karpechenko
(*c*) Nicotiana digluta	(*iii*) Nagashru, u
(*d*) Triangle of Brassica	(*iv*) Clausen & Goodspeed

Q. 62. Match the terms with symbols

Terms	Symbols
(*a*) Nullisomic	(*i*) $2n-1$
(*b*) Simple monosomic	(*ii*) $2n + 1$
(*c*) Simple trisomic	(*iii*) $2n-1-1$
(*d*) Double monosomic	(*iv*) $2n-2$

Q. 63. Match the terms with symbols

Terms	Symbols
(*a*) Basic number	(*i*) n
(*b*) Haploid number	(*ii*) X
(*c*) Nullisomic haploid	(*iii*) $n-1 + 1$
(*d*) Substitution haploid	(*iv*) $n-1$

Q. 64. Match the terms of group A with those of group B

Group A	Group B
(*a*) Autopolyploidy	(*i*) Hybrid polyploidy

(*b*) Allopolyploidy	(*ii*) Simple polyploidy
(*c*) Allotetraploid	(*iii*) Polysomics
(*d*) Hyperploids	(*iv*) Amphidiploid

Q. 65. Match the terms with symbols

Terms	**Symbols**
(*a*) Simple trisomic	(*i*) $2n + 2$
(*b*) Double trisomic	(*ii*) $2n + 1 - 1$
(*c*) Simple tetrosomic	(*iii*) $2n + 1$
(*d*) Quasidiploid	(*iv*) $2n + 1 + 1$

Q. 66. Match the terms with scientists

Terms	**Scientists**
(*a*) Nullisomic	(*i*) Hsu
(*b*) Hyperploid	(*ii*) Dermen
(*c*) Quasidiploid	(*iii*) Belar
(*d*) Colchiploidy	(*iv*) Blakeslee

Q. 67. Match the species with chromosome number

Species	**Chromosome number**
(*a*) Brassica juncea	(*i*) 34
(*b*) Brassica carinata	(*ii*) 36
(*c*) Brasica napus	(*iii*) 38
(*d*) Brassica nigra	(*iv*) 16

Q. 68. Match the terms with scientists

Terms	**Scientists**
(*a*) Alien addition line	(*i*) Kattermann
(*b*) Alien substitution line	(*ii*) Leightly and Taylor
(*c*) Cryptic structural hybridity	(*iii*) Anderson and Hubricht
(*d*) Introgression	(*iv*) Stebbins

Q. 69. Match the terms with scientists

Terms	**Scientists**
(*a*) Protoplast	(*i*) Larkin Scowcroft
(*b*) Somatic hybridization	(*ii*) Bunn et al.
(*c*) Cybrid	(*iii*) Henstein
(*d*) Somaclonal variation	(*iv*) Barki et al.

Q. 70. Match the terms of group A with group B

Group A	**Group B**
(*a*) Protoplast	(*i*) Meristem culture
(*b*) Somaclonal variation	(*ii*) Naked cells
(*c*) Gametoclonal variation	(*iii*) Protoplast culture
(*d*) Protochloral variation	(*iv*) Anther culture

Q. 71. Match the information of group A with group B

Group A	Group B
(*a*) Transgenic	(*i*) Biological method of DNA transfer
(*b*) Protoplast	(*ii*) Gordon & Ruddle
(*c*) Plasmid method	(*iii*) Particle acceleration
(*d*) Particle bombardment	(*iv*) Naked cell

Q. 72. Match the information of group A with group B

Group A	Group B
(*a*) Vertical resistance	(*i*) General resistance
(*b*) Horizontal resistance	(*ii*) Specifc resistance
(*c*) Non preference	(*iii*) Pathotype
(*d*) Physiological race	(*iv*) Antixenosis

Q. 73. Match the information of group A with group B

Group A	Group B
(*a*) Gene for gene hypothesis	(*i*) Sorghum
(*b*) Vertical resistance	(*ii*) Painter
(*c*) Mechanisms of insect resistance	(*iii*) Vander Plank
(*d*) Striga	(*iv*) Flor hypothesis

Q. 74. Match the resistant varieties with insects

Varieties	Resistant
(*a*) Will (Barley)	(*i*) Stemsawfly
(*b*) Rescue (wheat)	(*ii*) Greenbugs
(*c*) SAR1 (Sorghum)	(*iii*) Stem eelworm
(*d*) Greta (Oat)	(*iv*) Striga

Q. 75. Match the terms of group A with information of group B

Group A	Group B
(*a*) Xerophytes	(*i*) Acidity tolerant plants
(*b*) Oxylophytes	(*ii*) Moisture stress tolerant plants
(*c*) Halophytes	(*iii*) Salinity sensitive plants
(*d*) Glycophytes	(*iv*) Salinity tolerant plants

Q. 76. Match the information of group A with group B

Group A	Group B
(*a*) Lithophytes	(*i*) Shade loving plants
(*b*) Sociophytes	(*ii*) Plants which can grow on hills
(*c*) Hydroponics	(*iii*) Study of salt tolerance
(*d*) Lysimeter	(*iv*) Water Culture

Q. 77. Match the drought tolerant varieties with their crops

Varieties	Crops
(*a*) COS 510	(*i*) Rice

(*b*) NP 720	(*ii*) Sugarcane
(*c*) IGFRI 437	(*iii*) Wheat
(*d*) Akashi	(*iv*) Cowpea

Q. 78. Match varieties with their crops

Varieties	**Crops**
(*a*) C 281	(*i*) Barley
(*b*) CO 997	(*ii*) Wheat
(*c*) Motia	(*iii*) Sugarcane
(*d*) Karan 280	(*iv*) Durum wheat

Q. 79. Match drought tolerant varieties with their crops

Varieties	**Crops**
(*a*) K 46	(*i*) Sugarcane
(*b*) CO 1158	(*ii*) Wheat
(*c*) IGFRI 450	(*iii*) Groundnut
(*d*) Jyoti	(*iv*) Cowpea

Q. 80. Match salinity tolerant varieties with their crops

Varieties	**Crops**
(*a*) Ratana	(*i*) Rice
(*b*) COS 767	(*ii*) Barley
(*c*) Mohan	(*iii*) Okra
(*d*) Pusa Sawani	(*iv*) Sugarcane

Q. 81. Match salinity tolerant varieties with their crops

Varieties	**Crops**
(*a*) CO 1148	(*i*) Barley
(*b*) Karan 18	(*ii*) Sugarcane
(*c*) Hisar 2	(*iii*) Rice
(*d*) Mohan	(*iv*) Onion

Q. 82. Match the information of group A with group B

Group A	**Group B**
(*a*) Ideotype	(*i*) Donald (1962)
(*b*) Harvest Index	(*ii*) Donald (1968)
(*c*) Noirn 10	(*iii*) Rice
(*d*) Dee Geo Woogen	(*iv*) Wheat

Q. 83. Match information of group A with group B

Group A	**Group B**
(*a*) C3 plant	(*i*) Ideal plant type
(*b*) C4 plant	(*ii*) Wheat
(*c*) Biological yield	(*iii*) Total dry matter/plant
(*d*) Ideotype	(*iv*) Sugarcane

Q. 84. Match crop ideotypes with their scientists

Crop Ideotype	**Scientists**
(*a*) Wheat	(*i*) Rasmusson
(*b*) Six rowed barley	(*ii*) Donald
(*c*) Maize	(*iii*) Thurling
(*d*) Brassica napus	(*iv*) Mock & Pearce

Q. 85. Match the information of group A with group B

Group A	**Group B**
(*a*) Gossypol	(*i*) Soybean
(*b*) Neurotoxin	(*ii*) Repeseed & Mustard
(*c*) Erucic acid	(*iii*) Cotton
(*d*) Goitrogen	(*iv*) Lathyrus

Q. 86. Match the information of group A with group B

Group A	**Group B**
(*a*) Saponin	(*i*) Sorghum
(*b*) Coumarin	(*ii*) Alfalfa
(*c*) CN glucocides	(*iii*) Sweet clovin
(*d*) Steroidal alkaloids	(*iv*) Potato

Q. 87. Match the information of group A with group B

Group A	**Group B**
(*a*) Alkaloids	(*i*) Yam
(*b*) Resigns	(*ii*) Mango
(*c*) Anti Vitamin E factor	(*iii*) Safflower
(*d*) Polyphenolics	(*iv*) Field Pea.

Q. 88. Match the information of group A with group B

Group A	**Group B**
(*a*) Brittle 2	(*i*) Wheat
(*b*) Notch 2	(*ii*) Sorghum
(*c*) IS 11167	(*iii*) Barley
(*d*) Naphal	(*iv*) Maize

Q. 89. Match the information of group A with group B

Group A	**Group B**
(*a*) Riso 1508	(*i*) Sorghum
(*b*) Floury 2	(*ii*) Barley
(*c*) IS 11758	(*iii*) Maize
(*d*) Pusa 24	(*iv*) Lathyrus

Q. 90. Match the information of group A with group B

Group A	**Group B**
(*a*) P 721	(*i*) Wheat

(*b*) Opaque 2 (*ii*) Barley
(*c*) Notch 1 (*iii*) Sorghum
(*d*) Atlas 66 (*iv*) Maize

Q. 91. Match the varieties with crops

Varieties	Crops
(*a*) CO 671	(*i*) Maize
(*b*) Karan 92	(*ii*) Rapeseed
(*c*) K 88	(*iii*) Barley
(*d*) Protina	(*iv*) Sugarcane

Q. 92. Match the information of group A with group B

Group A	Group B
(*a*) Maize-Wheat	(*i*) Double Cropping
(*b*) Wheat + Raya	(*ii*) Monoculture
(*c*) Crops + Cattle	(*iii*) Mixed Cropping
(*d*) Rice-Rice-Rice	(*iv*) Mixed farming

Q. 93. Match the information of group A with group B

Group A	Group B
(*a*) Maize-Potato-wheat	(*i*) Triple cropping
(*b*) Rice-Rice regrowth	(*ii*) Mixed farming
(*c*) Wheat + Mustard	(*iii*) Rotoon Cropping
(*d*) Crops + Poultry	(*iv*) Mixed Cropping

Q. 94. Match the mother seed with their progeny

Mother seed	Progeny
(*a*) Nucleus seed	(*i*) Certified seed
(*b*) Breeder seed	(*ii*) Foundation seed
(*c*) Foundation seed	(*iii*) Breeder seed

Q. 95. Match the information of group A with group B

Group A	Group B
(*a*) Tetrazolum test	(*i*) Genetic purity
(*b*) Growout test	(*ii*) Inviable seeds
(*c*) Dead seeds	(*iii*) Seed viability
(*d*) Roguing	(*iv*) Off types

Q. 96. Match the physical purity with crops

Physical purity	Crops
(*a*) 95%	(*i*) Groundnut
(*b*) 96%	(*ii*) Carrot
(*c*) 97%	(*iii*) Wheat
(*d*) 98%	(*iv*) Soybean

Q. 97. Match isolation distance with crops

Isolation distance (m)	Crops
(*a*) 20	(*i*) Wheat
(*b*) 3	(*ii*) Chickpea
(*c*) 50	(*iii*) Pigenopea
(*d*) 200	(*iv*) Cotton

Q. 98. Match the isolation distance with crops

Isolation (m)	Crops
(*a*) 400	(*i*) Pearlmillet
(*b*) 1000	(*ii*) Mustard
(*c*) 800	(*iii*) Cabbage
(*d*) 1600	(*iv*) Cucurbits

Q. 99. Match the crops with germination percent

Crops	Germination (%)
(*a*) Rice	(*i*) 85
(*b*) Maize	(*ii*) 80
(*c*) Wheat	(*iii*) 75
(*d*) Pigeonpea	(*iv*) 90

Q. 100. Match crops with objectionable weeds

Crops	Weeds
(*a*) Wheat	(*i*) Hirankhuri
(*b*) Mustard	(*ii*) Sengi
(*c*) Methi	(*iii*) Cuscuta
(*d*) Lucern	(*iv*) Argemone

Q. 101. Match the institutes with their locations

Institutes	Locations
(*a*) IRRI	(*i*) Nigeria
(*b*) CIMMYT	(*ii*) Philippines
(*c*) CIAT	(*iii*) Mexico
(*d*) IITA	(*iv*) Colombia

Q. 102. Match the institutes with their locations

Institutes	Locations
(*a*) ICARDA	(*i*) Peru
(*b*) ICRISAT	(*ii*) Syria
(*c*) IPGRI	(*iii*) Rome
(*d*) CIP	(*iv*) India

Q. 103. Match the institutes with their year of establishment

Institutes	Year
(*a*) IRRI	(*i*) 1968

(*b*) CIMMYT	(*ii*) 1960
(*c*) CIAT	(*iii*) 1966
(*d*) CIP	(*iv*) 1972

Q. 104. Match the institutes with their year of establishment

Institutes	**Year**
(*a*) IITA	(*i*) 1972
(*b*) ICRISAT	(*ii*) 1968
(*c*) ICARDA	(*iii*) 1994
(*d*) IPGRI	(*iv*) 1976

Q. 105. Match ICAR institutes with their locations

Institutes	**Locations**
(*a*) CRRI	(*i*) Shimla
(*b*) CPRI	(*II*) Cuttack
(*c*) CTRI	(*iii*) Kasargod
(*d*) CPCRI	(*iv*) Rajahmundry

Q. 106. Match ICAR institutes with their locations

Institutes	**Locations**
(*a*) CTCRI	(*i*) Kanpur
(*b*) IIPR	(*ii*) Lucknow
(*c*) IISR	(*iii*) Bangalore
(*d*) IIHR	(*iv*) Trivandrum

Q. 107. Match ICAR institutes with their locations

Institutes	**Locations**
(*a*) CSSRI	(*i*) Mumbai
(*b*) CICR	(*ii*) Varanasi
(*c*) IIVR	(*iii*) Nagpur
(*d*) CIRCOT	(*iv*) Karnal

Q. 108. Match ICAR institutes with their locations

Institutes	**Locations**
(*a*) NDRI	(*i*) Mumbai
(*b*) IARI	(*ii*) Karnal
(*c*) IVRI	(*iii*) New Delhi
(*d*) CIFE	(*iv*) Izatnagar

Q. 109. Match ICAR institutes with their locations

Institutes	**Locations**
(*a*) DOR	(*i*) Karnal
(*b*) DWR	(*ii*) Hyderabad
(*c*) IGFRI	(*iii*) Barrackpur
(*d*) CRIJAF	(*iv*) Jhansi

Q. 110. Match the national research centres with their locations

National Research Centres	**Locations**
(*a*) NRC Soybean	(*i*) Hyderabad
(*b*) NRC Sorghum	(*ii*) Junagarh
(*c*) NRC Mustard	(*iii*) Indore
(*d*) NRC Groundnut	(*iv*) Bharatput

Q. 111. Match National Research Centres with their locations

NACs	**Locations**
(*a*) NRC Cashew	(*i*) Pune
(*b*) NRC Citrus	(*ii*) Puttar
(*c*) NRC Grapes	(*iii*) Trichanapalli
(*d*) NRC Banana	(*iv*) Nagpur

Q. 112. Match Agricultural Universities with their location

Institutes	**Locations**
(*a*) PAU	(*i*) Hisar
(*b*) HAU	(*ii*) Ludhiana
(*c*) RAU	(*iii*) Udaipur
(*d*) MPUAT	(*iv*) Bikaner

Q. 113. Match Agricultural Universities with their locations

Universities	**Locations**
(*a*) JNKVV	(*i*) Coimbatore
(*b*) GAU	(*ii*) Jabalpur
(*c*) TNAU	(*iii*) Dharward
(*d*) UAS	(*iv*) Dantiwoda

Q. 114. Match Agricultural Universities with their locations

Universities	**Locations**
(*a*) PDKV	(*i*) Assam
(*b*) MAU	(*ii*) Rahuri
(*c*) MPKV	(*iii*) Akola
(*d*) AAU	(*iv*) Parbhani

Q. 115. Match Agricultural Universities with their year of establishment

Universities	**Year**
(*a*) PAU	(*i*) 1970
(*b*) HAU	(*ii*) 1963
(*c*) TNAU	(*iii*) 1969
(*d*) PDKV	(*iv*) 1972

Q. 116. Match the Universities with their year of establishment

Universities	**Year**
(*a*) RAU	(*i*) 1987

(*b*) CAU	(*ii*) 1986
(*c*) IGKVV	(*iii*) 1972
(*d*) MAU	(*iv*) 1994

Q. 117. Match the information of group A with that of group B

Group A	Group B
(*a*) Z-test	(*i*) Paired observations
(*b*) t-test	(*ii*) Unpiared observations
(*c*) Student *t*	(*iii*) Small sample
(*d*) Fishers *t*	(*iv*) Large sample

Q. 118. Match the information of group A with that of group B

Group A	Group B
(*a*) X^2 test	(*i*) Fisher
(*b*) Augmented design	(*ii*) Karl Pearson
(*c*) F test	(*iii*) Galton
(*d*) Correlation	(*iv*) Federer.

Q. 119. Match the information of group A with that of group B

Group A	Group B
(*a*) 5 B system	(*i*) Cotton
(*b*) Nobilization	(*ii*) Wheat
(*c*) Pegs	(*iii*) Sugarcane
(*d*) Lint	(*iv*) Groundnut

Q. 120. Match the information of group A with that of group B

Computer Generations	Main Components
(*a*) First	(*i*) Transistor
(*b*) Second	(*ii*) Vacuum tubes
(*c*) Third	(*iii*) Large scale integrated circuits
(*d*) Fourth	(*iv*) Integrated circuits

ANSWERS MATCHING

1.	(*a*) (*ii*)	(*b*) (*i*)	(*c*) (*iv*)	(*d*) (*iii*)
2.	(*a*) (*iii*)	(*b*) (*i*)	(*c*) (*ii*)	(*d*) (*iv*)
3.	(*a*) (*iii*)	(*b*) (*iv*)	(*c*) (*ii*)	(*d*) (*i*)
4.	(*a*) (*ii*)	(*b*) (*i*)	(*c*) (*iv*)	(*d*) (*iii*)
5.	(*a*) (*ii*)	(*b*) (*iii*)	(*c*) (*iv*)	(*d*) (*i*)
6.	(*a*) (*ii*)	(*b*) (*iv*)	(*c*) (*i*)	(*d*) (*iii*)
7.	(*a*) (*ii*)	(*b*) (*iv*)	(*c*) (*i*)	(*d*) (*iii*)
8.	(*a*) (*iv*)	(*b*) (*i*)	(*c*) (*ii*)	(*d*) (*iii*)
9.	(*a*) (*iv*)	(*b*) (*iii*)	(*c*) (*ii*)	(*d*) (*i*)
10.	(*a*) (*iv*)	(*b*) (*i*)	(*c*) (*ii*)	(*d*) (*iii*)

11.	(*a*) (*iv*)	(*b*) (*i*)	(*c*) (*ii*)	(*d*) (*iii*)
12.	(*a*) (*ii*)	(*b*) (*iii*)	(*c*) (*i*)	(*d*) (*iv*)
13.	(*a*) (*iii*)	(*b*) (*iv*)	(*c*) (*i*)	(*d*) (*ii*)
14.	(*a*) (*ii*)	(*b*) (*iv*)	(*c*) (*i*)	(*d*) (*iii*)
15.	(*a*) (*ii*)	(*b*) (*iii*)	(*c*) (*iv*)	(*d*) (*i*)
16.	(*a*) (*iv*)	(*b*) (*i*)	(*c*) (*ii*)	(*d*) (*iii*)
17.	(*a*) (*ii*)	(*d*) (*iv*)	(*c*) (*iii*)	(*d*) (*i*)
18.	(*a*) (*iii*)	(*b*) (*iv*)	(*c*) (*ii*)	(*d*) (*i*)
19.	(*a*) (*iii*)	(*b*) (*iv*)	(*c*) (*i*)	(*d*) (*ii*)
20.	(*a*) (*ii*)	(*b*) (*iv*)	(*c*) (*i*)	(*d*) (*iii*)
21.	(*a*) (*iv*)	(*b*) (*i*)	(*c*) (*ii*)	(*d*) (*iii*)
22.	(*a*) (*iv*)	(*b*) (*iii*)	(*c*) (*i*)	(*d*) (*ii*)
23.	(*a*) (*ii*)	(*b*) (*i*)	(*c*) (*iv*)	(*d*) (*iii*)
24.	(*a*) (*iv*)	(*b*) (*i*)	(*c*) (*ii*)	(*d*) (*iii*)
25.	(*a*) (*ii*)	(*b*) (*iii*)	(*c*) (*iv*)	(*d*) (*i*)
26.	(*a*) (*ii*)	(*b*) (*iv*)	(*c*) (*i*)	(*d*) (*iii*)
27.	(*a*) (*iii*)	(*b*) (*iv*)	(*c*) (*i*)	(*d*) (*ii*)
28.	(*a*) (*ii*)	(*b*) (*iv*)	(*c*) (*i*)	(*d*) (*iii*)
29.	(*a*) (*iv*)	(*b*) (*iii*)	(*c*) (*i*)	(*d*) (*ii*)
30.	(*a*) (*ii*)	(*b*) (*i*)	(*c*) (*iv*)	(*d*) (*iii*)
31.	(*a*) (*iii*)	(*b*) (*ii*)	(*c*) (*iv*)	(*d*) (*i*)
32.	(*a*) (*ii*)	(*b*) (*i*)	(*c*) (*iv*)	(*d*) (*iii*)
33.	(*a*) (*ii*)	(*b*) (*iii*)	(*c*) (*iv*)	(*d*) (*i*)
34.	(*a*) (*iii*)	(*b*) (*iv*)	(*c*) (*i*)	(*d*) (*ii*)
35.	(*a*) (*iv*)	(*b*) (*i*)	(*c*) (*ii*)	(*d*) (*iii*)
36.	(*a*) (*iv*)	(*b*) (*i*)	(*c*) (*ii*)	(*d*) (*iii*)
37.	(*a*) (*iv*)	(*b*) (*i*)	(*c*) (*ii*)	(*d*) (*iii*)
38.	(*a*) (*ii*)	(*b*) (*i*)	(*c*) (*iv*)	(*d*) (*iii*)
39.	(*a*) (*ii*)	(*b*) (*i*)	(*c*) (*iv*)	(*d*) (*iii*)
40.	(*a*) (*iii*)	(*b*) (*i*)	(*c*) (*iv*)	(*d*) (*ii*)
41.	(*a*) (*ii*)	(*b*) (*iii*)	(*c*) (*iv*)	(*d*) (*i*)
42.	(*a*) (*ii*)	(*b*) (*i*)	(*c*) (*iv*)	(*d*) (*iii*)
43.	(*a*) (*iii*)	(*b*) (*i*)	(*c*) (*iv*)	(*d*) (*ii*)
44.	(*a*) (*iv*)	(*b*) (*i*)	(*c*) (*ii*)	(*d*) (*iii*)
45.	(*a*) (*iv*)	(*b*) (*iii*)	(*c*) (*i*)	(*d*) (*ii*)
46.	(*a*) (*iv*)	(*b*) (*ii*)	(*c*) (*iii*)	(*d*) (*i*)
47.	(*a*) (*iii*)	(*b*) (*i*)	(*c*) (*iv*)	(*d*) (*ii*)

48.	(*a*) (*ii*)	(*b*) (*iv*)	(*c*) (*i*)	(*d*) (*iii*)
49.	(*a*) (*iv*)	(*b*) (*iii*)	(*c*) (*ii*)	(*d*) (*i*)
50.	(*a*) (*ii*)	(*b*) (*i*)	(*c*) (*iv*)	(*d*) (*iii*)
51.	(*a*) (*ii*)	(*b*) (*i*)	(*c*) (*iv*)	(*d*) (*iii*)
52.	(*a*) (*ii*)	(*b*) (*iv*)	(*c*) (*i*)	(*d*) (*iii*)
53.	(*a*) (*ii*)	(*b*) (*iii*)	(*c*) (*iv*)	(*d*) (*i*)
54.	(*a*) (*ii*)	(*b*) (*i*)	(*c*) (*iv*)	(*d*) (*iii*)
55.	(*a*) (*ii*)	(*b*) (*i*)	(*c*) (*iv*)	(*d*) (*iii*)
56.	(*a*) (*ii*)	(*b*) (*i*)	(*c*) (*iv*)	(*d*) (*iii*)
57.	(*a*) (*iii*)	(*b*) (*i*)	(*c*) (*iv*)	(*d*) (*ii*)
58.	(*a*) (*iv*)	(*b*) (*iii*)	(*c*) (*ii*)	(*d*) (*i*)
59.	(*a*) (*iv*)	(*b*) (*i*)	(*c*) (*ii*)	(*d*) (*iii*)
60.	(*a*) (*iv*)	(*b*) (*iii*)	(*c*) (*i*)	(*d*) (*ii*)
61.	(*a*) (*ii*)	(*b*) (*i*)	(*c*) (*iv*)	(*d*) (*iii*)
62.	(*a*) (*iv*)	(*b*) (*i*)	(*c*) (*ii*)	(*d*) (*iii*)
63.	(*a*) (*ii*)	(*b*) (*i*)	(*c*) (*iv*)	(*d*) (*iii*)
64.	(*a*) (*ii*)	(*b*) (*i*)	(*c*) (*iv*)	(*d*) (*iii*)
65.	(*a*) (*iii*)	(*b*) (*iv*)	(*c*) (*i*)	(*d*) (*ii*)
66.	(*a*) (*iv*)	(*b*) (*iii*)	(*c*) (*ii*)	(*d*) (*i*)
67.	(*a*) (*ii*)	(*b*) (*i*)	(*c*) (*iii*)	(*d*) (*iv*)
68.	(*a*) (*ii*)	(*b*) (*i*)	(*c*) (*iv*)	(*d*) (*iii*)
69.	(*a*) (*iii*)	(*b*) (*iv*)	(*c*) (*ii*)	(*d*) (*i*)
70.	(*a*) (*ii*)	(*b*) (*i*)	(*c*) (*iv*)	(*d*) (*iii*)
71.	(*a*) (*ii*)	(*b*) (*iv*)	(*c*) (*i*)	(*d*) (*iii*)
72.	(*a*) (*ii*)	(*b*) (*i*)	(*c*) (*iv*)	(*d*) (*iii*)
73.	(*a*) (*iv*)	(*b*) (*iii*)	(*c*) (*ii*)	(*d*) (*i*)
74.	(*a*) (*ii*)	(*b*) (*i*)	(*c*) (*iv*)	(*d*) (*iii*)
75.	(*a*) (*ii*)	(*b*) (*i*)	(*c*) (*iv*)	(*d*) (*iii*)
76.	(*a*) (*ii*)	(*b*) (*i*)	(*c*) (*iv*)	(*d*) (*iii*)
77.	(*a*) (*ii*)	(*b*) (*iii*)	(*c*) (*iv*)	(*d*) (*i*)
78.	(*a*) (*ii*)	(*b*) (*iii*)	(*c*) (*iv*)	(*d*) (*i*)
79.	(*a*) (*ii*)	(*b*) (*i*)	(*c*) (*iv*)	(*d*) (*iii*)
80.	(*a*) (*ii*)	(*b*) (*iv*)	(*c*) (*i*)	(*d*) (*iii*)
81.	(*a*) (*ii*)	(*b*) (*i*)	(*c*) (*iv*)	(*d*) (*iii*)
82.	(*a*) (*ii*)	(*b*) (*i*)	(*c*) (*iv*)	(*d*) (*iii*)
83.	(*a*) (*ii*)	(*b*) (*iv*)	(*c*) (*iii*)	(*d*) (*i*)
84.	(*a*) (*ii*)	(*b*) (*i*)	(*c*) (*iv*)	(*d*) (*iii*)

85. (a) (iii) (b) (iv) (c) (ii) (d) (i)
86. (a) (ii) (b) (iii) (c) (i) (d) (iv)
87. (a) (i) (b) (ii) (c) (iv) (d) (iii)
88. (a) (iv) (iv)(iii) (c) (ii) (d) (i)
89. (a) (ii) (b) (iii) (c) (i) (d) (iv)
90. (a) (iii) (b) (iv) (c) (ii) (d) (i)
91. (a) (iv) (b) (iii) (c) (ii) (d) (i)
92. (a) (i) (b) (iii) (c) (iv) (d) (ii)
93. (a) (i) (b) (iii) (c) (iv) (d) (ii)
94. (a) (iii) (b) (ii) (c) (i)
95. (a) (iii) (b) (i) (c) (ii) (d) (iv)
96. (a) (ii) (b) (i) (c) (iv) (d) (iii)
97. (a) (ii) (b) (i) (c) (iv) (d) (iii)
98. (a) (ii) (b) (i) (c) (iv) (d) (iii)
99. (a) (ii) (b) (iv) (c) (i) (d) (iii)
100. (a) (i) (b) (iv) (c) (ii) (d) (iii)
101. (a) (ii) (b) (iii) (c) (iv) (d) (i)
102. (a) (ii) (b) (iv) (c) (iii) (d) (i)
103. (a) (ii) (b) (iii) (c) (i) (d) (iv)
104. (a) (ii) (b) (i) (c) (iv) (d) (iii)
105. (a) (ii) (b) (i) (c) (iv) (d) (iii)
106. (a) (iv) (b) (i) (c) (ii) (d) (iii)
107. (a) (iv) (b) (iii) (c) (ii) (d) (i)
108. (a) (ii) (b) (iii) (c) (iv) (d) (i)
109. (a) (ii) (b) (i) (c) (iv) (d) (iii)
110. (a) (iii) (b) (i) (c) (iv) (d) (ii)
111. (a) (ii) (b) (iv) (c) (i) (d) (iii)
112. (a) (ii) (b) (i) (c) (iv) (d) (iii)
113. (a) (ii) (b) (iv) (c) (i) (d) (iii)
114. (a) (iii) (b) (iv) (c) (ii) (d) (i)
115. (a) (ii) (b) (i) (c) (iv) (d) (iii)
116. (a) (ii) (b) (iv) (c) (i) (d) (iii)
117. (a) (iii) (b) (iv) (c) (ii) (d) (i)
118. (a) (ii) (b) (iv) (c) (i) (d) (iii)
119. (a) (ii) (b) (iii) (c) (iv) (d) (i)
120. (a) (ii) (b) (i) (c) (iv) (d) (iii)

APPENDIX—I

LIST OF JOURNALS OF PLANT BREEDING AND GENETICS

S. No.	Name of Journal	Published by/from
A. FOREIGN JOURNALS :		
1.	Plant Breeding	Berlin, Germany
2.	Netherlands Journal of Plant Breeding	Netherlands
3.	Japanese Journal of Breeding	Tokyo, Japan
4.	Plant Breeding Reviews	John Wiley & Sons Inc. New York
5.	Plant Breeding Abstracts	CAB
6.	Crop Science	Wisconsin, USA
7.	Theoretical and Applied Genetics	Springer-Verlag
8.	Japanese Journal of Genetics	Japan
9.	Journal of Heredity	Oxford Uni. UK
10.	Genetics	Maryland
11.	Genome	Canada
12.	Genomics	Academic Press
13.	Hereditas	Sweden
14.	Genetica	Kluwer Academic Pub.
15.	Heredity	Longmam Group Ltd.
16.	Cytologia	Tokyo, Japan
17.	Advances in Genetics	Academic Press NY
18.	Advances in Agronomy	Academic Press NY
19.	Annual Review of Genetics	Annual Rev. Inc.
20.	Biometrics	Virginia
21.	Barley Newsletter	Canada
22.	Pisum Newsletter	Sweden
23.	Current Genetics	Berlin, Germany

24.	Chromosoma	Berlin, Germany
25.	Australian Journal of Agricultural Research	Australia
26.	Advances in Radiation Biology	Academic Press NY
27.	Agronomy Journal	USA
28.	Journal of Agricultural Sciences	Cambridge Uni, UK
29.	Nature	MacMillan
30.	Science	New York USA
31.	Journal of Agricultural Research	Washington DC, USA
B. INDIAN JOURNALS :		
1.	Indian Journal of Genetics and Plant Breeding	IARI, New Delhi
2.	Indian Journal of Agricultural Sciences	ICAR, New Delhi
3.	Indian Journal of Experimental Biology	CSIR, New Delhi
4.	Journal of Genetics	IAS, Bangalore
5.	Journal of Cytology and Genetics	Bangalore Uni.
6.	Journal of oil seeds Research	DOR, Hyderabad
7.	Journal of Indian Society for Cotton Improvement	CIRCOT, Mumbai
8.	Journal of Cotton Research & Development	HAU, Hisar
9.	Annals of Agriculture Research	NAAS, IARI, New Delhi
10.	The Botanica	Delhi University
11.	Current Science	IAS, Bangalore

APPENDIX—II

SYLLABUS

New ARS Syllabus
(Genetics and Plant Breeding)

[Genetics & Plant Breeding w.e.f. 2013]

Unit 1: General Genetics and Plant Breeding

Mendelian inheritance. Cell structure and division, Linkage, its detection and estimation. Epistasis. Gene concept, allelism and fine structure of gene. Extra chromosomal inheritance. DNA – structure, function, replication and repair. Genetic code. Gene- enzyme relationship. Replication, Transcription and Translation. Gene regulation in prokaryotes and eukaryotes. Nuclear and cytoplasmic genome organization. Spontaneous and induced mutations and their molecular mechanisms. Crop domestication, evolution of crops and centres of diversity. Emergence of scientific plant breeding. Objectives and accomplishments in plant breeding and the role of National and International institutes. Gametogenesis and fertilization. Modes of sexual and asexual reproduction and its relation to plant breeding methodology. Apomixes, incompatibility and male sterility systems and their use in plant breeding.

Unit 2: Economics Botany and Plant Breeding Methods

Origin, distribution, classification, description and botany of cereals (wheat, rice, maize, sorghum, pearl millet, minor millets); pulses (pigeonpea, chickpea, black gram, green gram, cowpea, soyabean, pea, lentil, horse gram, lab-lab, rice bean, winged bean, lathyrus, Lima bean; oilseeds (groundnuts, sesamum, castor, rapeseed mustard, sunflower, Niger, linseed); fibers and sugar crops, fodder and green manures; Breeding methods for self-pollinated, cross-pollinated and clonally propagated crops.Component, recombinational and transgressive breeding. Single seed descent. Populations, their improvement methods and maintenance, Hybrid breeding and genetic basis of heterosis. Ideotype breeding. Mutation breeding.

Unit 3: Genome organization and Cytogenetics of Crop Plants

Chromosome structure, function and replication. Recombination and crossing over. Karyotype analysis. Banding techniques. *In situ* hybridization. Special types of chromosomes. Chromosomal interchanges, inversions, duplications and deletions. Polyploids, haploids, aneuploids and their utility. Wide hybridization and chromosomal manipulations for alien gene transfer. Pre-and post- fertilization barriers in wide hybridization. Genome organization and cytogenetics of important crop species- wheat, maize, rice, Brassica, cotton, Vigna, potato and sugarcane. Principles and procedures of genome analysis. Cytogenetic techniques for gene location and gene transfer, Construction and use of molecular marker based chromosome maps. Comparative mapping and genome analysis.

Unit 4: Quantitative and Biometrical Genetics

Quantitative characters. Multiple factors inheritance. Genetic control of polygenic characters. Genetic advance and types of selection and correlated response. Hardy Weinberg law. Linkage disequilibrium. Genetic load. Polymorphism. Breeding value, heritability. Response to selection, correlated response. Estimates of variance components and covariance among relatives. Mating designs with random and inbred parents. Estimation of gene effects and combing ability. Effects of linkage and epistasis on estimation of genetic parameters. Maternal effects. Genotype-environment interactions and stability of performance. Heterosis and its basis. Mating system and mating design- diallel, line X tester, NC-1, NC-II and NC-III designs, approaches to estimate and exploit component of self and cross pollinated crops. Genotype X environment interaction and stability analysis.

Unit 5: Genetic Engineering and Biotechnological Tools in Plant Breeding

Somatic hybridization, micropropagation, somachonal variation *in-vitro* mutagenesis. Artificial synthesis of gene. Genetic and molecular markers, generations of molecular markers and their application in genetic analyses and breeding. Molecular markers in genetic diversity analysis and breeding for complex characters. Gene tagging, QTL mapping and marker aided selection. Genome projects and utilization of sequence formation. Vectors. DNA libraries, DNA fingerprinting, DNA sequencing. Nuclei acid hybridization and immunochemical detection. Chromosome walking, Recombinant DNA technology, Gene cloning strategies. Genetic transformation and transgenics. Antisense RNA, RNAi and micro RNA techniques in crop improvement.

Unit 6: Plant Breeding for Stress Resistance and Nutitional Quality

Genetic basis and breeding for resistance to diseases and insect-pests. Breeding for vertical and horizontal resistance to diseases. Genetic and physiological basis of abiotic stress tolerance. Breeding for resistance to heat, frost, flood, drought and soil stresses. Important quality parameters in various crops, their genetic basis and breeding for these traits. Role of molecular markers in stress resistance breeding: MAS, MARS and MABB.

Unit 7: Plant Genetic Resources and their Regulatory System; Varietal Release and Seed Production

Plant exploration, germplasm introduction, exchange, conservation, evaluation and utilization of plant genetic resoures. Convention on Biological Diversity and International Treaty on Plant Genetic Resources for Food and Agriculture. Intellectual Property Rights. Biodiversity Act. Plant Variety Protection and Farmers' Rights Act. System of variety release and notification. Types of seeds and seed chain. Seed production and certification.

Unit 8: Statistical Methods and Field Plot Techniques

Frequency distribution. Measures of central tendency, probability theory and its applications in genetics. Probabaility distribution and tests of significance. Correlation, linear, partial and multiple regression. Genetic divergence. Multivariate analysis. Design of experiments- basic principles, completely randomized design, randomized block design and split plot design. Complete and incomplete block designs. Augmented design, Grid and honeycomb design. Hill plots, unreplicated evaluation. Data collection and interpretation.

APPENDIX—III

List of ICAR Crop Research Institutes

S. N.	Name of Research Institute	Location	State
A. Crop Research Institutes			
1.	Central Agricultural Research Institute	Port Blair	Andman & Nicobar
2.	Central Arid Zone Research Institute	Jodhpur	Rajasthan
3.	Central Institute for Cotton Research	Nagpur	Maharashtra
4.	Central Potato Research Institute	Shimla	Himachal Pradesh
5.	Central Research Institute for Dry land Agriculture	Hyderabad	Andhra Pradesh
6.	Central Research Institute for Jute & Allied Fibres	Kolkata	West Bengal
7.	Central Rice Research Institute	Cuttack	Odisha
8.	Central Soil salinity Research Institute	Karnal	Haryana
9.	Central Tobacco Research Institute	Rajamundry	Andhra Pradesh
10.	Directorate of Groundnut Research	Junagarh	Gujarat
11.	Directorate of Maize Research	Pusa	New Delhi
12.	Directorate of Oilseed Research	Hyderabad	Andhra Pradesh
13.	Directorate of Rapeseed Mustard Research	Bharatpur	Rajasthan
14.	Directorate of Rice Research	Hyderabad	Andhra Pradesh
15.	Directorate of Seed Research	Mau	Uttar Pradesh
16.	Directorate of Sorghum Research	Hyderabad	Andhra Pradesh

17.	Directorate of Soybean Research	Indore	Madhya Pradesh
18.	Directorate of Wheat Research	Karnal	Haryana
19.	Indian Agricultural Research Institute	Pusa	New Delhi
20.	Indian Grassland and Fodder Research Institute	Jhansi	Uttar Pradesh
21.	Indian Institute of Pulses Research	Kanpur	Uttar Pradesh
22.	Indian Institute of Sugarcane Research	Lucknow	Uttar Pradesh
23.	National Research Centre Plant Biotechnology	Pusa	New Delhi
24.	Sugarcane Breeding Institute	Coimbatore	Tamil Nadu
25.	National Bureau of Plant Genetic Resources	Pusa	New Delhi
26.	Vivekanand Parvatiya Krishi Anusandhan Sansthan	Almora	Uttarakhand
B. Horticulture Research Institutes			
1.	Central Institute for Arid Horticulture	Bikaner	Rajasthan
2.	Central Institute for Subtropical Horticulture	Lucknow	Uttar Pradesh
3.	Central Institute for Temperate Horticulture	Srinagar	Jammu & Kashmir
4.	Central Plantation Crops Research Institute	Kasargod	Kerala
5.	Central Tuber Crops Research Institute	Trivendrum	Kerala
6.	Directorate of Cashew Research	Puttur	Karnataka
7.	Directorate of Floriculture Research	Pusa	New Delhi
8.	Directorate of Medicinal And Aromatic Plants Research	Anand	Gujarat
9.	Directorate of Onion and Garlic Research	Pune	Maharashtra
10.	Directorate of Oil Palm Research	Pedavegi	A.P.
11.	Indian Institute of Horticulture Research	Bangalore	Karnataka
12.	Indian Institute of Vegetable Research	Varanasi	Uttar Pradesh
13.	National Research Centre for Citrus	Nagpur	Maharashtra

14.	National Research Centre for Grapes	Pune	maharashtra
15.	National Research Centre for Litchi	Muzaffarpur	Bihar
16.	National Research Centre on Pomegranate	Solapur	Maharashtra
17.	National Research Centre for Seed Spices	Ajmer	Rajasthan
18.	National Research Centre for Banana	Tiruchirapalli	Kerala
19	National Research Centre for Orchids	Gangtock	Sikkim
C. ICAR Research Complex and Others			
1.	ICAR Research Complex for NEH Region	Shillong	Meghalaya
2.	ICAR Research Complex for Eastern Region	Patna	Bihar
3.	ICAR Research Complex for Goa	Panaji	Goa
4.	National Bureau of Agriculturally Important Insects	Bangalore	Karnataka
5.	National Bureau of Agriculturally Important Micro-organism	Mau	Uttar Pradesh
6.	National Centre for Integrated Pest Management	Pusa	New Delhi
7.	National Institute of Abiotic Research management	Baramati	Maharashtra
8.	National Research Centre for Agro-forestry	Jhansi	Uttar Pradesh
9.	Project Directorate for Farming System Research	Meerut	Uttarpradesh

APPENDIX—IV

List of Agricultural Universities in India

I. Agricultural Universities

S. N.	Name of University	Location	State
1.	Acharya NG Ranga Agricultural University	Hyderabad	Andhra Pradesh
2.	Assam Agricultural University.	Jorhat	Assam
3.	Rajendra Agricultural University.	Pusa	Bihar
4.	Bihar Agricultural University.	Sabour	Bihar
5.	Birsa Agricultural University	Ranchi	Jharkhand
6.	Indira Gandhi Krishi Vishwavidyalaya.	Raipur	Chhatishgarh
7.	Anand Agricultural University.	Anand	Gujarat
8.	Junagadh Agricultural University.	Junagarh	Gujarat
9.	Navsari Agricultural University.	Navsari	Gujarat
10.	Sardarkrushinagar-Dantiwada Agricultural University.	Bansakantha	Gujarat
11.	Chaudhary Charan Singh Haryana Agricultural University.	Hisar	Haryana
12.	CSK Himachal Pradesh Krishi Vishvavidyalaya,	Palampur	Himachal Pradesh
13.	Sher-E-Kashmir Univ of Agricultural Sciences & Technology.	Jammu	Jammu & Kashmir
14.	Sher-E-Kashmir Univ of Agricultural Sciences & Technology of Kashmir.	Srinagar	Jammu & Kashmir
15.	University of Agricultural Sciences.	Bangalore	Karnataka
16.	University of Agricultural Sciences.	Dharwad	Karnataka
17.	University of Agricultural Sciences.	Raichur	Karnataka
18.	Kerala Agricultural University.	Vallanikara	Kerala
19.	Jawaharlal Nehru Krishi Viswavidyalaya.	Jabalpur	Madhya Pradesh

20.	Rajmata Vijayaraje Scindia Krishi Viswavidyalaya.	Gwalior	Madhya Pradesh
21.	Dr. Balasaheb Sawant Konkan Krishi Vidyapeeth.	Ratnagiri	Maharashtra
22.	Dr. Panjabrao Deshmukh Krishi Vidyapeeth.	Akola	Maharashtra
23.	Mahatma Phule Krishi Vidyapeeth.	Rahuri	Maharashtra
24.	Marathwada Agricultural University.	Parbhani	Maharashtra
25.	Odisha Univ. of Agriculture & Technology.	Bhubaneswar	Odisha
26.	Punjab Agricultural University.	Ludhiana	Punjab
27.	Maharana Pratap Univ. of Agriculture & Technology.	Udaipur	Rajasthan
28.	Rajasthan Agricultural University	Bikaner	Rajasthan
29.	Tamil Nadu Agricultural University.	Coimbatore	Tamil Nadu
30.	Chandra Shekar Azad University of Agriculture & Technology.	Kanpur	Uttar Pradesh
31.	Narendra Deva University of Agriculture & Technology.	Faizabad	Uttar Pradesh
32.	Sardar Ballabh Bhai Patel Univ. of Agriculture & Technology.	Meerut	Uttar Pradesh
33.	Manyavar Kashiram University of Agriculture & Technology.	Banda	Uttar Pradesh
34.	Govind Ballabh Pant University of Agriculture & Technology.	Pantnagar	Uttrakhand
35	Bidhan Chandra Krishi Viswavidyalaya.	Kalyani	West Bengal
36.	Uttar Banga Krishi Viswavidyalaya.	Cooch Bihar	West Bengal
37.	Central Agricultural University	Imphal	Manipur
	Deemed Universities		
38.	Indian Agricultural Research Institute.	Pusa	New Delhi
39.	Allahabad Agricultural Institute.	Allahabad	Uttar Pradesh

II. Animal Science Universities

S. N.	Name of University	Location	State
1.	Sri Venkateswara Veterinary University.	Tirupati	Andhra Pradesh
2.	Karnataka Veterinary, Animal and Fisheries Sciences University.	Bidar	Karnataka
3.	Maharashtra Animal Science & Fishery University.	Nagpur	Maharashtra
4.	Guru Angad Dev Veterinary and Animal Science University.	Ludhiana	Punjab
5.	Tamil Nadu Veterinary & Animal Science University.	Chennai	Tamil Nadu
6.	UP Pandit Deen Dayal Upadhaya Pashu Chikitsa Vigyan Vishwa Vidhyalaya evam Go Anusandhan Sansthan, Mathura	Mathura	Uttar Pradesh
7.	West Bengal University of Animal & Fishery Sciences.	Kolkata	West Bengal
8.	Rajasthan University of Veterinary and Animal Sciences	Bikaner	Rajasthan
	Deemed Universities		
9.	Indian Agricultural Research Institute	Pusa	New Delhi
10.	Indian Veterinary Research Institute.	Izatnagar	Uttar Pradesh
11.	National Diary Research Institute.	Karnal	Haryana
12.	Central Institute of Fisheries Education,	Mumbai	Maharashtra

III. Horticultural Universities

S. N.	Name of University	Location	State
1.	Andhra Pradesh Horticultural University	Tadepalligudem	Andhra Pradesh
2.	University of Horticultural Sciences	Bagalkot	Karnataka
3.	Dr. Yashwant Singh Parmar Univ of Horticulture & Forestry.	Solan	Himachal Pradesh

GLOSSARY

A

A LINE. The male sterile line.

ABNORMAL SEEDLINGS. Seedlings which are unable to develop into normal plants.

ABIOTIC STRESS. Adverse conditions for crop growth and production caused by environmental factors such as drought, water logging, frost, cold, heat, soil salinity and acidity etc.

ACCLIMATIZATION. The process of adaptation of a genotype to changed environment.

ADAPTABILITY. The capacity of a genotype or population to give stable performance over a wide range of environments.

ADAPTATION. The fitness of a genotype or population to a given environment. Also adjustment of organisms to environmental changes.

AFLP. Amplified fragment length polymorphism, a type of DNA marker.

ALLOGAMY. Development of seed by cross pollination.

ALLOPLOID. A polyploid individual which combines complete genomes from two or more species, also known as hybrid polyploid.

ANTHER. Pollen bearing portion of the stamen.

ANTHESIS. The process of pollen shed from the anthers.

APOMIXIS. Development of seed without sexual fusion.

ARITHMETIC MEAN. Sum of all observations of a sample divided by their number.

ARTIFICIAL SELECTION. The selection made by human. It favours those characteristics of plants which are related to yield and quality.

AUGMENTED DESIGN. An experimental design which is used to test a large number of germplasm lines in a limited area.

AUTOGAMY. Development of seed by self pollination.

AUTOPOLYPLOIDS. Polyploids which originate by mutiplication of chromosames of a single species, also known as simple polyploid.

AVERAGE HETEROSIS. Superiority of a hybrid over the mean value of its both parents.

B

BC$_1$ First backcross with recipient parent.

BC$_2$ Second backcross with recipient parent.

B LINE. Isogenic line of A line which is male fertile.

BALANCED POLYMORHISM. Regular occurrence of several phenotypes in a genetic population due to superirity of heterozygote over homozygotes.

BIOINFORMATICS. The computer aided study of biology especially genetics, molecular genetics and biotechnology.

BACK CROSS. Crossing of F$_1$ with either of its parents.

BASIC NUMBER. The gametic chromosome number of a true diploid species.

BETA PARTICLES. Radioactive particles which are generated from radioactive decay of heavier elements such 3H, 32 P, 32 S. These are negatively charged, sparsely ionizing and more penetrating.

BIOMETRICAL GENETICS. A branch of genetics which utilizes various statistical concepts and procedures for the study of genetical principles, also known Statistical or Mathematical Genetics.

BIOMETRY. The science which deals with application of statistical concepts and procedures to the study of biological problems, also known as biometrics.

BIOTIC STRESS. Adverse conditions for crop growth and production caused by biological factors such as diseases, insects and parasitic weeds.

BIOTYPE. Variant or new strain of an insect.

BIPARENTAL CROSS. Crossing of randomly selected plants in F$_2$ or subsequent generation of a cross between two pure lines, in a definite fashion.

BISEXUALITY. Presence of male and female organs in the same flower.

BREEDER SEED. Progeny of nucleus seed or breeder seed produced under the strict supervision of original or sponsoring plant breeder.

C

CERTIFIED SEED. The progeny of foundation or registered or certified seed produced under the supervision of National or State Seed Corporation.

CHARACTER. Any property of an individual showing heritable variation. It includes morphorogical, physiological and biochemical properties. Also called trait.

CHASMOGAMY. In plants, fertilization after opening of flower or floret.

CHIMERA. Presence of more than one genotype of cells in an individual.

CHISQUARE TEST. A test of statistical significance which is used to test the significance of difference between observed and expected frequencies or ratios.

CLEISTOGAMY. Completion of pollination and fertilization in unopened flower bud.

CLONE. Progeny of a single plant obtained by asexual reproduction.

CLONING. Making inentical copies of individuals, cells and genes by asexual process.

CLUSTER DIAGRAM. In D^2 statistics, a linear diagram which is prepared with the help of D^2 values.

COMPLETELY RANDOMIZED DESIGN. A statistical design which is used when experimental plot is limited and homogeneous.

COMPOSITE VARIETY. In cross pollinated species, a variety developed by mixing the seed of various genotypes which are similar in height, maturity, seed size, colour etc.

COMPUTER. An eletronic device that can transmit, store and process information or data.

CONTRASTING CHARACTERS. Characters with marked phenotypic differences.

CORRELATION. The statistics which measures the degree and direction of association between two or more variables. It is represented by *r*.

CRITICAL DIFFERENCE. Least significant difference, greater than which all differences are significant.

CROSS POLLINATION. Pollination of a flower by pollen of another flower.

CROSSING. Artificial pollination between genetically dissimilar individuals.

CULTIVAR. Cultivated variety of a crop species, also called cultigen.

CYTOPLASMIC GENIC MALE STERILITY. Pollen sterility which is controlled by both cytoplasmic and nuclear genes.

CYTOPLASMIC INHERITANCE. Inheritance which is governed by cytoplasmic genes found in chloroplasts and mitochondria.

CYTOPLASMIC MALE STERILITY. Pollen sterility which is caused by cytoplasmic genes.

D^2 STATISTICS. A biometrical technique used for the study of genetic divergence in a population.

DEFECTIVE SEEDS. Seeds which are broken, disease infested, insect damaged, undeveloped and unfit for germination.

DEGREE OF FREEDOM. The number of independent comparisons.

DIALLEL ANALYSIS. Analysis of a set of diallel cross.

DIALLEL CROSS. Mating of selected parents in all possible combinations *i.e.* $n(n\text{-}2)/2$.

DICHOGAMY. Maturation of anthers and stigma of a flower at different times.

DIOECY. Existence of male and female flowers on different plants.

DIPLOID. An individual with two basic sets of chromosomes.

DIPLOIDIZATION. The process by which a polyploid species behaves like a diploid species.

DIPLOID NUMBER. The somatic chromosome number of a species.

DIRECT CROSS. A cross between two individuals.

DIRECT INTRODUCTION. Introductions which are immediately adapted to the changed environment.

DIRECTIONAL SELECTION. Selection in favour of an extreme phenotype such as earliness or lateness, dwarfness or tallness.

DISCRIMINANT FUNCTION. A statistical approach used in the construction of selection index.

DISEASE. Disorders of crop plants caused by pathogens.

DISRUPTIVE MATING. Mating between two extreme types for a character in segregating population.

DNA MARKERS. A unique sequence of nucleotides found on a strand of DNA; also called genetic markers or molecular markers.

DISTANT HYBRIDIZATION. Crossing between two different species of a genus or between different genera of a family, also called wide crossing.

DOMESTICATION. The process of bringing wild species under cultivation.

DOMINANCE. Masking effect of one allele of a gene over the other. Condition in which F_1 resembles one of the its parents.

DOMINANT. The character which expresses in F_1.

DONOR PARENT. The parent which donates desirable genes, also called non-recurrent parent.

DOUBLE CROSS HYBRID. The hybrid progeny from a cross between two single crosses *i.e.* (A × B) × (C × D).

DOUBLE HAPLOID. The diploid lines which are obtained by doubling the chromosome number of a halpoid line by colchicine treatment.

DOUBLE TOP CROSS. A cross between a single cross and an open pollinated variety *i.e.* (A × B) × OP variety.

DROUGHT. The condition of soil moisture deficiency or water scarcity.

ECONOMIC HETEROSIS. Superiority of F_1 hybrid over the best commercial variety of a region, also called useful heterosis.

EMASCULATION. Removal of anthers from a flower.

EMBRYO. Rudimentary plant in a seed.

ENVIRONMENT. External conditions that influence expression of genes in an individual.

EPIDEMIC. Wide spread uncontrolled incidence of a disease, also called disease outbreak or epiphytotic.

EPISTASIS. Interaction of genes at different loci affecting the same character.

EXPERIMENT. A scientifically planned method.

EXPERIMENTAL DESIGN. Various types of plot arrangements which are used to test a set of treatments to draw a valid conclusion about a problem.

EXPERIMENTAL ERROR. Variation due to environmental factors or uncontrolled factors.

EXOTIC COLLECTION. The germplasm which is collected or received from other countries.

EXOTIC VARIETY. A foreign variety which is directly recommended for commercial cultivation.

F_1. The progeny of a cross between two inbreds or pure lines.

F_2. Progeny of F_1 plants obtained by selfing or intermating.

F_3. The selfed progeny of F_2 plants.

F TEST. A test of statistical significance which is used to compare the difference among several means.

F VALUE. The ratio of treatment variance to the error variance.

FAST NEUTRONS. High velocity, densely ionizing particles generated from radioactive decay of heavier elements in atomic reactors or cyclotrons.

FIELD RESISTANCE. Resistance which provides an effective control of a parasite under field conditions.

FOUNDATION SEED. The progeny of breeder seed produced by National Seed Corporation.

FULL DIALLEL. All possible both way (direct and reciprocal) crosses among *n* genotypes *i.e.* $n\ (n\text{-}1)$

GAMETE. A sexual unit or a cell produced by female and male reproductive organs.

GENE FLOW. Mov..ment of various alleles from one genotype to another due to open or cross pollination.

GENE REVOLUTION. Significant increase in the productivity due development of transgenic crop cultivars.

GENE BANKS. Various organizations where genetic variability is maintained in living state ; also called germplasm banks.

GENE POOL. Whole library of different alleles of a species, also called germplasm, genetic stock, genetic resources.

GENE SANCTUARIES. Protected areas of great genetic diversity under natural conditions.

GENERAL COMBINING ABILITY. Average performance of a genotype in a series of hybrid combinations. It is a measure of additive gene action.

GENETIC BASE. The quantum of genetic diversity in a breeding population.

GENETIC DIVERSITY. The variety of genes and genotypes found in a particular crop species.

GENETIC ENHANCEMENT. Transfer of usuful genes fron exotic or wild typed in to agronomically acceptable genotypes; also called pre-breeding.

GENETIC POLYMORPHISM. Regular occurrence of several phenotypes in a genetic population.

GENETIC ENGINEERING. Isolation, introduction and expression of foreign DNA in plants and animals.

GENETIC MALE STERILITY. Pollen sterility which is controlled by nuclear genes.

GENETIC PURITY. Absence of seeds of other variety of the same crop as well as of other crops.

GENETIC RESISTANCE. Ability of some genotypes to give higher yield of good quality than other varieties at the same initial level of disease or insect infestation under similar environmental conditions.

GENETIC VARIANCE. Heritable portion of total (phenotypic) variance.

GENOME. A basic set of chromosomes in which each type of chromosome is represented only once.

GENOMICS. The study of structure and function of entire genome of a living organism.

GENOTYPE. Genetic constitution of an individual viz. AA, Aa or aa.

GERMINATION. Emergence of normal seedlings from the seed under ideal conditions of light, temperature, moisture, oxygen and nutrients.

GLOBALIZATION. The process of organizing the whole world in to a single integrated marketing unit.

GREEN REVOLUTION. Revolution (quantum jump) in the productivity of cereals (wheat and rice) by evolving dwarf high yielding varieties.

GERMPLASM. Sum total of genes in a species.

GENERAL RESISTANCE. Host plant resistance against all prevalent races of a pathogen.

GENERAL COMBINING ABILITY. Average performance of a strain in a series of cross combinations.

GYNODIOECY. Coexistence of female (male sterile) and hermaphrodite individuals in a population.

HALF DIALLEL. All possible one way (direct) crosses among *n* genotypes.

HARD SEEDS. Seeds which do not absorb water.

HAPLOID. An individual with half of somatic chromosome number.

HAPLOID NUMBER. The gametic chromosome number of a species.

HARVEST INDEX. The ratio of economic yield to biological yield.

HERITABILITY. The ratio of genotypic variance to phenotypic variance (broad sense) or the ratio of additive variance to phenotype variance (narrow sense).

HERMAPHRODITE. A flower with both male and female reproductive organs.

HETROBELTIOSIS. Superiority of F_1 over its better parent.

HETEROSIS. Superiority of F_1 in one or more characters over its parents, also called hybrid vigour.

HETEROZYGOTE. An individual having dissimilar alleles on the corresponding locus of homologous chromosomes.

HETEROZYGOUS. Same as heterozygote.

HETEROZYGOSITY. The proportion of heterozygous individuals in a segregating population.

HOMEOLOGOUS PAIRING. Pairing between the chromosomes of two different species.

HOMEOSTASIS. The buffering capacity of a genotype to environmental changes.

HOMOGAMY. Maturation of anthers and stigma of a flower at the same time.

HOMOGENEOUS POPULATION. A population of genetically similar plants such as pure line, F_1 between two pure lines and progeny of a clone.

HOMOLOGOUS PAIRING. Intra-genomal pairing between homologous chromosomes.

HOMOZYGOUS. An individual having similar alleles at the corresponding locus of homologous chromosomes.

HOMOZYGOUS POPULATION. A true breeding population on selfing.

HOMOZYGOSITY. Proportion of homozygous individuals in a segregating population.

HORIZONTAL RESISTANCE. Resistance of a host to all the prevalent races of a pathogen. Also called general resistance, polygenic resistance, minor gene resistance and nonspecific resistance.

HOST. The plant attacked by a disease, insect or parasitic weed.

HYBRID. The progeny of a cross between genetically dissimilar parents.

HYBRID VARIETY. The F_1 population which is released and notified for commercial cultivation.

HYBRIDIZATION. Crossing between genetically dissimilar genotypes or individuals.

HYPOSTATIC GENE. A non-allelic gene whose effect is masked.

IDEOTYPE. A plant model which is expected to yield greater quantity of grains, fibre, oil or other useful product when developed as a cultivar.

IMPROVED SEED. Seed of a popular and released variety produced by scientific method.

INBRED. In cross pollinated species, a true breeding line obtained by continuous inbreeding.

INBREEDING. Mating between closely related individuals such as self pollination.

INBREEDING DEPRESSION. Reduction in vigour, fitness and productivity due to fixation of unfavourable recessive genes in F_2 as a result of inbreeding.

INCOMPLETE DOMINANCE. Partial resemblence of F_1 with one of its parents.

INDIGENOUS COLLECTION. Germplasm which is collected within the country.

INTELLECTUAL PROPERTY RIGHTS. The rights granted to a person for his creations of mind that can be used on commercial scale.

INDUCED MUTATIONS. Mutations which are produced by the treatment of mutagenic agents.

INERT MATTER. Non-living materials such as sand, pebbles, soil particles, straw etc. in the seed.

INTERSPECIFIC CROSS. A cross between two different species of the same genus.

INTERGENERIC CROSS. A Cross between two genera of the same family.

INTRASPECIFIC CROSS. A cross between two genotypes of the same species.

INTROGRESSION. Incorporation of some genes of one species into the genetic background of another species through interspecific hybridization and backcrossing.

IONIZING RADIATIONS. Radiations which produce ions in the medium through which they pass.

ISOGENIC LINES. Genotypes having one gene difference.

ISOLATION. Separation of field of a variety from another variety of same crop to the prescribed distance to avoid contamination.

KARYOTYPE. Phenotypic features of the chromosomes of a Species.

LAND RACES. Primitive cultivars which were selected and cultivated by farmers for many generations.

LATIN SQUARE DESIGN. The experimental design which simultaneously controls fertility variation in two directions.

LATTICE DESIGN. Incomplete block design in which the number of varieties or treatment forms a square.

LINE x TESTER CROSS. Mating scheme in which some genotypes are used as females and others as males.

LOCAL CONTROL. The principle of making use of greater homogeneity in groups of experimental units (*e.g.*, land) for reducing experimental error. Also called error control.

MAINTENANCE BREEDING. A branch of plant breeding which deals with principles and methods of breeder seed production.

MAJOR GENES. Genes with distinct and easily identifiable effect.

MALE GAMETOCIDES. Chemicals which are used for induction of male sterility.

MAPPING POPULATIONS. Various types of plant materials which are used for gene mapping.

MALE STERILITY. A condition in which either pollen is absent or nonfunctional in flowering plants.

MASS SELECTION. A method of crop Improvement in which individual desirable plants are selected on the basis of phenotype from a mixed population, their seeds are bulked and used to grow the next generation.

METABOLOMICS. The study of all metabolic path ways in a living organism.

MATING SYSTEM. The method by which individuals are paired for crossing.

METROGLYPH ANALYSIS. A semigraphic method used for assessment of morphological variation in the germplasm.

MOLECULAR BREEDING. Genetic improvement of crop plants for various economic characters using molecular marker and trasformation technology.

MID-PARENT HETEROSIS. Heterosis over the mean value of both the parents.

MINOR GENES. Genes having small and invisible effect on a character.

MONOECY. Separate existence of male and female flowers on the same plant in the same inflorescence or different inflorescence.

MONOGENIC TRAIT. A character which is governed by single gene.

MULTILINES. The deliberate seed mixtures of isogenic lines, closely related lines or unrelated lines.

MULTIPLE CROSS. A cross involving more than four inbred lines, also known as composite cross.

MULTIPLE CORRELATION. Joint influence of two or more independent variables on a dependent variable. It is denoted by R.

MUTATION BREEDING. Improvement of crop plants by induced mutagenesis.

MUTATOR GENE. A gene which enhance the mutation rate of other gene in the same genome.

NATURAL POLYPLOID. Naturally occurring polyploids.

NATURAL SELECTION. The process of evolution in nature in which fittest individuals survive and rest are wiped out. It favours those characters which are advantageous for survival.

NON-PREFERENCE. Various features of host plant that make the host undesirable or unattractive to insects for food, shelter or reproduction, also called Non-acceptance or Antixenosis.

NON RECURRENT PARENT. See donor parent.

NORMAL SEEDLINGS. Seedlings with normal growth of root and shoot.

NUCLEUS SEED. The initial seed of a variety produced in limited quantity by originating plant breeder.

NULLISOMIC. An individual lacking one pair of chromosome from a diploid set ($2n$-2).

OBSOLETE CULTIVARS. Improved cultivars of the recent past.

OLIGOGENES. One or few genes governing the same qualitative character.

OLIGOGENIC CHARACTER. A character which is governed by few major genes.

OLIGOGENIC INHERITANCE. Inheritance of oligogenic or qualitative characters.

ORGANIC FARMING. Crop cultivation without the use of any chemical.

ORGANIC PLANT BREEDING. Development of new crop cultivars without the use of any chemical.

OLIGOGENIC MUTATION. A mutation in ologogenic character.

OPEN-POLLINATION. Natural cross pollination.

ORTHODOX SEED. Seed which can be dried to low moisture content and stored at low temperature without losing their viability.

OUTBREEDERS. Cross pollinated species.

OUTBREEDING. Mating between unrelated individuals.

OUTCROSS. A natural cross to a plant with different genotype.

OVER-DOMINANCE. Superiority of F_1 heterozygote for one or more characters over its both the parents (homozygotes).

OVER DOMINANCE HYPOTHESIS. Heterosis due to superiority of heterozygote over both the homozygotes.

PARTIAL DIALLEL. A modified form of diallel in which selected crosses are used for analysis.

PARASITE. An organism or virus which lives upon or within another living organism.

PARTICIPATORY PLANT BREEDING. Plant breeding programmes in collaboration with farmers and consumers.

PARTHENOGENESIS. Development of embryo from the egg cell without fertilization.

PARTICULATE RADIATIONS. Raditions which emit particles when pass through a medium.

PATH ANALYSIS. A statistical procedure which splits the corretation co-efficient into direct and indirect effects and measures direct and indirect contribution of each independent character on the dependent variable.

PATHOGENS. Various disease causing organisms such as fungi, viruses and bacteria.

PHATOTYPE. Strain of a pathogen differing in pathogenicity, also called physiological race.

PEDIGREE. Record of the ancestry of an individual selected plant for its various generations.

PEDIGREE BREEDING. A method of breeding in which pedigree records of selected plants are maintained in each generation.

PEST. An animal or higher plant which parasitises crop plants such as insects, birds, parasitic weeds etc.

PHENOTYPE. Morphological appearance of an individual.

PHENOTYPIC CORRELATION. The observable correlation between two variables.

PHYSICAL PURITY. Freedom of seed from inert matter and defective seeds.

PLANT BREEDERS' RIGHTS (PBR). Legal rights provided to the originating plant breeder or owner of a variety to regulate production and marketing of his variety.

PLANT BREEDING. A science, art and technology which deals with genetic improvement of crop plants in relation to their economic use for mankind.

PLEIOTROPY. A gene having manifold effects.

POINT MUTATION. A mutation in the structure of base sequence of DNA.

POLYGENES. Several genes governing the same character.

POLYGENIC CHARACTER. A character which is governed by several genes.

POLYGENIC INHERITANCE. Inheritance of quantitative characters.

POLYPLOID. An individual having more than two basic sets of chromosomes.

POLLINATION. Transfer of pollen from anther to stigma.

POPULATION. A group of interbreeding members of one kind of organism.

PREDATOR. An animal naturally preying on others, an insect naturally preying on other insect.

PROGENY TEST. Evaluation of genotypic value of an individual based on its progency performance.

PROTANDRY. Maturation of anthers before the pistil.

PROTEOMICS. The study of structure and function of all proteins in a organism.

PROTOGYNY. Maturation of pistil before anthers.

PURE LINE. Progeny of a self pollinated homozygous plant obtained by selfing.

QTL. A quantitative trait locus, a region of DNA containing genes controlling expression of a quantitative trait.

QTL MAPPING. The technique of locating genes which control expression of quantitative traits.

QUADRIALLEL ANALYSIS. Analysis of double cross hybrids.

QUALITATIVE TRAIT. Same as aligogenic character.

QUALITY BREEDING. Genetic improvement of crop plants for various quality attributes.

QUANTITATIVE GENETICS. A branch of genetics which deals with the inheritance of quantitative characters.

QUANTITATIVE TRAIT. Same as polygenic character.

QUASI-DIPLOID. An individual with trisomic-monosomic constitution (2n+1–1)

RAPD. Random amplified polymorphic DNA, a molecular DNA marker technique used in genetic engineering.

RFLP. Restriction fragement length polymorphism, a molecular DNA marker technique.

RECESSIVE ALLELE. The allele whose expression is masked by a dominant allele of the same gene.

RECIPIENT PARENT. The parent which receives a desirable gene; also called recurrent parent.

RECIPROCAL CROSS. A cross in which order of male and female parents is reversed.

RECURRENT SELECTION. Reselection generation after generation with intermating of selected plants to provide for genetic recombination.

REPLICATION. Repetition of treatments under investigation.

REPRODUCTION. The process by which living organisms give rise to offspring of similar kind (species).

R LINE. Line which restores fertility when crossed with cytoplasmic genic male sterile line.

REVERSE MUTATION. A change from mutant allele to wild type.

ROGUING. The process of removal of off type plants from the field of an improved variety to avoid contamination.

SEED. The product of fertilized ovule or any plant part which is used for mass multiplication of a crop.

SEED CERTIFICATION. A legal system which ensures production of high quality seed in terms of genetic purity and germination.

SEED PRODUCTION TECHNOLOGY. A branch of plant breeding which deals with principles and methods of improved seed production.

SEED TESTING. The process of evaluation of seeds in terms of purity and germination.

SEEDLING RESISTANCE. Resistance exhibited by young seedlings.

SELF INCOMPATIBILITY. Inability of a plant with functional pollen to set seed when self pollinated.

SELFING. The process of artificial self pollination.

SELF POLLINATION. Pollination of a flower by its own pollen.

SEXUAL REPRODUCTION. Multiplication of plants through fertilized embryos.

SIMPLE CORRELATION. The association between two variables.

SNP. Single nucleotide polymorhism, a type of molecular marker used in genetic engineering.

SINGLE CROSS HYBRID. The hybrid progeny of a cross between two inbreds or varieties, viz., (A × B).

SOMATIC HYBRIDIZATION. Crossing of plants through protoplast fusion.

SOMATIC MUTATION. A mutation in somatic tissues or cells.

SPECIES. A group of interbreeding natural population which is reproductively isolated from other such group.

SPECIFIC COMBINING ABILITY. The performance of a parent in specific cross in relation to general combining ability.

SPERM. The male gamete.

SSCP. Single strand conformational polymorphism, a type of DNA marker.

SSR. Simple sequence repeat, a type of molecular marker.

SPLIT PLOT DESIGN. Experimental design having main plots, subplots and ultimate plots and can study several factors simultaneously.

SPONTANEOUS MUTATIONS. Naturally occuring mutations.

STABILITY. The suitability of a variety for general cultivation over a wide range of environmental conditions.

STANDARD HETEROSIS. The superiority of a hybrid over the local commercial hybrid of a crop.

SYNTHETIC VARIETY. A variety which is developed by intermating in all possible combinations a number of inbreds with good general combining ability and mixing the seed of crosses in equal quantity.

TERMINATOR GENE. A gene which allows germination of seeds for one generation only in transgenic plants. It leads to embryo abortion in the next generation.

TEST CROSS. Crossing of F_1 with its homozygous recessive parent.

THREEWAY CROSS HYBRID. The hybrid progeny between a single cross and an inbred *i.e.*, (A × B) × C

TRANS GENES. A foreign genes which are used for developing transgenic individuals through genetic cengineering.

TRANS GENES. A foreign genes which are used for developing transgenic individuals through genetic engineering.

TRANSGENIC. Genetically engineered organisms, may a plant, an animal or a microorganism.

TRANSGENIC PLANT BREEDING. Genetic improveent of crop plants through genetic engineering.

TOP CROSS. A cross between an inbred line and an open pollinated variety.

TRANSGENIC PLANTS. Plants which are developed by the process of genetic engineering.

TRANSGRESSIVE SEGREGANTS. Segregants which fall out side the limits of both the parents for a character.

TREATMENTS. Various objects which are used for comparison in an experiment.

TRIALLEL ANALYSIS. Analysis of a set of three way crosses among *n* parents.

TRIPLE TEST CROSS. Crossing of randomly selected F_2 plants with both the parents involved in the cross and their F_1 hybrid.

USEFUL HETEROSIS. Same as economic heterosis.

VARIABILITY. The existence of phenotypic differences among the genotypes of a population.

VARIETY. A genotype released and notified for commercial cultivation.

VARIETAL DETERIORATION. Permanent reduction either in the genetic or agronomic value of a released variety.

VERTICAL RESISTANCE. Resistance of a host to the particular race of a pathogen, also called major gene, oligogenic and qualitative resistance.

VIRULENT BACTERIA. Disease causing bacteria leading to death of the host plant.

WORKING COLLECTIONS. Germplasm which is most commonly used in crossing programmes.

WTO. World trade organization, an international body which deals with the rule of trade between nations.

X-RAYS. Sparsely ionizing radiations used for induction of mutation.

ZYGOTE. A product of the union of male and female gametes. It gives rise to adult organisms.

KEY REFERENCES

1. **ALLARD, R. W.** 1960. Principles of Plant Breeding. John Wiley and Sons Inc. New York.
2. **ALLARD, R. W.** 1999. Principles of Plant Breeding 2nd edn. John Wiley and Sons Inc. New York.
3. **BOROJEVIC, S.** 1990 Principles and Methods of Plant Breeding. Alsevier, Amesterdam.
4. **BRIGGS, F. N. AND KNOWLES, P. F.** 1967. Introduction to Plant Breeding.Reinhold, New York.
5. **CHOPRA, V. L. (ED.)** 2000. Plant Breeding: Theory and Practice 2nd edn. Oxford and IBH Publishing Company, Pvt. Ltd. New Delhi.
6. **FEHR, W. R.** !987. Principles of Cultivar Development Volume 1. Theory and Techniques. Macmillan Publishing Company, New York.
7. **FREY, K. J.** (ed.) 1966. Plant Breeding. Iowa State University, Press, Ames, Iowa, USA.
8. **FREY, K. J.** (ed.) 1981. Plant Breeding II. Iowa State University, Press, Ames, Iowa, USA.
9. **HARTEN, A. V.** 1998. Mutation Breeding: Theory and Practical Applications. Cambridge Uni. UK.
10. **KUCKKUCK, H., KOBADE, G. AND WENGEL G.** 1993. Fundamentals of Plant Breeding. Narosa Publishing House, New Delhi.
11. **MAYO, O. 1984**. The Theory of Plant Breeding. Clarendon, Oxford.
12. **POEHLMAN, J. M.** 1987. Breeding Field Crops 3rd edn.AVI Publishing Co. Inc.West Port, Connecticut, USA.
13. **POEHLMAN, J. M. AND BORTHAKUR, D. N.** 1969. Breeding Asian Field Crops. Oxford and IBH Publishing Company, New Delhi.
14. **PHUNDAN SINGH** 2013. Essentials of Plant Breeding 5th edn. Kalyani Publishers, New Delhi.
15. **PHUNDAN SINGH** 2013. Elements of Genetics 4th edn. Kalyani Publishers, New Delhi.

16. **PHUNDAN SINGH** 2012. Cotton Breeding 3rd edn. Kalyani Publishers, New Delhi.

17. **PHUNDAN SINGH AND SANJEEV SINGH** 2010. Breeding Hybrid Cotton 3rd edn. Kalyani Publishers, New Delhi.

18. **SHARMA, J. R.** 1984. Principles and Practice of Plant Breeding.Tata McGraw-Hill Publishing Company Ltd. New Delhi.

19. **SIMMONDS, N. W.** 1979. Principles of Crop Improvement. Longman, London.

20. **SNEEP, J.AND HENDRIKSON, A. J. T.** (eds) 1979. Plant Breeding Perspectives. Pudoc, Wageningen, Netherlands.

21. **VOSE, P. B. AND BLIXT, S. G.** (eds) 1984. Crop Breeding : A Contemporary Basis. Pergamon Press, London.

22. **WILLIAMS, W.** 1964. Genetic Principles and Plant Breeding, Oxford, Blackwell.